Fluid Phenomena

African and Asian Anthropocene: Studies in the Environmental Humanities

VOLUME 2

The titles published in this series are listed at *brill.com/aaeh*

Fluid Phenomena

The River and Anthropocene Life-World in the Asia-Pacific

Edited by

Iftekhar Iqbal
Noor Hasharina Hassan
Asiyah Kumpoh

BRILL

ISSN 2772-8536
ISBN 978-90-04-75809-4
ISBN 978-90-04-75810-0 (PDF)
DOI https://doi.org/10.1163/9789004758100

Library of Congress Control Number: 2026005206

Brill and Koninklijke Brill BV are part of De Gruyter Brill.
www.degruyterbrill.com

Questions about General Product Safety Regulation: productsafety@degruyterbrill.com

Cover illustration: On the Sungai Sikuati in Kudat in Northern Sabah, where mangroves meet timber piers and a village leans into the tide—boats awaiting the turn—lives, livelihoods, and weather move with the river. Photo credit: Paul J. Carnegie.

Contents

PART 2
Everyday Life and Livelihoods

PART 3
Development and Governance

Acknowledgements

The journey of this book began with a hybrid-format conference held at Universiti Brunei Darussalam (UBD) in February 2022. All contributors brought a rich array of interdisciplinary insights centered on the unifying theme of the river. The conference itself, along with post-conference discussions and successive rounds of internal peer review, fostered a remarkable sense of intellectual camaraderie—indeed, the contributors became something of a scholarly family, bound by the fluid force of the river. We extend our heartfelt thanks to each contributor. Special thanks are due to John Charles Ryan for his invaluable guidance and support, and to Rohan D'Souza, whose generosity in writing the book's epilogue is deeply appreciated.

We are grateful to the Faculty of Arts and Social Sciences at UBD for a generous two-year research grant, which made possible the organization of the initial conference that laid the foundation for this volume. We also wish to acknowledge our colleagues who served as session moderators and provided early, constructive feedback on the draft chapters: Stephen Druce, Sarah Ghazali, Muhammad Arafat bin Muhamad, Angela Tritto, and Philippe Forêt. Our sincere thanks to Paul Carnegie for providing the beautiful photograph used on the book's cover. We thank Md Mamunur Rashid and Ilyas Abdul Manaf for their help in reviewing the indexing and stylistic issues, and Fatin Razak for her secretarial assistance.

Subsequent stages of review—both at the internal Series editorial level and through external peer review—were instrumental in shaping the final manuscript. We are especially thankful to Meng Yue, Howard Choy, Ignasi Ribó, and three anonymous reviewers for their insightful comments and feedback. We are indebted to Stephanie Carta at Brill, whose early engagement with the book project was immensely helpful, and to Iulia Ivana, whose support during the peer review process and the final stages of production contributed greatly to the smooth and rewarding completion of this project.

Finally, we thank all members of the series editorial board for *African and Asian Anthropocene: Studies in the Environmental Humanities* for their enthusiastic yet critical endorsement of this volume.

Bandar Seri Begawan
January 2026

Figures and Tables

Figures

Tables

Notes on Contributors

Maria Carinnes P. Alejandria
is an anthropologist specializing in disaster governance and health justice in Southeast Asia. She is an assistant professor and the Programme Coordinator of the Sociology and Anthropology Programme at the Universiti Brunei Darussalam. She held fellowship positions with Brown University and Loughborough University for her work on disaster studies and human rights. Her publications include the edited volumes *Disaster Archipelago: Locating Vulnerability and Resilience in the Philippines* and *Aging in the Global South: Challenges and Opportunities*. She has also authored journal articles and edited special issues addressing complex disasters, humanitarian coordination, and social health.

David Arnold
is a historian of modern South Asia and a Fellow of the British Academy. Formerly Professor of Asian and Global History at the University of Warwick, he has written extensively on the history of medicine, science, technology and environment. Published work includes *Colonizing the Body: State Medicine and Epidemic Disease in Nineteenth-Century India* (1993); *The Problem of Nature: Environment, Culture and European Expansion* (1996); *The Tropics and the Traveling Gaze: India, Landscape, and Science, 1800–1856* (2006); *Everyday Technology: Machines and the Making of India's Modernity* (2013) and *Toxic Histories: Poison and Pollution in Modern India* (2016).

Paul J. Carnegie
is Associate Professor of Political Science at the Institute of Asian Studies, Universiti Brunei Darussalam. His interests focus on the politics of development, state formation, marginality, and precarity in Southeast Asia, especially Indonesia. He is the author of *The Road from Authoritarianism to Democratization in Indonesia* (Palgrave Macmillan, 2010) and co-edited *Human Insecurities in Southeast Asia* (Springer, 2016) and *(Re)Presenting Brunei Darussalam: A Sociology of the Everyday* (Springer, 2023). He is also a section editor for the *Palgrave Handbook of Ethnicity* and sits on the editorial board of the IAS-Springer, *Asia in Transition Series*. His research articles have appeared in leading journals such as *Pacific Affairs*, *Australian Journal of Politics and History*, *Journal of Population Research* and *Australian Journal of International Affairs*.

Ta-Wei Chu
is an assistant professor in the Department of Social Science and Development, Chiang Mai University. His research interests are in security studies, hydropolitics in the Mekong Basin and transdisciplinarity. His articles have appeared in leading international journals including *Asian Survey*, *Asian International Studies Review*, *Journal of Current Southeast Asian Affairs* and *Wiley Interdisciplinary Reviews Water*. His current research project focuses on Belt and Road Initiative (BRI) in Myanmar.

Rommel A. Curaming
is a senior assistant professor at the History and International Studies Programme at the Universiti Brunei Darussalam (UBD). He obtained a PhD at the Australian National University (ANU) and was a Postdoctoral Fellow at La Trobe University and the National University of Singapore (NUS). The empirical and thematic areas of his research focus on the analysis of the politics of the (supposedly) non-political, particularly historical knowledge production, consumption and distribution in Indonesia, Malaysia and the Philippines. His book *Power and Knowledge in Southeast Asia* was published by Routledge (2020).

Maya Dania
is an assistant professor of Sociology at the School of Social Innovation, Mae Fah Luang University. Her research integrates the Anthropocene and Posthumanism into environmental sustainability, addressing critical social and environmental challenges in Southeast Asia. She is a Fellow of the UK Professional Standards Framework (UKPSF) and a UNESCO Chair-Kobe University Fellow on Gender, Vulnerability and Well-Being in Disaster Risk Reduction. She has received several prestigious grants, including funding from the Sumitomo Foundation, and has published extensively in leading academic outlets such as Springer and Brill. Additionally, she is a core member of the Disaster Resilience and Environmental Sustainability Research Program (DRES-ARCID) at Mae Fah Luang University.

Heather Downey
is a social work academic at La Trobe University's Albury-Wodonga campus in Victoria, Australia. Her research concerns social and environmental justice in the rural context. As a regional academic, her interest lies in the economic, social, cultural, recreational, and environmental meaning of water for all rural residents. She is currently collaborating with multidisciplinary researchers in the Murray Darling Water and Environment Research Program to explore how

communities have navigated environmental, water policy, and social change. She has published widely in international journals, most recently on the impacts of climate change on Australian freshwater tourism.

Rohan D'Souza

is an environmental historian specializing in South Asia, with research interests in colonial water governance, technological transformations, and ecological change. He is based at the Graduate School of Asian and African Area Studies, Kyoto University. His works include *Drowned and Dammed: Colonial Capitalism and Flood Control in Eastern India*, and numerous articles on the political ecology of rivers. His scholarship critically examines the intersections of colonialism, capitalism, and environmental change. He has also contributed to debates on the Anthropocene and environmental futures, positioning his work within global environmental history and political ecology frameworks.

Sk. Tawfique M. Haque

is a professor of Public Policy and Governance and Director of the South Asian Institute of Policy and Governance (SIPG) at North South University (NSU), Bangladesh. His research interests include migration, Rohingya crisis, organizational culture, local governance, globalization, international trade regime, development management, geopolitics, and climate change. He has authored seven books, and fourteen book chapters from leading international publishing presses, including Springer, Routledge, Ashgate, Palgrave Macmillan, in addition to several articles in nationally and internationally indexed journals. His most recent book, *The Displaced Rohingyas: A Tale of a Vulnerable Community*, was published by Routledge in 2024. He serves as member of the editorial board of South East Asia: A Multidisciplinary Journal (SEAMJ).

Mohammad Tareq Hasan

is an associate professor of Anthropology at the University of Dhaka (DU), Bangladesh. After the initial training in Anthropology from DU, he pursued MPhil in Anthropology of Development (2014) and PhD in Social Anthropology (2018) from the University of Bergen (UiB), Norway. He also completed a postdoctoral research fellowship (2021) at the International Institute for Asian Studies (IIAS), Leiden University, the Netherlands. Besides teaching at the undergraduate and graduate levels, his research revolves around anthropology of work, state formation, political economy, and egalitarianism. His recently authored book is "Everyday Life of Ready-made Garment Kormi in Bangladesh: An Ethnography of Neoliberalism" (Palgrave Macmillan, 2022).

Noor Hasharina Hassan
is a senior assistant professor in the Faculty of Arts and Social Sciences, Universiti Brunei Darussalam. She was previously a Postdoctoral Visiting Research Fellow at the Department of Geography at King's College London and a Visiting Scholar at the East West Centre, Hawaii. Her research falls under the umbrella of urban landscape and consumption, where her research interests include retail and consumption dynamics; housing development; landscape studies and tourism; sustainable development and welfare. She is the Head of Borneo Studies Network Secretariat Office, UBD and a Research Associate at the Institute of Asian Studies, UBD.

Iftekhar Iqbal
is a historian at Universiti Brunei Darussalam, where he serves as an associate professor and programme leader for the History and International Studies Programme. His research interests lie in environmental humanities and intellectual history, and his works include *The Bengal Delta: Ecology, State and Social Change, 1840–1943* (Palgrave Macmillan, 2010) and *The Range of the River: A Riverine History of Empire across China, India, and Southeast Asia* (Stanford University Press, 2025). He is currently working on a book project, "Anthropocene South Asia: A Short History." Iqbal is an editorial board member with *South Asia: Journal of South Asian Studies* and *Conservation and Society*. At UBD, he leads the research cluster "The Environment and Transformations."

Andrew Alan Johnson
PhD, is Associate Professor at Stockholm University and a writer at Firaxis Games since 2020. He was previously Visiting Scholar at the Centre for Southeast Asian Studies, UC Berkeley (2019–2022), and held faculty positions at Princeton, Yale-NUS College, and postdoctoral fellowships at Columbia and the National University of Singapore. He earned his PhD in Anthropology from Cornell University in 2010. His research explores how people rebuild life amid economic and environmental crises. He is the author of *Ghosts of the New City* (2014), on urban decay in Chiang Mai, and *Mekong Dreaming* (2020), which examines the transformations wrought by dams on the Mekong river. He has also published widely in journals and edited volumes.

Magne Knudsen
is an anthropologist specializing in environmental anthropology and political ecology, with a focus on livelihood transitions in fishing and farming communities across Southeast Asia. Since his doctoral project in social anthropology at the Australian National University in 2010, his work has spanned studies of fishing families' livelihood challenges and land rights, disaster response,

agrarian relations, traditional knowledge and human uses of river across Southeast Asia, including the Philippines and Brunei. Following a postdoctoral fellowship in the Department of Sociology at the National University of Singapore (NUS), he served as an assistant professor of Sociology and Anthropology at Universiti Brunei Darussalam from 2015 to 2024. Beyond academia, he has worked as a consultant on environmental issues and as a livelihood and land tenure expert for humanitarian crisis response efforts. Currently, he is an independent researcher based in Oslo, Norway.

Asiyah Kumpoh
(PhD, Leicester), is a senior assistant professor at the Faculty of Arts and Social Sciences, Universiti Brunei Darussalam. Her current research and publications focus on Brunei's history, conversion narratives, the Brunei Dusuns, and diasporas in Brunei. Recent publications include "Exclusion and Inclusion: Melayu Islam Beraja in Bruneian Nationalism and National Identity," [pp. 549–560] (co-authored with Nani Suryani Abu Bakar) in the *Routledge Handbook of Nationalism in East and Southeast Asia* (Routledge, 2023) and "Can Cultural Homogenization be an Open-Ended Process? Reconstructing the Narratives of Brunei's Homogenization Process" in the *Journal of Ethnic and Cultural Studies*, 10, 2 (2023).

David Ludden
is an emeritus professor of Political Economy and Globalization at the Department of History at New York University. He has held the position of Senior Fellow at the Institute of Public Knowledge; Chairman of History Department at NYU; Fulbright Fellow; Fellow at the John Simon Guggenheim Memorial Fund; and a member of American Council of Learned Societies. He is the Editor-in-Chief of *Oxford Research Encyclopedia of Asian History*. He was the President of Association for Asian Studies (2002–3). He has published more than 10 books and more than 65 articles and book chapters. He is currently working on a monograph: *Global Asia: A Spatial History*. His major monographs and edited volumes include: *India and South Asia: A Short* History (Oxford 2002); *Peasant History in South India* (Oxford, 1985, 2005); *An Agrarian History of South Asia* (Cambridge, 1999); *Capitalism in South Asia* (Association for Asian Studies, 2004) and *Contesting the Nation: Religion, Community, and the Politics of Democracy in India* (Philadelphia, 1996).

Misa Juliana Minggu
holds an undergraduate degree in Sociology and Anthropology. A young scholar from the Iban community of Temburong, she has worked as a research assistant on several projects on indigenous cultures and ecological knowledge on Borneo.

Kathrina Mohd Daud
is a senior assistant professor in the English Studies programme at the Faculty of Arts and Social Sciences, Universiti Brunei Darussalam. Her research focuses on the intersections of popular fiction, Bruneian fiction, and representations of religion in literature. Her work has appeared in several edited volumes published by Routledge and Springer, as well as in the *Journal of Commonwealth Literature* and *World Englishes*. Most recently she co-edited *The Literature of Brunei: History, Culture and Challenges* with Ooi Keat Gin, Routledge 2025.

Izni Azrein bin Noor Azalie
teaches at the University of Brunei Darussalam's Faculty of Arts and Social Sciences. He earned his BA Hons Major in Geography and Minor in History at Universiti Brunei Darussalam. At Loughborough University in the UK, he received his Master of Science (MSc) in Global Transformations. After obtaining his PhD in Development Policy from University College London (UCL), his research interests now focus more on halal industrial governance and economic growth, urban economies and development, and how human-environmental dynamics affect communities and nature. His other research interests also cover environmental monitoring and assessment; globalisation; sustainable development; creative cities and urban planning; issues and politics of development, and geopolitics.

John Charles Ryan
is a researcher in literary studies, creative writing, and environmental humanities. He is Adjunct Senior Research Fellow at the Nulungu Research Institute, University of Notre Dame, Australia, and Visiting Researcher at the Arctic Centre, University of Lapland, Finland. Funded by the Kone Foundation, his current project examines possibilities for communication and collaboration between people and trees in Northern Finland. His books include *Environment, Media and Popular Culture in Southeast Asia* (2022, Springer, co-edited) and *Introduction to the Environmental Humanities* (2022, Routledge, co-authored). He is the Chief Editor of the journal *Plant Perspectives* and Managing Co-Editor of *The Trumpeter*. For more information, see www.johncharlesryan.com.

Md Mamunur Rashid
(PhD, Universiti Brunei Darussalam) is a professor of History at Jagannath University, Dhaka. His research interests include, among other areas, historical relations between the economy and environment in South and Southeast Asia. In addition to publishing in several leading journals, Rashid is currently working on a book project that explores the tripartite relationship between economy, transport, and ecology in colonial Bengal Delta and northern Borneo.

Ryoji Soda
is a professor of Geography at the Graduate School of Literature and Human Sciences, Osaka Metropolitan University, Japan. He has conducted extensive field research in Malaysia, Japan, and other Asian countries, focusing on human-nature interactions and the environmental humanities. His recent research interests include the expansion of oil palm smallholdings in Southeast Asia and the politics surrounding dam construction in East Asia. His notable publications include: *People on the Move: Rural-Urban Interactions in Sarawak* (2007); "The Diversity of Small-Scale Oil Palm Cultivation in Sarawak, Malaysia," *The Geographical Journal*, 182 (2015) and *Anthropogenic Tropical Forest: Human-Nature Interfaces on the Plantation Frontier* (2020).

Willem van Schendel
(University of Amsterdam and International Institute of Asian Studies) works in the fields of history, sociology and anthropology of Asia. He served as Professor of Comparative History and then as Professor of Modern Asian History. Recent books include *Entangled Lives: Human-Animal-Plant Histories of the Eastern Himalayan Triangle* (2022, with J. Pachuau); *Flows and Frictions in Trans-Himalayan Spaces: Histories of Networking and Border Crossing* (2022, ed. with G. Cederlöf); *A History of Bangladesh* (2015); *The Camera as Witness: A Social History of Mizoram, Northeast India* (2015, with J. Pachuau) and *The Bangladesh Reader: History, Culture, Politics* (2013, ed. with M. Guhathakurta). For more publications, see uva.academia.edu/WillemVanSchendel.

Gabriel Yong
is a lecturer in the Faculty of Arts and Social Sciences at Universiti Brunei Darussalam. Currently, he is the programme leader for Geography and Environmental Studies. His research interests include sustainability issues from a complex living systems perspective; human-environment evolution in Brunei & Brunei Bay; Arcology; heritage conservation, focusing on Kampong Ayer; data capture methodologies in an information-rich but noisy environment; and learning & knowledge development through experiential, project/field-based dialectic learning & use of social networks.

Md. Parvez Hasan Yousuf
is a research associate at the South Asian Institute of Policy and Governance (SIPG), North South University, Bangladesh. Trained in Disaster and Human Security Management at the Bangladesh University of Professionals (BUP), his interest lies in disaster management, environment, climate change, rivers, refugee crisis, displacement, human security, public policy, and governance. His work on the social cohesion of the host and Rohingya communities in

Bangladesh has appeared in *Displacement and Refugee Issues in South Asia-Uncovering the Contested Realities'* (Lexington Books, 2024). His study of Bangladesh's climate policy was published recently (*The Elgar Encyclopedia of Climate Policy*, 2024). He is a member of the Center for Migration Studies (CMS), and the China–South Asia Center for Sociocultural Studies (CSCSS) of SIPG.

Introduction

Iftekhar Iqbal, Noor Hasharina Hassan and Asiyah Kumpoh

> We say: water is the temperament of the spirit,
> and delight of the soul, and the constitution of the
> body of all men and animals, since it resembles
> them and conforms to them. One of the virtues of
> water is that no drink, be it pure, clear, agreeable,
> or sweet as it can be, has any substitute for water.
> Without water any living creature would die, as
> water is the stamina in its body, its energy, its nourishment,
> and is always within it. Watching flowing
> water brings joy to the soul, pleasure to the eyes and
> solace to the heart. When water is salty, it produces
> ambergris, pearls and gems. When it is clear and
> pleasant, it revives the soul. There is nothing that
> does not contain water, or is touched by water or
> created out of water, as water is another word for
> sperm (nuṭfah) and nuṭfah another name for water.
>
> RAPOPORT AND SAVAGE-SMITH (2014, 504)

∵

This book explores the "life-world" of the riverspace that has shaped the ecological, economic, cultural, and governing processes affecting both humans and non-humans in the Asia-Pacific region, including parts of South Asia, Southeast Asia, Japan, and Australia. It employs interdisciplinary conceptual and methodological tools for river studies, focusing on three broad thematic areas: environmental humanities, livelihood research, and development studies. In other words, the book seeks to tie together three strands of human engagements with rivers: the representation of the river as an other-than-human entity in intellectual discourse, indigeneity, and creative imagination; the layered livelihood activities as supported and endangered by the river; and the governance of the river from the vantage point of development projects and popular responses.

 | HTTPS://DOI.ORG/10.1163/9789004758100-002

Each of the three areas has seen remarkable scholarship in its respective field. However, given the planetary urgency for transcending disciplinary boundaries across environmental humanities and social sciences, a key question driving this book relates to how these three thematic areas speak to each other. Is the river an ontologically static entity whose autonomous essence needs to be revered and imagined in its own aura, or is it an interactive site that is constructed by layers of power, participation, and political protocols? Can the debates around a hydroelectric dam on a river, for example, be explored from the vantage point of literary imagination of the same? Can the study of everyday life and livelihood at the societal peripheries help forge conversations between non-modern practices and modern development needs on the site of the river? In exploring these questions, the contributions in the book, based on seventeen research chapters and an epilogue, suggest that the river cuts across disciplines in more ways than previously acknowledged. Thus, the book on riverine life-worlds contributes to the burgeoning interest in rivers by demonstrating how environmental humanities, livelihood research and development studies form an interrelated content.

We aspire to demonstrate the centrality of the river in these disparate areas by using the lens of the Anthropocene. The term has been in use in the field of natural sciences since the late nineteenth century and received wider currency only at the start of this millennium. Paul Crutzen and his colleagues in the scientific community used the term to mark the violent power of humans at the end of the Holocene epoch to bring changes to the planetary system that non-human natural forces were previously unable to accomplish. Although the term evolved in the specific circuit of geophysical sciences, it quickly gained traction in the social sciences and humanities.

However, the integration of the scientifically informed concept of the Anthropocene into the social sciences and humanities has not resulted in a seamless assimilation of these broader disciplinary domains. The idea of the Anthropocene continues to present an epistemic duality. On one hand, it identifies humans as the prime force for destabilizing the balance of the planetary system, especially through prompting global warming—a rational, scientific appreciation of the human agency driving toward potential self-extinction. On the other hand, the Anthropocene has, particularly in the past two decades, emerged as a formative trope for sensitivities to non-human or other-than-human actors, consigning humans to be just another species among millions who have evolved through mutual interactions—an understanding that has triggered the emergence of the field of environmental humanities. Therefore, the Anthropocene is as much about destruction as it is about healing.

The question remains whether this duality within the Anthropocene could be reconciled. What is increasingly becoming clear is that the exclusively scientific discourse of human-induced ecological precarity and decline also reflects an indifference to humans' deep-time, affective, and creative power of connecting to the pulse of the planet Earth. This book suggests that environmental humanities possess the tools and trajectory to engage both the science and arts of existence and thereby narrow the gap between the epistemic duality of the Anthropocene.

Why river? Because among all bodies of water, the river offers the fullest scope for multimodal relationality. Water in its glacial, salty oceanic, or evaporated form—all have useful properties. But because of its sweet water properties and location to terrestrial ecology and fluxes, the river does the following for human and non-humans: quenches thirst; helps vegetation flourish; nurtures fishes for biodiversity and human nutrition; offers energy in the forms of electricity and irrigation; enables micro navigability for daily movement across the land; acts as rainmaker via evapotranspiration, provides accessible material bases of livelihood, enriches soil fertility; facilitates transport and communication; nourishes the surrounding underground moisture level and broader biodiversity. No wonder, at least 90 percent of the world's population lives within 10 kilometers of a river system. This materiality of the river and the science that celebrates it also creates space for the human mind, for culture, imagination, and spirituality where the river is not merely a body of water that exists to serve human's material need. In this context, the river is human's perennial interlocutor with an agency of its own. If water is 'radical' with an agency, as Irine J. Klaver (2021) poignantly suggests, the river is then the most radical of them.

But the river is as radical as it is relatable. This relationality, embedded in what we call "fluid", has a long traceable history. "Fluid Phenomena" marks the river as a site that demonstrates that the Anthropocene brings not only decline and despair but also announces the ability and need for human introspection regarding their relation to nature (Khan 2023). Heidegger's fourfold relations of "dwelling" between sky and earth and among mortals and the divine—as mediated by the ontopoetics of existence—inform the concept of what he terms "Unity of Being." In the context of this book, we suggest that in recent times phenomenological ideas have flowered into Bruno Latour's "Parliament of things," which seeks to restore the formal political agency of non-human entities squarely into the world of human politics. Even if not directly linking the question of phenomenology in its Heideggerian manifestation, each of the chapters in this book connects to the river through exploring the imagined, felt, lived, and governmentalized life-world on a relatable scale. A poet, a farmer or a hydraulic engineer—all are conscious rational human beings, but each is also, in one way or the other, an agent that dwells, imagines, and feels in encountering the river. The scholarly, cutting-edge, and passionate contributions in this book seek to grasp this relational engagement with the river at a time when anthropogenic encounters with nature have not only become overwhelming but also diverse. In this larger context of the diverse experience, *Fluid Phenomena* uses as a site that demonstrates that the Anthropocene is not only a marker of human capacity to harm natural balance but also a reminder of human capacity to relate to nature through embodied encounters in this time of increasing planetary anxiety.

The invocation of phenomenology in exploring the relationality between humans and the waterworld in the Asia-Pacific has a spatial connotation. Phenomenology is neither originally Western nor exclusively modern. The spiritual dimension of subject-object transfusion is ingrained in Sufi Islamic tradition. Twelfth-century Spanish commentator Ibn ʿArabi's idea of "Unity of Existence," which posits that the entire creation is the actualization of God's knowledge of Himself, is an enduring theme in Sufi literature. Ibn Sina's contributions to the field are also considered a precursor to Heidegger's (El-Bizri 2000). The influence of ancient Indian and Chinese thoughts, especially Daoist and Buddhist classics, on Heidegger has also surfaced in recent research (May 1996). This book's focus on the Asia-Pacific region and the experience of riverine water spaces as explicated in each chapter, returns our attention to a cultural milieu where phenomenology is no stranger.

David Arnold sets the tone for the book in two connected ways. One relates to the ambivalent site of disjuncture between the indigenous and colonial perception and practices regarding riverspaces. Such disjuncture manifested in the dualistic manner in which India's tropicality was perceived and engaged with by the colonial power. It symbolized a return to pristine nature, untouched by modernity, while also representing an opportunity to apply Western science and technology to unlock India's potential, ushering anthropogenic change and increased productivity for the market and revenue. In this temporal juncture, the rivers served as a "marker of what was and what might yet be." Arnold observes, "Transitions seemed more rapid and more drastic in and alongside rivers than in the slow-moving expanses and circulating currents of the oceans. Conflict, and what it connotes for the exercise or contestation of power, appears more pervasive, more ingrained, in fluvial studies than in its oceanic counterpart." This insight about the river as a site of study of the past, present and future leads Arnold to proclaim a new historiographical area of "fluviology," in contrast to highly invested oceanic studies recently termed "Thalassology."

While Arnold delves into the anthropogenic protocol of engagement with rivers, the inescapable flux of co-constitutive relationships between human society and rivers finds apt empirical windows in the chapters by Willem van Schendel and David Ludden. Van Schendel considers how different groups of people have imagined and used the Karnaphuli river in the wettest corner of the Monsoon Asia, flowing from the Indo-Burma Mountain ranges to the Bay of Bengal. Ancient and monsoon-fed, the Karnaphuli has evoked diverse imaginations among precolonial and colonial-era inhabitants, European missionaries and administrators, and contemporary policymakers and environmentalists. With the thriving social existence of the Hindus, Buddhists, Christians, and Muslims in different spatial nodes of the river and along its abiotic substance and range of organisms, van Schendel understands it as "a cultural layer cake as well as ecological layer cake." He suggests that although the colonial partition of the subcontinent in 1947 fractured this diverse but seamless hydrosocial space of the river, the most impactful anthropogenic intervention has been in place in more recent decades under the watch of the postcolonial states. The age-old hydrosocial dynamics of plurality and fluvial lifelines that Karnaphuli had shaped are now shattered by the singular force of the Anthropocene, which van Schendel considers a "process" rather than merely a concept. These dynamics that mark a "hydrosocial universe," call for a "water optic" to explore further the stories like that of the Karnaphuli river.

The "water optic" gets another use in David Ludden's chapter. He shows how the watery places span very small spaces of river drainage, "dammed with dirt and stone to train moving water into paddy fields to feed rice to feed people who praised kings who deployed labor to water the land." Various renditions of that dynamic interplay of rain, water, soils, plants, food, culture, and power have formed a vast diversity of river territories in Monsoon Asia. For Ludden, the river is not merely a node to relationality in an abstract sense; but it informs a material circularity that touches all conceivable corners of life. Spatial crossings, large-scale migrations, and the making and remaking of places throughout Monsoon Asia's long historical trajectory were the hallmark of rivers, giving rise to cities and settlements. This riverine process of territorial place-making is often embodied, as beautifully expressed in a Tamil poem that Ludden quotes: "Food is first for all bodies made of food/and food is but soil and water mixed/So those who unite water and soil/create living bodies and life in this world." The resultant "river territory spatial histories," as Ludden calls it, are, however, upended by social power relations that "pile the worst climate suffering onto people least responsible for the pollution which is aggravating Anthropocene extremes."

The multitude of human entanglements to the riverspace, both historically and contemporarily, as outlined by Arnold, van Schendel and Ludden, underscores the possibility of a finer creative appreciation of the deeper meaning of relationality—a topic taken up by Kathrina Daud and John Charles Ryan. The question of relationality takes centre-stage in Kathrina Daud's excursion into the creative imagination of the natural world of Brunei, a watery-forested slice of the Borneo Island. Through contemporary visual and written narratives, Daud demonstrates how the fictive spaces shape and govern the Bruneian imagination about the environment and national identity. She argues that the river is imagined as a gestational space onto which a nationally predominant value is projected—a site and symbol for "traditional" wisdom and power—and as an unknowable yet active, and agentic space from which values are enforced. Conversely, land-based rainforests are imagined as a site of modernization, progress, and multiethnic plurality that define the nation's cosmopolitan ethos. This raises the question of "what is to be understood as the space between the water and the rainforest, between the fickle unknowability of the sea, the mysterious governing of the jungle and the possibility and price of urban built environments." In either case, the question of the human relation to nature, mediated by political and cultural factors, remains latent in Daud's chapter—yet is gravely at stake in the overwhelming impact of the Anthropocene, which affects both the forest and the water.

If the river is crucial to the functioning of the planetary body that we all share for survival in a world marked by power politics and economic vocations, John Charles Ryan places it in the inner core of society and family, nurturing human relations with the river through the cultivation of its corporeality and sentience. Concerning Australian Aboriginal people's shared "custodial ethics of rivers shaped by ancestral connections to water," Ryan views the river not merely as relational, but as inter-relational too. It may not be far-fetched to suggest that this inter-relational engagement with the river is possible not just because of human capacity to empathize with it, but because the river itself has a voice and language that call for recognition. Using the lens of indigenous hydropoetics to examine several rivers in Western Australia, New South Wales and Queensland, Ryan argues that rather than presuming to give voice to rivers—and therefore inscribing an anthropocentric regard for rivers and water more generally—his conceptual framework acknowledges the inherent language(s) of rivers and, as a consequence, proffers a medium to listen to their voice(s), which transpires across time and space. Ryan's chapter, therefore, flags the deeper elements of the river's agency that cogently speak to that of the human.

As the imaginative and creative understanding of the human and non-human entanglement with riverspaces unfolds in the above-mentioned chapters, Rommel Curaming raises useful questions about the posthuman turn. Drawing from Dipesh Chakrabarty and Astrida Neimanis, he interrogates the complexities inherent in the posthuman trope. He observes that invoking a deep-ecological paradigm to explore the collective existence of humans and other-than-humans often fails to address anthropogenic destruction of nature. With reference to the Ganges, he notes that although many people hold a reverential attitude towards nature, aligning with planetary or deep historical perspectives, it appears insufficient to prevent ecologically-insensitive behaviour. Moreover, the formative role that humans play in confronting the desacralization of the sacred sites of nature like Ganges—for example in the form of offering legal personhood to rivers in different parts of the world—introduces further contradictions. Curaming asks: "If the posthumanist position needs human laws to instantiate and enforce, does it suggest the inutility of the posthumanist stance?"

The chapters in the first section take us to a realm of human-nature relationality as manifested in timeless, creative and embodied entanglements with river spaces. However, underlying this felt relationship, as these chapters show, lies a disconnect or rather "irrelationality" that has developed in the shadow of the Anthropocene. It is not to suggest that before the current spell of the Anthropocene human relationship with nature was pristine as there was considerable damage to the rivers since at least the mid-Holocene era (Roe and Iqbal 2025). But the level of disconnect affecting a large section of the riverine human and other-than-human population and their life and livelihoods has been unprecedented over the past few centuries, especially since the mid-twentieth century (McNeill 2000). How these seeds of irrelationality affect the mundane world of life and livelihoods is the subject of the second section of the book.

Md Mamunur Rashid's chapter offers a historical glimpse into the utilization of river networks in Northwest Borneo by riverine ethnic groups who developed a complex network between upriver inland ecology and the coastal trading centres as part of their life and livelihood. These river networks served as vital conduits for socio-economic development in the pre-colonial and colonial eras when roads and railways were absent. Consequently, the extensive riverbanks evolved into agroecological zones, spurred by ethnic mobility within the river-based commercial networks. Throughout the second half of the nineteenth century, cash crops began to be cultivated in river valleys initially by the inland ethnic groups and later by the coastal inhabitants such as the Melanaus in Sarawak. Upriver trading networks were established by the Malays and later superseded by Chinese traders. Due to the vigorous involvement of the Chinese in the timber trade, industrial zones and steam sawmills were established in Sandakan Bay and Rejang. However, Rashid also observes that, alongside the economic gains derived from rivers, gradual adverse impacts emerged. In the early twentieth century, extensive forest logging resulted in soil erosion, landslides, and river pollution in the region. As Rashid suggests, ecological livelihood gradually became dependent on market entitlements, which foreshadowed a more complex and precarious livelihood scenario in the postcolonial era.

Maria Alejandria's exploration of the marginalized community of Pasig river in Manila highlights their ongoing struggle to secure land titles and residency rights. The community faces major threats from flooding and river pollution, with women and children being particularly vulnerable to violence and food insecurity. Despite governmental recognition of these issues, root causes like excessive groundwater extraction and industrial projects remain unaddressed. Instead, slum settlements are often wrongly blamed for blocking tributaries and polluting the river. Government initiatives, including rehabilitation projects, flood risk management systems, and access to evacuation centres, often serve more to enhance politicians' public image than to genuinely benefit the community. Alejandria demonstrates that despite their vulnerabilities, the Pasig river community has adapted to the changing environment by developing individualized evacuation protocols based on perceived risks. Residents rely on personal networks to address safety concerns, ensure food security, and mitigate income disruptions. Flooding is no longer viewed as a threat by the community: children have become desensitized to the dangers, reducing their fear of floods, and residents perceive a "graysa" (blessing) in flooding, as debris collected can be turned into meals and income. This is a poignant testament to how people at the margin are resiliently negotiating the Anthropocene for their quest for life and livelihood, while the neoliberal logic of the state pushes its citizens further away from social safety nets.

Similarly, Mohammad Tareq Hasan unpacks multifaceted relationships between humans and the Chandpur river port in Bangladesh, a confluence of the Ganga and Meghna rivers. Like those on the Pasig river, inhabitants of this fluvial frontier encounter perennial precarity and marginalization in their pursuit of livelihoods, exacerbated by the gradual rise in sea levels.

Certain government-sponsored development projects have aggravated rather than mitigated these existential challenges because they are often designed to enrich the elites only. Nevertheless, the river port and its surrounding areas have evolved into a dynamic space where human mobilities embody myriad meanings, encompassing shelters, hopes, and aspirations for a living. For Hasan, the river by the port is a subject in its own right. Rivers epitomize a holistic social phenomenon integral to economic activity, livelihoods, social sentiments, occupational patterns, governance, and politics. The intertwined relationships between people and the river contribute to the construction of a "space" imbued with a profound sense of place, belonging, and interconnectedness.

Maya Dania proposes a revaluation of human-nature dynamics amidst the complex anthropogenic ecological disruptions in the Chiang Khong district in northernmost Thailand. Here, the Mekong river serves as vital transboundary water resources between Thailand and Laos, hosting the critically endangered giant Mekong catfish in the ancestral community lands. However, Chiang Khong district is also a nodal point of China's Belt and Road Initiative's (BRI) investments, exemplified by the Xayaburi dam in northwest Laos. The dam, the first operating hydropower energy project constructed in northwest Laos, has worsened the environmental consequences, including the loss of flood pulses, floodplain habitats, wetlands, and crucial sediment transport. Amid these disruptions, Dania explores the concept of tentacular thinking in "Making Kin," inspired by Donna Haraway, emphasizing interdependencies in a "natureculture" narrative that transcends anthropocentrism. Dania underscores the urgent need to revive both material and spiritual cosmos, such as Naga mythology and cultural folklore, to sustain ecological relationships.

If the river spaces in the Philippines, Bangladesh and Thailand exemplify the dynamics of subaltern adaptability to new political-ecological challenges, how do some of the spaces that are still relatively less impacted by the Anthropocene shape livelihood efforts? Focusing on the river spaces cutting through one of the remaining rainforests in Temburong district of Brunei on the Borneo island, Magne Knudsen and Misa Juliana Minggu offer insight into that issue. They explore the livelihood, food security, and socio-cultural importance of freshwater river fisheries within the Iban community. They illustrate that Ibans, as the main resource users of the Temburong river, possess specialized knowledge about the riverine environment and skills crucial for sustainably navigating and accessing remote fishing grounds such as Wong Wan. This specialized knowledge includes an excellent command of the temporal aspects of freshwater river fisheries, cognitive maps of the river system, key locations, and topographical features. Knudsen and Minggu also highlight that the ability to catch fish species such as Ikan Semah or Tor Tambra, which holds significant cultural value among the Ibans, is reserved for those with specialized knowledge of fish, fisheries and river ecology of the Temburong river. The exclusivity of catching Ikan Semah in Wong Wan elevates the fisher's prestige. Hence, a bountiful catch of Ikan Semah is easily translated as a successful mudik expedition, bringing much prestige to the fishers. Given this, the habitat of Ikan Semah is significant in defining the Ibans' relationships with their riverine environment as well as their cultural practices. The relationship between human and nature is thus recognized by the sense of prestige that it brings as a way of livelihood. Unmediated by extreme anthropogenic hazards, livelihood is not just bare existence; it is a process that attains legitimacy by the non-human subject on which it depends.

In Australia, within a socio-economic context that is distinct from those of the Philippines or Bangladesh, the river's connection to everyday life and livelihoods remains equally important. Through the lens of green social work and drawing on the Indigenous worldview, Heather Downey analyses river imaginaries and diverse aspects of hydro-social relationships within the rural communities of Mildura and Albury/Wodonga and their connection with the Murray river in Southeastern Australia. One new area that Downey explores is the community's deep engagement with the river through recreational activities. In all other respects, the Australian experiences reflect similar sense of attachment to and anxiety about riverspace. Downey navigates the community's concerns amid the impacts of La Niña and El Niño, anthropogenic climate change, blue-green algae, overuse of water for irrigation, fish population health, and commodification. Yet, despite current anxieties and future uncertainties in the riparian communities, Downey sees the water-space providing a "psychological buffer in a rapidly changing environment, opportunities for incidental social interactions, capable of binding communities together."

The chapters in this section illustrate a constant tension between the riverine life-world and its anthropogenic nemesis that has gathered force in recent times. The perennial bond that has marked human-river relationships and the persistent longing for a return to its mystery and material endowment cannot be fully severed. However, the strains of this relationship are too evident to be ignored in our time. An urgent question, therefore, is how to navigate the forces beyond the abstract trope of the "Anthropocene" that trigger and perpetuate irrelationality in nature. The next set of chapters aspires to address this question.

The first chapter in the third section by Ta-Wei Chu and Paul Carnegie reflects fieldwork experiences studying dam-induced challenges in northeast Cambodia at the Sesan river site. It highlights how representatives of a country's state-capital-development nexus can significantly influence the primary aims and rationale of conducting transdisciplinary projects in the Global South. The chapter employs a "social conflict-cum-politics of scale" lens to detail how eliciting and integrating knowledge co-production into project aims and outcomes is especially vulnerable to a state's political-economic dynamics. It foregrounds the political-economic agency of actors in a riverine livelihood study and the substantial influence certain actors can exert on project outcomes. The problem of governance here is exposed in the way state-capital-development nexus entangles with the local knowledge base around such major water-related projects as dam construction. Perspectives from the local knowledge base are often lost in the discussions about development projects affecting water spaces. The result is, as Chu and Carnegie conclude: "When dominant ruling coalition actors, as evidenced in Cambodia, gain ascendancy, it can severely constrain *bona fide* attempts at knowledge co-production for problem-solving."

The exclusion of the knowledge base of primary stakeholders can result in major policies and governance frameworks that are insensitive to the broader ecology of water space, as shown in the next chapter. The chapter by Sk. Tawfique M. Haque and Md. Parvez Hasan Yousuf explores the idea of "ecological governance" of rivers within both national and transnational contexts of India and Bangladesh, focusing on Padma (Ganges) and Teesta rivers. By "ecological governance," the authors refer to a comprehensive approach to managing water resources that considers the needs of both human and other-than-human entities that define and depend on these water resources. The primary concern is the water supply that meets the needs of both countries. Regarding the sharing of transboundary water space, they address issues such as the Farakka Barrage, which has deprived Bangladesh of its legitimate share despite a treaty, causing significant ecological impact on the country. This critique of the transboundary water situation is complemented by a discussion on the national legal framework of water governance in Bangladesh. They highlight the flaws in the Bangladesh Water Act 2013 and National Water Policy (1999), attributing these deficiencies to excessive bureaucratic power, corruption, lack of public engagement and inter-institutional cooperation, limited access to river use by certain stakeholders, poor management, and a scarcity of studies or empirical assessments. While the chapter critiques the inherent deficiencies in the governance of rivers in both international and national contexts, it also underscores the inadequacy of the governance framework in addressing the roles of other-than-human actors that shape the waterspace. Haque and Yousuf thus probe how "ecological governance" remains detached from anthropogenic governance.

The disconnect between "ecological" and "anthropogenic" governance that characterize development projects like large dams is further explored in the chapter by Andrew Alan Johnson. Johnson describes the drastic alteration of the hydrology of the Mekong river due to dams upstream in China and Laos, including Tonle Sap Lake and the deltaic region further downstream. Fish and plant species have declined, soil nourishing silt being trapped by the dams, leading to erosion and increased salinity due to the fall of fresh water level. But beyond these hydrological and ecological issues, what drives Johnson's exploration is the pluriverse of the people facing these challenges more intimately. In the experience of those living alongside the great river, these impacts amount to an existential crisis—personally, as one's position vis-à-vis the landscape and one's kin, both human and nonhuman, is altered. Johnson charts the intimacies of these changes, how declines in fisheries lead to greater importance on cultivating networks of kin living abroad, or in the introduction of new spiritual forces that govern power and potency on the great river. In the wake of the river's decline, where the potential that it once held is gone, how can new potencies be unveiled?—asks Johnson.

Johnson's question is, in some ways, addressed in Ryoji Soda's chapter, which explores the Japanese river improvement strategies on the traditional craftsmanship and technology used for flood management. The term "traditional" implies a sustainable approach that does not involve perceiving people as having absolute power over the natural environment, especially the rivers. Soda explores the evolution of technologies and craftsmanship in river improvement, aiming to understand how local and context-specific attributes shape river engineering. Soda argues that there is no binary distinction between technology and craftsmanship. He also notes that while rivers have agency, they are also affected by the people who depend on the river and their unpredictable behaviour. For a more effective river rehabilitation or engineering, understanding the relationship and interaction between people and the river is important. Soda explores traditional construction methods and the various technologies and craftsmanship used in these techniques. He emphasizes viewing the river as having unique characteristics based on their different locations, necessitating localized river engineering as part of best practices for river management and governance. For example, Soda argues that transplanting Dutch engineering methods or technology to Japan's river issues or flood control techniques is not part of best practice. Marrying tradition and modernity in river technology and the need for place-dependent localized technologies and craftsmanship appears as the best solution for the sustainable life of rivers.

The concluding chapter by Gabriel Yong, Izni Azrein bin Noor Azalie, and Noor Hasharina Hassan explores the interplay of tradition and contemporary ecological challenges in Brunei Darussalam. They trace the nation's history from its riverine roots to recent environmental challenges. The authors describe how Brunei's "Water Village" evolved, emphasizing its vital role in daily life, commerce, and settlements. They argue that modernization has shifted Brunei's landscape and culture from its riverine networks to land-based urbanization driven by the oil and gas industry. This shift has diminished the significance of traditional water networks, now reduced to storm drains and inconspicuous channels. The impact of climate change, including sea level rise and unusually heavy rainfall, is leading to the inundation of riparian zones. However, ironic as it is, the authors argue that climate change, via sea-level rise, appears to pave the way to a return to the water world that people had long left behind in their quest for land-based progress and prosperity. The authors seize this opportunity to propose a "blue urban development" plan, merging traditional and modern practices, and envisioning a society readapting to life with water. For the authors, thus, adaptation is not merely a policy trope, but a process that speaks to the deeper human cultural, social, and economic entanglement with the riverspace.

Rohan D'Souza's epilogue encapsulates the book's themes, emphasizing the need to transcend the conceptual obsession with the river as merely a resource or an object of scientific and technical experiments. He suggests that the Anthropocene demands a departure from conventional environmentalism focused solely on achieving "sustainable development." Instead, we must reconsider what "environmental conditions enable human survivability in an altered earth system." To this end, he echoes the broader call of the other authors to explore the river's nuanced and profound relationship with human imagination, culture, and daily life.

The chapters in the book cover a large space spanning the Asia-Pacific region, focusing on areas that have experienced different social, political, and economic processes, including Bangladesh, India, Cambodia, Vietnam, Japan, and Australia. Despite these varied spatial and temporal contexts, anthropogenic footprints have bound these regions together, representing a general scenario across the region that includes climatic issues, scarcity, drought, and a range of other precarity. Each chapter mentions the trope of the Anthropocene in its specific context, transcending individual differences. These disparate places are also bound by the memories and history of intimate entanglements, as seen in creative imaginations, myths, and embodied experiences in the aqueous spaces—what we have termed the "fluid phenomena." How riverine livelihoods, governed by elites and bureaucracy empowered by codes of capitalist praxis, speak to these phenomena—and are responded to by an array of existentially alive and place-bound human and other-than-human agencies—has been a significant part of the book's exploration. These experiences in the Asia-Pacific are as deeply rooted locally as they are globally resonant within the many meanings of the Anthropocene.

Interweaving conversations about the Anthropocene through the varied manifestations of the riverworld, the book invites meaningful reflection on the place and predicaments of the Anthropocene. Each chapter is not merely situated within its own geographical, intellectual, or methodological comfort zone; rather, they come together in a shared conversation, connected through multiple nodes provided by the concept of the Anthropocene.

In Chapter 5, John Ryan explores "two aspects of an Indigenous hydropoetics—corporeality and interrelationality"—which, he argues, are especially vital for articulating the concept's potential in responding to the urgencies of the Anthropocene. For Ryan, Indigenous approaches to river management are grounded in "reciprocal relationships" between people and the other-than-human world. His contribution, embedded within cultural studies of rivers, thus connects directly to the question of river governance.

Willem van Schendel, in Chapter 2, describes the Anthropocene as a "very influential notion," albeit a "quite unstable" one. Still, he finds it a useful lens through which to examine how "non-participatory development" has contributed to the degradation of riverscapes. This observation applies as much to the Karnaphuli river as it does to numerous rivers across South Asia and beyond.

In Chapter 3, David Ludden links the Anthropocene to the "spatial inequities of capitalist economic development," showing how its effects take root unevenly across regions. He further explores how poetry offers a voice to issues of food, livelihood, and ecological dispossession under Anthropocene conditions.

The first section thus, while grounded in the environmental humanities, engages with the disciplinarily distant questions of water management, food and livelihood, spatial injustice, and issues of development and governance.

Similarly, the chapters in the second Part, 'Everyday Life and Livelihood,' finds nuanced resonance with the chapters in Part 3, 'Development and Governance.'

For example, whereas in Chapter 7 Alejandria highlights how residents of a Manila riverspace rely on grassroots resilience to navigate an evolving urban environment in the absence of robust institutional support, she also points to other chapters in Part 3 (Yong, Azalie, and Hassan; Chu and Carnegie) that reveal contrasting dynamics of the state involvement. In doing so, she underscores the dual possibilities of the Anthropocene—as both a reductive force that marginalizes communities and a constructive one that enables new forms of engagement between state and society.

While Downey, in Chapter 9, in discussing the wellbeing of humans under the fluid umbrella of rivers in Australia, refers to the similarities in which colonialism reached the interior in other regions via rivers (Arnold; van Schendel), she also connects to the chapter by Ryan from Part 1 by suggesting that the river is central to "ways of being," and therefore, river health and the health of Indigenous peoples are intertwined. Noting that "mobility is integral to place," particularly in the context of river-places, Hasan's chapter finds resonances with Maya Dania's chapter, which explores the web of relationships that the mobile riverscapes forge. In both cases, the question of livelihood is central to these relationships.

Grappling with the hydropower question that transforms river ecologies and alters phenomenological pathways of 'becoming', Andrew Alan Johnson connects his analysis to the ways in which Daud and Ryan, in their respective chapters, foreground imaginations of an other-worldly world within the unstable chasm of the Anthropocene. Through his interlocutor, Johnson thus discerns how the question of hydropower may meaningfully intersect with that of hydropoetics.

The material question of river improvement—and even river engineering—according to Soda, "requires a heightened level of consideration for human–nature interaction compared to other infrastructure development, fostering a fluid relationship between the two." This effort to engage a formative aspect of Anthropocene awareness links Soda back to the conceptual question of the river's agency, as suggested in the Introduction and explored in several chapters across the sections.

These cross-references among the chapters across three disparate thematic clusters suggest that dialogue between the environmental humanities and the social sciences is not only possible, but also carries practical implications for how the Anthropocene reveals itself in the context of the riverworld—both as a conceptual category and as a methodological approach. This book is an exercise in bridging these disciplinary divides, offering new insights into both the fractures and interconnectedness of the fluid phenomena at the intersections of rivers, humans, and other-than-human entities.

In shaping the conceptual framework of this volume, we take seriously the multiple and sometimes competing valences of the term *Anthropocene*. While the term threads through the chapters in diverse ways, we recognize the need to clarify its role as more than a passing reference to the environmental crisis. For this collection, we treat the Anthropocene as encompassing at least two distinct, yet interrelated, orientations. The first (*Anthropocene 1*) is grounded in the recognition of humanity as a dominant physical force in shaping planetary systems, a condition evidenced in phenomena as varied as altered hydrological cycles, accelerated species loss, and breaches of planetary boundaries. The second (*Anthropocene 2*) frames the Anthropocene as a provocation to self-critique, exposing the paradoxes of modernity in which humans—especially in many parts of the Asia-Pacific—are at once agents and casualties of ecological transformation. This second orientation invites responses that foreground repair, care, and the renewal of relations among human and more-than-human actors.

By distinguishing these dimensions, we hope to avoid a reductive binary between science/reason and sensitivity/affect, and instead offer the riverspace as a conceptual and material terrain through which both orientations can be examined. Rivers, in their physical and cultural life-worlds, embody the entanglements of Anthropocene crisis and critique: as sites of extraction, control, and ecological rupture, and as corridors of memory, sustenance, and resistance. They render visible the forces that drive environmental change while also hosting possibilities for alternative ways of living, sensing, and governing.

In this light, "riverspace" is not only a descriptive category but also a prototype for rethinking the more-than-human world. It accommodates varied intellectual traditions—indigenous cosmologies, eco-Marxist critiques, posthumanist theory—while remaining attentive to the political economy of water and the historical specificities of riverine societies (Iqbal 2025). By situating the contributions in dialogue with ongoing debates over the Anthropocene, the Capitalocene (Moore 2015) and the Chthulucene (Haraway 2016), we aim for a framing that is both regionally grounded and planetarily resonant.

To reiterate the central argument of *Fluid Phenomena*, our recognition of the human propensity to unleash destruction upon the river spaces—and other waterbodies—is matched by our enduring faith in the human capacity to heal and restore.[1] However, in the past few centuries, often characterized as 'modern,' the force of healing and restoration has largely capitulated to the force of unsustainable interventions. The failure of this formative encounter with nature lies not always necessarily in modernization, development, or the neoliberal protocols of governance per se, but in the tendency of these forces to fear or disregard the boundless creative and cultural repertoire of humanity, which we would name "Environmental Humanities" in the context of the book. The chapters in the book show that it is possible to foster conversations between material growth and ecological sensibilities, between institutional rationalities and cultural imaginations, thereby opening up space for more sustainable engagements with the riverine and aquatic world, where an Anthropocene of hope prevails over an Anthropocene of destruction. If *Fluid Phenomena* serves this purpose even in the smallest measure, we would regard our efforts as meaningful—and our collective journey worthwhile.

References

Armstrong, Christopher, Matthew Evenden, and H. V. Nelles. 2009. *The River Returns: An Environmental History of the Bow*. Montreal and Kingston: McGill-Queen's University Press.

El-Bizri, Nader. 2000. *The Phenomenological Quest between Avicenna and Heidegger.* Global Academic Publishing.

Haraway, Dona J. 2016. *Staying with the Trouble: Making Kin in the Chthulucene*. Duke University Press.

Iqbal, Iftekhar. 2025. *The Range of the River: A Riverine History of Empire across China, India, and Southeast Asia*. Stanford University Press.

Khan, Naveeda. 2023. *River Life and the Upspring of Nature*. Duke University Press.

Klaver, Irene J. 2021. "Radical water." In *Hydrohumanities: Water Discourse and Environmental Futures*, edited by Kim De Wolff, Rina C. Faletti and Ignacio Lez-Calvo. University of California Press.

Knoll, Martin, Uwe Lübken, and Dieter Schott, eds. 2017. *Rivers Lost, Rivers Regained: Rethinking City-River Relations*. University of Pittsburgh Press.

---

1 Two important examples of such interplay of decline and revival of river spaces are: Christopher Armstrong, Matthew Evenden, and H. V. Nelles, *The River Returns: An Environmental History of the Bow* (Montreal and Kingston: McGill-Queen's University Press, 2009) and Martin Knoll, Uwe Lübken, and Dieter Schott, eds., *Rivers Lost, Rivers Regained: Rethinking City-River Relations* (Pittsburgh: University of Pittsburgh Press, 2017).

May, Reinhard. 1996. *Heidegger's Hidden Sources: East Asian Influences on His Work.* Translated by Graham Parkes. Routledge.

McNeill, John Robert. 2000. *Something New under the Sun: An Environmental History of the Twentieth-Century World.* ww Norton.

Moore, Jason W. 2015. *Capitalism in the Web of Life: Ecology and the Accumulation of Capital.* Verso Books.

Rapoport, Yossef, and Emilie Savage-Smith, eds. and trans. 2014. *An Eleventh-Century Egyptian Guide to the Universe: The Book of Curiosities.* Brill.

Roe, Alan, and Iftekhar Iqbal. 2025. "Riverine Environments." In *A Companion to Global Environmental History*, 2nd ed., edited by John Robert McNeill and Erin Stewart Mauldin. Wiley Blackwell.

PART 1

Reimagining the Other-Than-Human

CHAPTER 1

Rivers as Representational Tropes in the Tropicalization and Modernization of India

David Arnold

Recent decades have seen a dramatic rise in the historical study of seas—the "new thalassology" as it has been described, or the "oceanic turn." From the Mediterranean to the Atlantic to the Indian Ocean and the Pacific, seas have generated an ambitious and dynamic history that reaches far beyond the bounds of conventional maritime studies. In this innovative historiography, seas have become connecting bridges instead of barriers or boundaries; they present the possibility of movement over vast distances and fluidity between places and cultures; they are the bearers (but also the re-shapers) of commodities, people, and ideas; they serve as a means to explore climatic influence and track environmental change. In the now distant wake of Fernand Braudel, the new scholarship of the seas propels us beyond superficial, short-term histories; it questions our obsessions with national histories and with individual imperial regimes and land-based governance (Ghosh and Muecke 2007; Gupta, Hofmeyr, and Pearson 2012; Horden and Purcell 2006; Vink 2007). But, in the context of this exploratory volume, one might wonder whether the historical and cultural study of fluviology has kept pace with the innovations and challenges of the new thalassology. And, more specifically, has our understanding of Asian river history undergone a comparable re-imagining?

While rivers played little part in Braudel's magisterial account of the Mediterranean, first published in 1949, his later work on the identity of France afforded rivers greater prominence. There they illustrate the impact of the physical environment on human activities across an extended time period; they help build the geopolitical character of France, but also constitute obstacles to trade, communications, and administrative integration (Braudel 1978, 1989). Rivers divide as well as unify and hinder movement as well as facilitate it. Their physical character and political significance can shift over time, so that they become, especially in the modern period, makers and markers of anthropogenic environmental change.

This chapter addresses the question of rivers and their historical significance in a very different context from that explored by Braudel. It focuses on the Indian subcontinent during the period of British colonial rule from the

© 2026 THE AUTHOR(S) | HTTPS://DOI.ORG/10.1163/9789004758100-003

mid-eighteenth to the mid-twentieth centuries. It is concerned with the riverine world not of temperate, continental Europe but that of the tropics and sub-tropics and with tropicalization and modernization as salient tropes in the colonial engagement with, and anthropogenic reconfiguration of, that vast and diverse region. The example of South Asia, specific though it might be in many respects, can help raise questions and suggest lines of analysis of wider value, not least to the fluvial history of Southeast Asia that is this volume's primary focus.

There are, of course, many different ways of addressing and assessing the Anthropocene, its nature, meaning, and consequence; but rivers and river systems surely constitute one of the most momentous illustrations. In South Asia the human impact on riverine geography and ecology is as ancient, extending back over millennia, as it has been profound and far-reaching. But—to an extent that the Braudelian model fails to recognize—the human-wrought transformations of the past two centuries, corresponding to the colonial era and after, have been singularly rapid, intense, and, to a large degree, irreversible. Physical change—the rerouting and canalization of rivers, the building of dams and bridges, the harvesting or persecution of riverine life, the spread of riverside cities and the reckless pollution of waterways—has been a major part of this riverine transformation, but so too have ideology, greed and ambition, and the urgent desire for anthropogenic change, often articulated through governing notions of tropicality, "improvement", and modernity, that have so often been the driver behind material change.

1 River History and Symbolic Landscapes

Rivers have long served to articulate India's identity and antiquity. The Ganges, Indus, Brahmaputra, Narmada, Mahanadi, Krishna, Godavari, and Kaveri are names that resonate as much for India as the Loire, Seine, or Rhône do for France; the cultural, political, economic, and ecological axes that they represent have long shaped the conceptualization and narrativization of India (Coningham and Young 2015). The archaeological discovery of the Indus Valley Civilization in the 1920s opened up a new perspective on India's fluviocentric history, but even before that, the Indus had long held significance as marking the topographical and geopolitical boundary between India and its western neighbours, even endowing India (as the land beyond the Indus) with its historic name. The Ganges, too, has been seen from remote antiquity as the central strand in the making of Hindu civilization just as reverence for this and other sacred rivers has played an essential role in Hindu rites and sacred geography (Eck 2012).

The writing of history through the story of a region's paramount river and its environs is hardly a novelty, the fluvial flow giving direction and form to the narrative stream of history itself (Albinia 2008). Indeed, a great deal of colonial as well as precolonial history focused on what W. W. Clemesha (1912, 2) dubbed the "leviathan" rivers of South Asia. Early in the twentieth century, the imperial geographer, T. H. Holdich, paid tribute to the Ganges, observing that its valley was in certain respects "the most important physical feature of India. Within the basin of the Ganges have ever been founded the chief kingdoms of the plains; the most ancient cities; the earliest centres of civilization, of industry, and of wealth" (Holdich 1909, 22). "There is not a river in the world", he further claimed, "which has influenced humanity or contributed to the growth of material civilization, or of social ethics, to such an extent as the Ganges" (26). This emphasis on the historical centrality of South Asian rivers has been reiterated many times, for instance by Sudipta Sen in his recent account of the Ganges and the merits of "thinking with a river". He argues that the Ganges was "indispensable to thinking about the history and culture of the Indian subcontinent", "a reflexive extension of something akin to a uniquely Indian consciousness" (Sen 2019, 6).

The Ganges, like many other Asian (and European) rivers, has given rise to a "rich and boisterous mythology" (Sen 2019, 8), a "hydro-mythology" in Simon Schama's felicitous phrase (1995, 277). But in their cultural significance, as in their physical form, rivers are far from timeless and immutable. Braudel, who died in 1985, did not live long enough to engage with the Anthropocene debate that has now become so central to environmental history. And yet even Braudel's rivers, which seem at first to be locked into the slow evolutionary flows of long-term symbiosis between peoples and places, are seen to change significantly with the coming of canalization, new modes of river transport, and, latterly, the building of dams and power stations along the Rhône (Braudel 1989, 281, 297). Likewise, but arguably more radically and rapidly, from the late eighteenth century, British colonialism ushered in a new era for India's principal rivers, novel in its degree of mastery, repurposing, and exploitation. In a development that paralleled European intervention in the Americas, Africa, Southeast Asia, and China, Indian rivers were viewed by colonizers as a vital means of "opening up" a continental interior, facilitating access to its material and human resources, its centres of wealth and power, even its history. Early colonial mapmaking in India, notably by James Rennell, reflected the perceived centrality of the subcontinent's major river systems and their importance to the delineation and appropriation of the surrounding terrain (Edney 1997, 98–104).

This was especially so for Bengal, the principal site of colonial ingress into India. Lacking roads and other distinctive topographical features, the Ganges and its distributaries provided the incoming power with an indispensable cartographic grid (Seton-Karr 1848, 1–3). Madras and Bombay lacked significant river access and had to await roadbuilding and railway construction to extend and consolidate their subregional dominance. But, as the epicentre of British rule in India until 1912 (its loss of capital status a mark of how river power had yielded to a new land-based geopolitics), Calcutta was the main port of disembarkation for arriving Europeans, a global trading hub and Indian Ocean entrepot, and a vital centre for British administration and political control. The Hooghly, which flowed past the city, facilitated entry upriver to the vast and populous Gangetic basin. At a time when transport by road was slow and haphazard, the Ganges remained, as it had been in precolonial times, a strategic axis, linking the Bay of Bengal to the interior towns and cities of northern India and beyond into Central Asia.

Unlike some Asian regions, such as Burma (Headrick 1981, 20–21) or Vietnam, India's rivers played little part in colonial conquest and territorial acquisition, though they helped shape the British preference for the fertile and productive river valleys and coastal plains over the dry, famine-prone, interior districts. But there were other kinds of conquest, "deep conquest" in Neeladri Bhattacharya's apt phrase (2019, 1), that rivers enabled, including colonization by technological means. In the mid-nineteenth century, the three main rivers of India—the Indus, Ganges, and Brahmaputra—opened up the prospect of steam navigation, with vessels operating from coastal or deltaic ports to distant upcountry locations (Bernstein 1960; Dewey 2014). For a while, in pursuit of cotton or coal, other river routes were also investigated: in the 1870s, for example, steamers even plied the Godavari during the monsoon months when the river ran high (Madras Revenue Proceedings 1877). However, despite initial optimism, even the Ganges (with 700 miles of theoretically navigable river between Calcutta and Mirzapur) proved difficult for steamboats to ply profitably and reliably, given the many obstructions to the river's course and the wide variation between dry-season and monsoon river levels. Instead, the railways stole much of the erstwhile river traffic along the Ganges and its main tributaries (Derbyshire 2022). Further, rather than following the familiar and long accustomed course of the rivers, many railways in northern and eastern India, aided by ambitious bridge-building, boldly cut across the fluvial axes (Berridge 1969, 43–53), forging new patterns of cross-country trade, labour migration, and political control. Steam navigation on the Brahmaputra proved more durable, however. With the development of tea estates in Assam from the 1860s onward, migrant labour, largely recruited from central India, moved upriver just as the harvested tea travelled downstream on steamers to Calcutta.

In another, more culturally accented, mode of conquest, it was also along the Ganges, its tributaries and distributaries, that many European travellers moved in the late eighteenth and early nineteenth centuries. Some were engaged in artistic reconnaissance or scientific exploration, others in writing travelogs, in missionizing, or simply sampling the sights and pleasures of upcountry life. The cities that they visited or passed through—Patna, Benares, Allahabad, and ultimately, via the Jumna, Agra, and Delhi—formed the essence of the Indian "Grand Tour" (Heber 1828; Roberts 1835; Tennant 1804). If the river afforded safe and privileged access to the observer, there was also a sense of political and cultural progression to this upriver voyaging—a sense of moving (as Europeans construed it) from the "civilized" world of Calcutta to the more alien and unfamiliar interior. This reading of South Asia upstream, from the sea to the source, was the opposite of the dominant indigenous idiom, for precolonial regimes whose power extended down the Ganges valley, ultimately to Bengal, and for Hindus who traced the descent of the Ganges from its Himalayan origins to the distant ocean and whose downstream passage was celebrated through a series of holy sites from Hardwar to Sagar Island. Hindu sacred geography and colonial penetration travelled in opposing directions, though modern writers, eschewing colonial proclivities, have tended to reinstate the downstream narrative (Newby 1983; Sen 2019).

2 Tropical Landscape and Scenery

The riverine orientation of early colonialism gave rise to an aquatic equivalent of the symbolic landscapes discussed by geographer Denis Cosgrave (1984). Riverscapes, like landscapes, were a culturally distinctive way of seeing and interpreting the observed world. Between the 1760s and 1860s the upriver scenario helped to frame the way in which Europeans conceptualized eastern and northern India, especially the passage through the sluggish waterways and dense jungles of the Sundarbans *en route* to Calcutta. Even while not scenically and topographically representative of the drier, cultivated plains of the interior, this visual antechamber informed a typification of lowland India as "tropical", with all the exotic allure and lurking danger, the pleasure, and the pestilence such an idea increasingly connoted in the Western imagination (Arnold 2006; Driver and Martins, 2005; Stepan 2001). Early in the nineteenth century, R. H. Colebrooke described the riverbanks of deltaic Bengal as being "over-run with thick jungles of long grass which are the usual haunts of tygers, wild buffaloes, and other animals" (Colebrooke 1808, 29). In a similar vein, in 1846, the Sundarbans were described as being "in their present wild, jungly state, the abode of tigers, rhinoceroses and alligators" and "the seat of malaria" (Long 1846, 249). "Jungle", wild beasts, and "malaria" were among the many prejudicial tropes used to typify this seemingly morbid, featureless, and "primitive" landscape.

Once beyond this gloomy tropical portal (Duff 1839, 200–205), impressions of riverine India might change significantly. It was not uncommon, for example, for visitors to remark on how the European villas that lined the banks of the Hooghly on the approach to Calcutta resembled riverside dwellings near London, or how, through a common association, the river at this point was "like the Thames at Westminster". Extending an English landscape aesthetic to India, pictorial views of the Hooghly and Ganges formed an essential ingredient in the romantic scenes and "picturesque" representation of eastern and northern India. An influential example of this was the Daniells' aquatints of the 1780s and 1790s, showing the water teeming with indigenous rivercraft, the banks lined with temples, mosques, forts, and palaces (Archer 1980, 57–71). This was a celebration of the aesthetic appeal of water rather than a demonstrably tropical trope, but, even by the 1830s, it was being supplanted by an increasing insistence on India's overarching tropicality. Thereafter, and for more than a century, the trope of the tropics, that expansive, colonizing toponym, informed Western perceptions of almost every aspect of Indian life. Rivers were no exception.

In the 1870s, following his long sojourn in Amazonia and Southeast Asia, the naturalist, Alfred Russel Wallace, wrote of the tropical world: "The luxuriance and beauty of Tropical Nature is a well-worn theme, and there is little new to say about it. The traveller and the naturalist have combined to praise, and not infrequently to exaggerate the charms of tropical life—its heat and light, its superb vegetable forms, its brilliant tints of flower and bird and insect" (Wallace 1878, VII). Like many contemporary naturalists, Wallace was a tropical enthusiast, yet he was also aware of the less attractive side of "tropical nature"—the oppressive heat and humidity, the violent rainstorms, the gloomy forests, the irritating insects, the deadly fevers. This negative representation of tropical difference was arguably not without precedent, even for India. Muslim writers had long suggested that moving down the Ganges into Bengal represented a marked shift in topography, vegetation, and climate. This was a land of tigers, pestilence, and swamps, far more inimical to outsiders from the north and west than the more arid and temperate lands to which they were historically accustomed and culturally attuned (but for the political and religious incorporation of deltaic Bengal, see Eaton 1993). What is evident, however, is how important the idea of the tropics became not just to Western perceptions of an alien environment but to the transformative agenda of "improvement" that accompanied the establishment of colonial rule.

There was a significant duality to tropicality. On the one hand, for Europeans, it signalled a return to a primordial world, in which nature still ruled, in which modern agriculture and urban civilization barely impinged. But it also represented a programme for change, an opportunity to apply, under Western tutelage, the benefits of modern science, medicine, and technology to unlock the potential wealth and productivity of the tropical domain. In this latter sense, tropicality stood on the cusp of anthropogenic change: it looked back, sometimes wistfully, often in dismay, at a kind of Edenic primitivism; but it also looked forward, prospectively, to the coming of the modern, anthropocentric age. Rivers were a marker of what was and what might yet be.

John M'Cosh's *Topography of Assam* embodies the growing perception of India's scenic and pestilential tropicality and the centrality of the river, here the Brahmaputra, as the primary topographical feature of the region and the key to its scientific investigation and imperial incorporation. In travelling upriver—a slow, tedious, and dreary journey by his account—he observed a landscape covered by "a jungle of gigantic reeds, traversed only by the wild elephant or the buffaloe; where a human footstep is unknown, and the atmosphere even to the natives themselves is pregnant with febrile miasmas and death" (M'Cosh 1837, 13). Of such agriculture as he was able to observe, the author mused: "Assam, with all its wastes and jungles, however much neglected and abused by man, has not been altogether forgotten by nature in her distributing of the good things in life" (31). He had little positive to remark, however, either on "the natives" or the violence of the seasons. As to salubriousness, or rather the want of it, this, too, was read from a European perspective and with a view, prospectively, to environmental transformation and "material improvement". "It ought not to be forgotten," M'Cosh observed, "that all new countries in India, when first taken possession of by the British, have been found unhealthy; nor have they ceased to be otherwise, till a settled state of the country allowed them to abandon these sinks of human life ...; till too often dearly-bought experience enabled them to select stations more congenial to European constitutions; till the new cantonments were cleared of their native pestilential jungle, the swamps drained, and comfortable houses erected" (98–99). Meanwhile, the mournfulness of the landscape, as he saw it, was reflected in the melancholy sight of the tombs of the Britons who had died taking possession of this febrile, inhospitable country.

3 Engineering the Tropics

If ever there was a colonial "project" deserving of all that term might convey in terms of intentionality and scale, resource deployment, and consistency of purpose, it lay in schemes for the remaking of India's rivers. Under colonial rule, the tropical landscape, broadly defined, became a vast scheme of "improvement", aimed at converting what had previously, in the colonial estimation, been untamed land and underutilized fluvial resources into modern, productive commodities. There were, of course, abundant precolonial antecedents for a strategy of environmental management as it concerned South Asia's rivers and deltas, especially through the construction of anicuts, dams, and other irrigation works; but in the colonial narrative, these were often ignored or relegated to a more primitive technological plane. Control of water, especially given India's erratic monsoon regime, was already seen by the 1840s as essential to the "progress of tropical agriculture" in India (Smith 1849, 79). The disciplining of unruly rivers and the creation of vast irrigation schemes were among the things that defined British India as a "scientific empire" (Gilmartin 1994) and so, in Western eyes at least, set it apart from pre-colonial regimes. Engineering was the harbinger, the facilitator, of momentous change. Crossing the bed of the Soane river, the Ganges' second largest tributary, more than two miles wide, with an iron-girder bridge was indicative of the formidable scale of such transformative undertakings but one that was deemed essential for railway construction to proceed (Theobald 1847, 335).

Beginning in the 1830s, but with gathering momentum from the 1870s, the colonial project of hydraulic transformation further entailed the regulation and canalization of the Ganges, Indus, Kaveri, Godavari, Krishna, Mahanadi, and other river systems, the creation of canal colonies in Punjab, and the wholesale extension of rice, wheat, sugar, and cotton cultivation (Amrith 2018; Bhattacharya 2019; D'Souza 2006; Headrick 1998). Hydraulic engineering for the tropics became a highly technical enterprise. The history of South Asian rivers and deltas was increasingly one of regulation and containment, of ever-expanding extraction and widening utility (Walch 1899), an agenda accelerated and intensified in the years after Independence. Mastering the tropics through water management on this scale might draw upon techniques in part familiar in Europe, North America, and elsewhere (cf. Worster 1985). But it also had its regional or pantropical specificities, as in the indigenously derived (but greatly modified) technique of well-sinking to erect bridge piers across broad, unstable floodplains and shifting river courses, or in dealing with the extreme variations in river levels caused by monsoonal floods and the melting of the

Himalayan snows (Berridge 1969, 43–52). Such wholesale environmental intervention, such momentous exercises in the "science of river-training", resulted in the loss of riverine habitats, the precipitous decline of aquatic and riverbank flora and fauna, and the shrinking of rivers whose dwindling waters were now diverted to fed irrigated land (Berridge 1969, 52, 76). The Anthropocene had begun to bite.

Not only did rivers become capitalist commodities, resources conceived of and managed on a colossal scale; they also became the theatre in which colonialism staged some of its grandest works—the exemplary sites of engineering ingenuity and the riverine performances intended to curb famine, drive agrarian change, and fuel industrial transformation. There was, though, much that could (and did) go wrong in tropical hydraulic engineering. Canals might be rendered ineffective by scouring or heavy silt loads and require substantial remodelling to redress earlier mistakes (Crofton 1865). Aqueducts failed; floods destroyed embankments or swept away bridges and dams (Anon. 1888). Still more negative and of long-term consequence was the spread of malaria and malaria-carrying anopheles mosquitoes due to leaking canals or disrupted drainage (it was not only in "primitive" jungles that malaria abounded), and the large areas of land rendered uncultivatable by *reh*, the subsurface salts brought to the surface in perennially waterlogged soils (Klein 2001; Whitcombe 1995). By the interwar years, India seemed already to be entering an age of environmental crisis, not least with respect to its rivers and water resources (Klingensmith 2014).

The transformation of the tropics required more than grand schemes of physical engineering. It also called for social engineering on an almost equal scale of magnitude. In a development that both reflected and compressed Europe's fluvial history, the liminal terrain between land, river, and sea was seen as fit for productive reformation through dykes, embankments, canals, roads, railways, and harbour works. This entailed the displacement or annihilation of existing trades and occupations and the people (pastoralists, fishermen, boatmen, fowlers, river bandits) who pursued them and their replacement by cultivators from outside social and ethnic groups (Bhattacharya 2019; Bhattacharyya 2018). In deltaic India, as elsewhere, this particularly connoted the creation of new or extended rice lands, a monoculture designed to produce a commodity as much for export as for local consumption. Rivers might play a less direct role in the development of other tropical monocultures—tea, coffee, rubber, cinchona—but they were still instrumental in facilitating their export, the importation of outside labour, and even the use of rivers to extract otherwise inaccessible timber or float felled trees downstream to awaiting sawmills and ports.

Few regions of the colonized tropics were, in however abbreviated a form, entirely without their "white highlands". But, in general, the lowland tropics were seen as inimical to European settlement—whether for reasons of climate and disease, from unfamiliarity with the crops that such regions produced, or from distaste for the arduous labour their cultivation required. If the tropical "estate" were to be developed, it needed non-white labour in abundance—for massive construction works, for policing and administration, for trade and, above all, for cultivation itself. From early in the colonial era, prompted by the formal abolition of slavery in the 1830s, the British and other colonial powers began to recruit "tropical races" for this purpose (Indians, Chinese, Javanese) and to shuffle ethnically defined cohorts of labour within and between tropical territories in the service of land clearance, settled agriculture, plantation economies, even penal settlements. The tropicalization of labour was accompanied by often contradictory conceptual tropes—of laziness and industry, of resistance to disease and pathogen transmission, of the poverty or utility of tropical, especially rice-based, diets. Little of this was an exclusively riverine discourse, but rivers were a catalyst, a connecting thread, a central component in accessing and transporting labour and in the wholesale movement and transregional management of people, labour practices, diets, and pathogens.

It was one of the self-appointed tasks of colonial science to identify and classify the characteristic features of the Asian tropical environment, and rivers provided one of the primary sites or routes for doing so. The cartographic imperative of tracing the routes of major rivers and placing them on maps or using watersheds as a means demarcating territory, helped to create a visual grid onto which the essential concerns of colonial governmentality (rainfall, disease, crops, livestock, and land use) could be superimposed. What rivers contained might be as important as what rivers did. Here, too, was a place for tropical taxonomies. From early in the colonial era, there was an interest, driven partly by scientific curiosity and partly by economic opportunity, in exploring riverine ecologies and their constituent elements—the fish and animals that inhabited India's waterways, the rivers' salinity and silt loads, and their shifting courses and significance in terms of climatology, geology, and hydrology (on fish, see Day 1873; Hamilton 1822). By the late nineteenth century, Indian rivers were subjected to close and repeated scientific examination to determine their chemical content and bacteriological loads (Clemesha 1912). This was notably so on the lower Ganges and in the environs of an increasingly industrial Calcutta, to assess the extent of pollution and the dwindling suitability of river water for human consumption (Arnold 2016; Chakrabarti 2015; Wilhelm 2016).

In the new thalassology, the movement of disease has become a critical demonstration of oceanic interconnectedness (Campbell and Knoll 2020);

such might also be the case in riverine studies. In the Western perception, tropical rivers harboured pathogenic dangers that were different from (or more extreme than) those of rivers in more temperate locations. The explanations offered for this difference were both environmental and cultural. From as early as 1817, cholera, the ur-pandemic of the nineteenth century, was identified with deltaic Bengal, and from there it was seen to move, aided by monsoon winds and human migration routes, along the line of the Ganges valley to northern India, and to circulate around maritime Asia, ultimately becoming near-global in its reach (Arnold 2022). River routes could themselves become pathogenic highways, as with the epidemic cholera that repeatedly moved upstream along the Brahmaputra with the "coolie" steamer traffic from Calcutta to Assam. One could further extend this association of Asia's tropical rivers and their deltas with other diseases, too—with the incidence and spread of malaria, for example, or even the beriberi that accompanied the consumption of heavily milled deltaic rice (M'Cosh 1837, 99–102; Taylor 1840, 330).

4 Conclusion

In the age of the new thalassalogy, two kinds of aquatic metaphors inform the debate over rivers and seas. On the one hand, there is the idea of fluidity (Gupta 2010, 3–4), the notion that people, objects, and ideas move, relatively unhindered but sometimes undergoing subtle processes of change, between different times, places, and cultures. Things flow like rivers and circulate like ocean currents. History is itself, like the landmass of Eurasia or the vastness of the Indian Ocean World, a stream of narrativized connectivity (Subrahmanyam 1997). And yet, on the other hand, there is a sense of history as disjuncture, of moments of rapidly accelerated, perhaps irreversible, change, captured by the idea of an "ecological watershed" (Gadgil and Guha 1992, 116), dividing off the colonial from the pre-colonial world, or heralding the abrupt and irredeemable arrival of the Anthropocene (Chakrabarty 2021; Ghosh 2016). How we date the coming of the Anthropocene and the criteria by which we judge its arrival and impact remain matters of debate and contention. But one marker of profound and relatively abrupt ecological change, one index of the deeply transformative effects of anthropogenic intervention, might surely be found in the riverine history of South Asia in the watershed moment of transition from precolonial to colonial rule.

This discussion has suggested some of the ways in which fluvial studies might connect with, or learn from, the new thalassalogy, or at the least develop an enlarged historical understanding of the rivers of South Asia and elsewhere in the Global South. There are connections and insights to be gained from such a comparative perspective, aligning seas and their burgeoning literature alongside what we increasingly know of rivers and changing fluvial environments. But it is striking how different, even divergent, these two species of hydraulic studies might prove to be in practice. Fluvial studies, and especially what we might in this context term the study of tropical fluviology, address more directly issues of colonial governance and the performative shifts evident in the move from precolonial times to colonial histories and the transformative projects of colonial "improvement" and hydraulic engineering. From the perspective of the Anthropocene, the triumphalist narratives of hydraulic engineering written in, or in the immediate aftermath of, the imperial past now seem more questionable in their assumptions of progress and the benefits of industrial civilization (cf. Buckley 1905). Certainly rivers, like oceans, can speak to historical and spatial connectivity, to the Braudelian *longue durée*; but theirs is also a more entangled history, one more intimately and intricately bound up with shifting patterns of land use, resource extraction, and human migration, with environmental change, and the often rapidly shifting patterns of disease ecology. In colonial times, tropical rivers like the "leviathans" of South Asia became, to a quite unprecedented degree, a project—an agenda for science, technology, medicine, and engineering—in a way in which, or to a degree in which, seas have only recently become. Rivers serve as a measure of what the coming of the Anthropocene might mean and when (in relation to colonialism and the tropical world) that epoch-changing event might truly be said to have started. There are, of course, other ways and other timescales on which we can plot environmental change, not least with respect to the long-term uses of water in South Asia (Morrison 2014); but this chapter has tried to argue that the period of colonial rule brought about unprecedented change at both a cultural and conceptual level as well as in material practice. People changed rivers; rivers changed people. Transitions seemed more rapid and more drastic in and alongside rivers than in the slow-moving expanses and circulating currents of the oceans. Conflict, and what it connotes for the exercise or contestation of power, appears more pervasive, more ingrained, in fluvial studies than in its oceanic counterpart.

References

Albinia, Alice. 2008. *Empires of the Indus: The Story of a River*. John Murray.

Amrith, Sunil. 2018. *Unruly Waters: How Mountain Rivers and Monsoons Have Shaped South Asia's History*. Penguin.

Anon. 1888. *Failure of the Kali Nadi Aqueduct, Lower Ganges Canal*. Superintendent of Government Printing.

Archer, Mildred. 1980. *Early Views of India: The Picturesque Journeys of Thomas and William Daniell, 1786–1794*. Thames & Hudson.

Arnold, David. 2022. *Pandemic India: From Cholera to Covid-19*. Hurst.

Arnold, David. 2016. *Toxic Histories: Poison and Pollution in Modern India*. Cambridge University Press.

Arnold, David. 2006. *The Tropics and the Traveling Gaze: India, Landscape, and Science, 1800–1856*. University of Washington Press.

Bernstein, Henry T. 1960. *Steamboats on the Ganges: An Exploration in the History of India's Modernization through Science and Technology*. Orient Longmans.

Berridge, P. S. A. 1969. *Couplings to the Khyber: The Story of the North-Western Railway*. David & Charles.

Bhattacharya, Neeladri. 2019. *The Great Agrarian Conquest: The Colonial Reshaping of a Rural World*. Permanent Black.

Bhattacharyya, Debjani. 2018. *Empire and Ecology in the Bengal Delta: The Making of Calcutta*. Cambridge University Press.

Braudel, Fernand. 1978. *The Mediterranean and the Mediterranean World in the Age of Philip II, Volume 1*. Translated by Sian Reynolds. Fontana.

Braudel, Fernand. 1989. *The Identity of France, Volume 1: History and Environment*. Translated by Sian Reynolds. Fontana.

Buckley, Robert Burton. 1905. *The Irrigation Works of India*. 2nd ed. E. & F. N. Spon.

Campbell, Gwyn, and Eva-Maria Knoll, eds. 2020. *Disease Dispersion and Impact in the Indian Ocean World*. Palgrave Macmillan.

Chakrabarti, Pratik. 2015. "Purifying the River: Pollution and Purity of Water in Colonial Calcutta." *Studies in History* 31 (2): 178–205.

Chakrabarty, Dipesh. 2021. *The Climate of History in a Planetary Age*. University of Chicago Press.

Clemesha, William Wesley. 1912. *The Bacteriology of Surface Waters in the Tropics*. Thacker, Spink.

Colebrooke, R. H. 1808. "On the Course of the Ganges through Bengal." *Asiatic Researches* 8: 1–31.

Coningham, Robin, and Ruth Young. 2015. *The Archaeology of South Asia: From the Indus to Ashoka, c. 6500 BCE–200 CE*. Cambridge University Press.

Cosgrave, Denis E. 1984. *Social Formation and Symbolic Landscape*. Croom Helm.

Crofton, J. 1865. *Report on the Ganges Canal*. Public Works Department.

Day, Francis. 1873. *Report on the Freshwater Fish and Fisheries of India and Burma*. Office of the Superintendent of Government Printing.

Derbyshire, Ian D. 2022. *Railways' Economic Impact on Uttar Pradesh and Colonial North India (1860–1914): The Iron Raj*. Cambridge Scholars.

Dewey, Clive. 2014. *Steamboats on the Indus: The Limits of Western Technological Superiority in South Asia*. Oxford University Press.

Driver, Felix, and Luciana Martins, eds. 2005. *Tropical Visions in an Age of Empire*. University of Chicago Press.

D'Souza, Rohan. 2006. *Drowned and Dammed: Colonial Capitalism and Flood Control in Eastern India*. Oxford University Press.

Duff, Alexander. 1839. *India, and India Missions*. John Johnstone.

Eaton, Richard M. 1993. *The Rise of Islam and the Bengal Frontier, 1204–1760*. University of California Press.

Eck, Diana. 2012. *India: A Sacred Geography*. Harmony.

Edney, Matthew H. 1997. *Mapping an Empire: The Geographical Construction of British India, 1765–1843*. University of Chicago Press.

Gadgil, Madhav, and Ramachandra Guha. 1992. *This Fissured Land: An Ecological History of India*. Oxford University Press.

Ghosh, Amitav. 2016. *The Great Derangement: Climate Change and the Unthinkable*. University of Chicago Press.

Ghosh, Devleena, and Stephen Muecke, eds. 2007. *Cultures of Trade: Indian Ocean Exchanges*. Cambridge Scholars.

Gilmartin, David. 1994. "Scientific Empire and Imperial Science: Colonialism and Irrigation Technology in the Indus Basin." *Journal of Asian Studies* 53 (4): 1127–49.

Gupta, Pamila. 2010. "Introduction." In *Eyes Across the Water: Navigating the Indian Ocean*, edited by Pamila Gupta, Isabel Hofmeyr, and Michael Pearson, 3–4. Unisa Press.

Gupta, Pamila, Isabel Hofmeyr, and Michael Pearson, eds. 2010. *Eyes Across the Water: Navigating the Indian Ocean*. Unisa Press.

Hamilton, Francis. 1822. *An Account of the Fishes Found in the River Ganges and its Branches*. 2 vols. A. Constable.

Headrick, Daniel R. 1988. *The Tentacles of Progress: Technology Transfer in the Age of Imperialism, 1850–1940*. Oxford University Press.

Headrick, Daniel R. 1981. *The Tools of Empire: Technology and European Imperialism in the Nineteenth Century*. Oxford University Press.

Heber, Reginald. 1828. *Narrative of a Journey through the Upper Provinces of India, from Calcutta to Bombay, 1824–25*. 3rd ed., 3 vols. John Murray.

Holdich, T. H. 1909. "Physical Aspects." In *The Imperial Gazetteer of India: The Indian Empire*. Vol. 1. Oxford at The Clarendon Press.

Horden, Peregrine, and Nicholas Purcell. 2006. "The Mediterranean and 'the New Thalassology'." *American Historical Review* 111 (3): 722–40.

Klein, Ira. 2001. "Development and Death: Reinterpreting Malaria, Economics and Ecology in British India." *Indian Economic and Social History Review* 38 (2): 147–79.

Klingensmith, Daniel. 2014. "Nature and Politics at the End of the Raj: Environmental Management and Political Legitimacy in Late Colonial India, 1919–47." In *Shifting Ground: People, Animals, and Mobility in India's Environmental History*, edited by Mahesh Rangarajan and K. Sivaramakrishnan. Oxford University Press.

Long, James. 1846. "The Portuguese in North India." *Calcutta Review* 5 (10): 243–92.

M'Cosh, John. 1837. *Topography of Assam*. G. H. Huttmann.

Madras Revenue Proceedings, 25 February 1877, no. 474. India Office Records, British Library, London.

Morrison, Kathleen D. 2014. "Conceiving Ecology and Stopping the Clock: Narratives of Balance, Loss, and Degradation." In *Shifting Ground: People, Animals, and Mobility in India's Environmental History*, edited by Mahesh Rangarajan and K. Sivaramakrishnan. Oxford University Press.

Newby, Eric. 1983. *Slowly Down the Ganges*. Picador.

Roberts, Emma. 1835. *Scenes and Characteristics of Hindostan with Sketches of Anglo-Indian Society*. 3 vols. W. H. Allen.

Schama, Simon. 1995. *Landscape and Memory*. Harper Collins.

Sen, Sudipta. 2019. *Ganges: The Many Pasts of an Indian River*. Yale University Press.

Seton-Karr, W. S. 1848. "The Plains of the Lower Ganges." *Calcutta Review* 9 (17): 1–28.

Smith, R. Baird. 1849. "Canals of Irrigation in the N. W. Provinces." *Calcutta Review* 12 (23): 79–183.

Stepan, Nancy Leys. 2001. *Picturing Tropical Nature*. Reaktion Books.

Subrahmanyam, Sanjay. 1997. "Connected Histories: Notes Towards a Reconfiguration of Early Modern Eurasia." *Modern Asian Studies* 31 (3): 735–62.

Taylor, James. 1840. *A Sketch of the Topography and Statistics of Dacca*. G. H. Guttmann.

Tennant, William. 1804. *Indian Recreations*. 2nd ed., 2 vols. Longman, Hurst, Rees, & Orme.

Theobald, W. 1847. "Our Indian Railways." *Calcutta Review* 7 (14): 321–71.

Vink, Marcus P. 2007. "Indian Ocean Studies and the 'New Thalassology'." *Journal of Global History* 2 (1): 41–62.

Walch, George T. 1899. *The Engineering Works of the Kistna Delta: A Descriptive and Historical Account*. 2 vols. Government Press.

Wallace, Alfred R. 1878. *Tropical Nature and Other Essays*. Macmillan.

Whitcombe, Elizabeth. 1995. "The Environmental Costs of Irrigation in British India: Waterlogging, Salinity, Malaria." In *Nature, Culture, Imperialism: Essays on the Environmental History of South Asia*, edited by David Arnold and Ramachandra Guha. Oxford University Press.

Wilhelm, Janine. 2016. *Environment and Pollution in Colonial India: Sewerage Technologies Along the Sacred Ganges*. Routledge.

Worster, Donald. 1985. *Rivers of Empire: Water, Aridity, and the Growth of the American West*. Oxford University Press.

CHAPTER 2

A Monsoon River and Its Humans: the Karnaphuli in India and Bangladesh

Willem van Schendel

Life on earth depends on an endless "water cycle" of evaporation, condensation, and precipitation. Water runs downwards, and rainfall on land launches innumerable streams that seek lower ground. This has created our "rivered earth" (Seth 2011). Where downpours are seasonal, as in Monsoon Asia, watercourses can develop from trickles in the dry season to torrents in the wet season.

Rivers and river basins have their own histories that long predate *Homo sapiens*. But when humans appeared, rivers became crucial co-constituents of human social life. And rivers were no longer purely "natural" but began to be co-constituted by humans. Flowing water, sediments, animals, plants, microorganisms, and human societies became permanently entangled, a connection that is currently imagined as the "hydrosocial cycle … a socio-natural process by which water and society make and remake each other over space and time" (Linton and Budds 2014, 170; also Wesselink, Kooy and Warner 2017; Tvedt 2021b). By focusing on this fundamental relationship, water historians contribute to a comparative approach to the "life-worlds and webs of significance people have spun around water" in the past and the present (Tvedt and Oestigaard 2006, xv; also Amrith 2018).

Rivers differ in how they shape humans, however, and humans differ in how they view and exploit rivers. By taking a single river as a springboard, an entire hydrosocial universe of "riparian relations" can be explored (Wilson 2021, 25–36). Biographies of rivers are not uncommon, but this chapter merely seeks to present a sketch of hydrosocial change (Saikia 2019; Tvedt 2021a; Archer 2015; Cioc 2002). We consider the evolving relationship between humans and a middling river in one of the wettest corners of Monsoon Asia.

This tropical river is located just east of the Bengal Delta, the largest delta on earth, which is formed by two gigantic waterways—the Ganges and the Brahmaputra—and dozens of smaller ones. Most of these originate in the glaciers of the Himalayas. The Bengal delta has attracted much scholarly attention because it discharges 650 billion m^3 of water and a billion metric tons

 | HTTPS://DOI.ORG/10.1163/9789004758100-004

of sediment yearly (mainly between May and October), is seasonally flooded, features wandering riverbeds, sustains very dense human populations, and is especially vulnerable to rising sea levels (Bhattacharyya 2021).

1 A Monsoon River

The river that takes centre stage in this chapter has other dynamics, however, and its relationship with humans has always been different. It is hard to give this waterway a name because sections and tributaries bear various designations and it is important to acknowledge the cultural and linguistic histories behind each of these.[1] For brevity's sake we will stick with its most-used, downstream name: Karnaphuli.[2] In this region the Karnaphuli does not stand out as a particularly long or large river. So why does it matter? And to whom?

The Karnaphuli originates in a range of mountains that run south from the Eastern Himalayas.[3] Known as the Indo-Burma Arc, this 1,200-kilometres-long range is made up of steep parallel mountains stretching down the full length of western Myanmar (Figure 2.1). Many rivers begin here but, unlike rivers originating in Himalayan ice, these rivers are purely rain-fed. The Karnaphuli is the only one to break through the range to make its way west and empty directly into the Indian Ocean (here known as the Bay of Bengal). Today, its upper course is in India and its mouth in Bangladesh (Figure 2.2).

1 Upstream it is known as Khawthlangtuipui and another name is Kaincha or Kansa khyong (Sen 1928, 32–34, 95). The dozens of streams that feed the main channel often have different names in different languages, for example, Kawrpui Lui/Thega Khal and Sazuk Lui/Bara Harina Chhara. Some other tributaries are known as Kau, Deh, Phairuang, Tuichawng, Tuilianpui, Chota Harina Chhara, Kassalong, Subalong, Chengi, Kaptai, Rainkhiang, Ichamati, and Halda (see Figure 2.3).

2 Over time this word has been spelled in many other ways, from eighteenth-century Gurumfallee, Curnafoolee, Curnafooly, and Currumfullee to twenty-first-century Karnafuli and Kornofuli. Early-colonial sources sometimes referred to the Karnaphuli as "the Chittagong River" or "the Islamabad River." Plaisted 1764; Pemberton 1835, 5–6; Barbe 1845.

3 Iftekhar Iqbal suggests the concept of "water-world" to characterize the region formed by the major rivers that originate in the Himalayan-Tibetan uplands and "spread out like a necklace around Yunnan and cut across parts of the territories that are known as South, Southeast and East Asia." The Karnaphuli can be understood as a minor river belonging to the margins of that "water-world." Iqbal 2020.

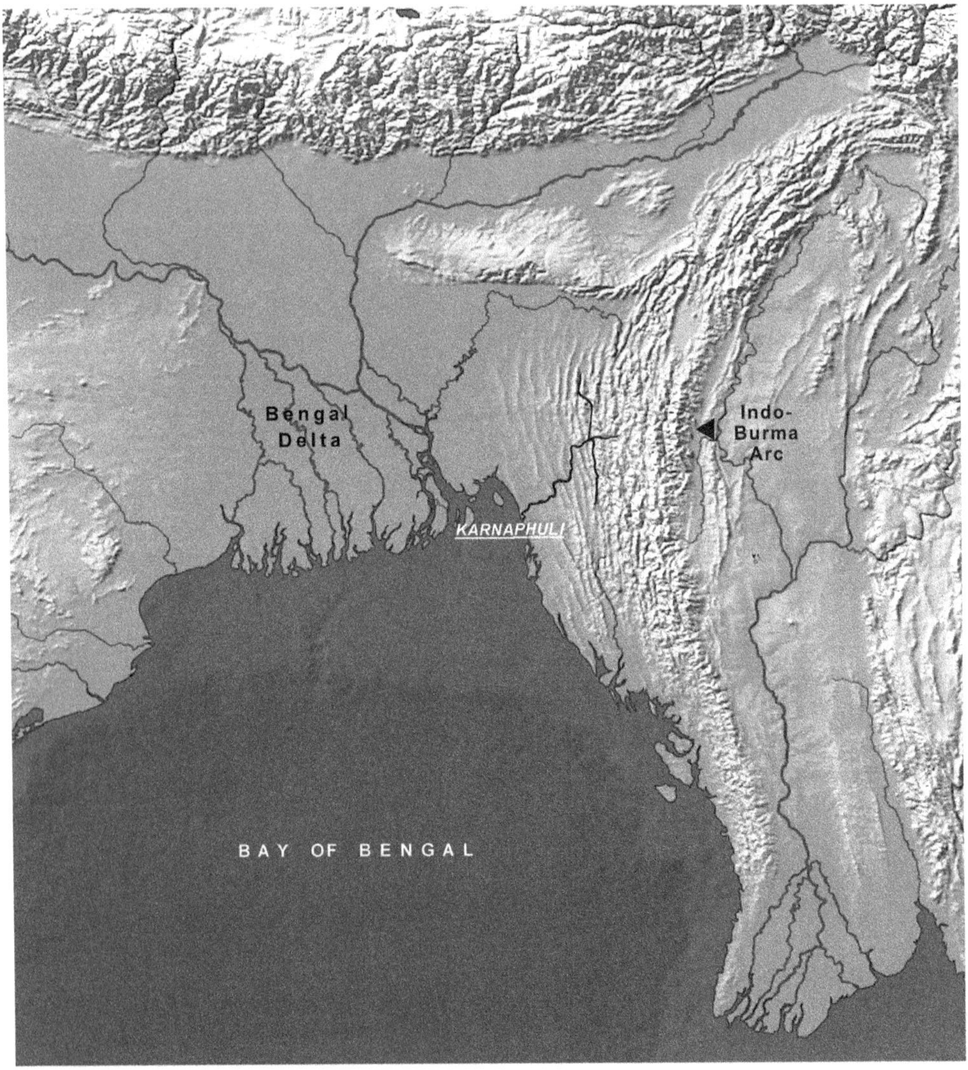

FIGURE 2.1 The Karnaphuli river

Fed by numerous tributaries, it is some 270 kilometres long and, in the lowlands, about 600 metres wide. The impact of the sea tides used to reach 50 kilometres upstream (Account 1798; Van Schendel 1992, 105, 120; Figure 2.3). Although the Karnaphuli usually is a predictable river, it can be dangerous: it is prone to irregular flash floods, tidal flooding, and storm surges that destroy human lives and property (Sen 1932, 34–35; Adnan, Dewan, Zannat and Abdullah 2019).

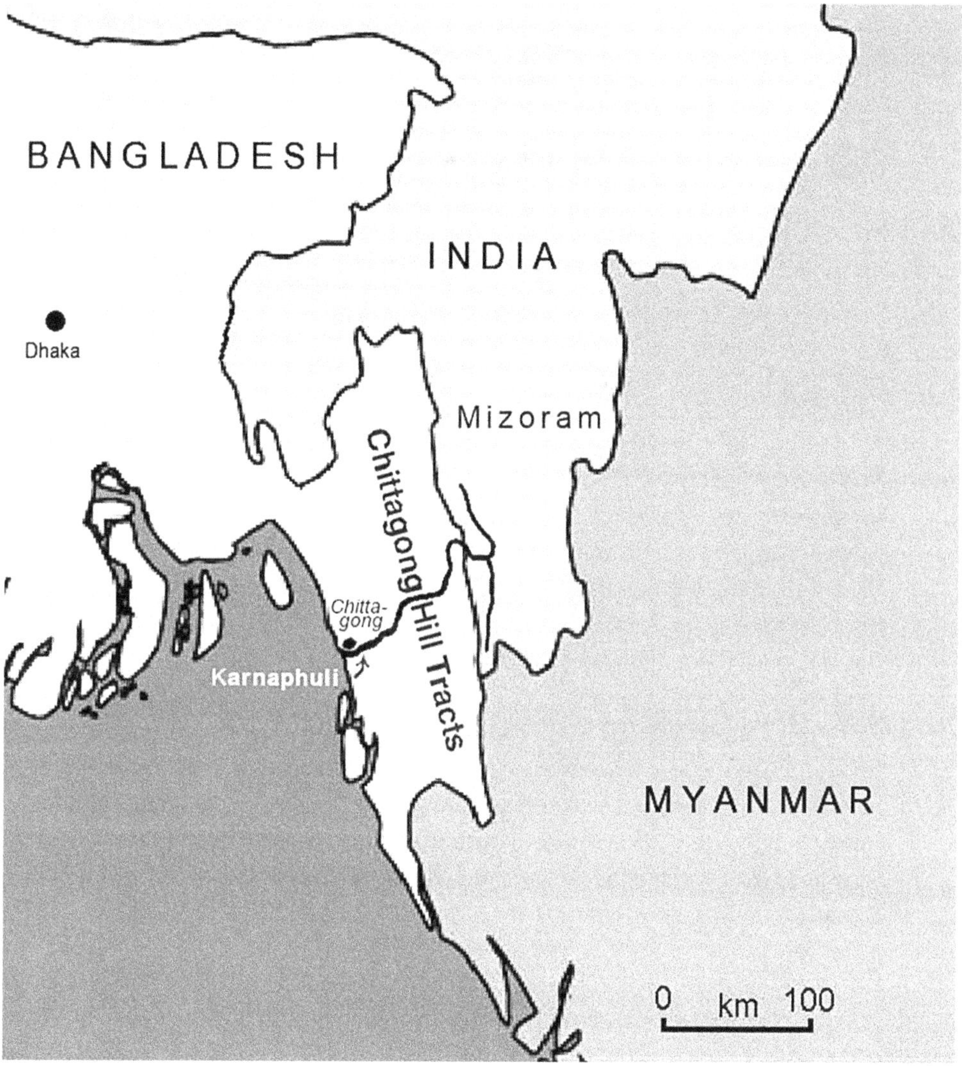

FIGURE 2.2 The Karnaphuli river and international borders today.

The river stands out in the region for being non-wandering and ending in a natural harbour, shielded from the sea.[4] This allowed for urban settlement right on the coast in a region where all major towns are established inland for protection against cyclones and tsunamis. This coastal harbour settlement has

4 The mouth of the river has been somewhat more flexible, moving very gradually southwards, but the Karnaphuli does not exhibit the sudden shifts of many of the Bengal Delta's rivers.

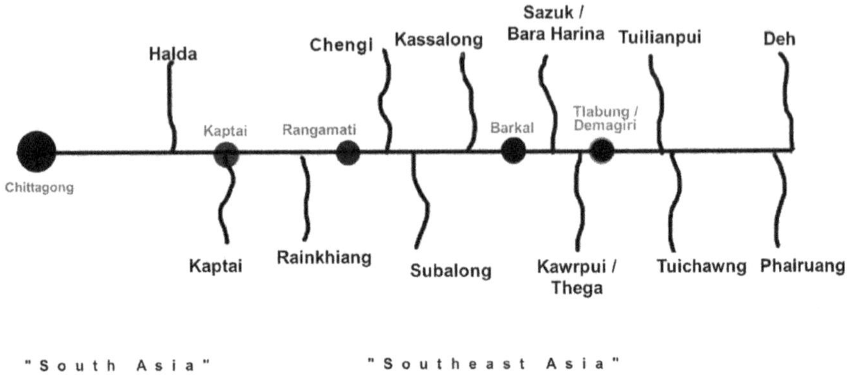

FIGURE 2.3 The Karnaphuli river system, pre-1860
Note: This figure shows only major tributaries – the Karnaphuli is fed by many more rivulets and streams.

had many names in the past.[5] Today it is known as Chittagong (or Chattogram). It grew from a small town into the busiest seaport on the northern Bay of Bengal, with a population of over five million. Today, Chittagong is the prime port of Bangladesh (Figure 2.2).

In this chapter we look at several ways in which this river and humans have made and remade each other between the eighteenth and the twenty-first centuries. The first part considers *shifting human power*: how some humans have used the river to exert power over other humans. The second part considers *shifting nonhuman power*: how humans have sought to control nonhuman "nature."

2 Shifting Human Power

As the only direct connection between the uplands and the sea, the Karnaphuli has always been of great cultural and economic importance. Little is currently known about early human mobilities along the river, about the meanings that humans gave to the Karnaphuli, about how they valued it, or how their activities created vernacular geographies (Jordan 1978). An eighteenth-century visitor noted that locals of various religious persuasions made offerings of grain, flowers, and eggs to deities at an upriver place of pilgrimage today known as

5 For example, Pentapolis, Sudkawan, Porto Grande, Xatigam, Bengala, Chottogram, Saitagong, and Islamabad. See Van Schendel 2015.

Chitmaram (Account 1798; Van Schendel 1992, 102–05).[6] He also learned that both the ritual and linguistic dimensions of the river were highly complex.[7]

2.1 *Four Cultural Zones*

It is essential to disaggregate the Karnaphuli waterscape and to understand it as a cultural layer cake.[8] Populations may have been mobile over time, but four distinct cultural zones—across only 100 kilometres—have been in place for centuries. They are, from east to west, the uplands, the foothills, the coastal plains, and the river mouth.

The inhabitants of the uplands formed numerous ethnic groups speaking languages of the Trans-Himalayan (Tibeto-Burman) language family. Their systems of rule were local and often transient. They practised shifting cultivation and were believers in local religions. From the late nineteenth century onwards, they adopted Christianity. Today the largest group is known as Mizo.

The inhabitants of the foothills were speakers of both Trans-Himalayan and Indo-European languages. Their systems of rule were local but occasionally tributary to lowland states. They also practised shifting cultivation, and they followed local religions and forms of Buddhism. Today the largest groups are known as Chakma and Marma.

Both the uplands and the foothills are usually thought to be part of the world region of Southeast Asia. By contrast, the narrow coastal plains are usually seen as belonging to South Asia.[9] The inhabitants of these plains spoke Indo-European languages (Chittagonian, Bengali), and practised irrigated rice cultivation. They regularly lived under established state systems whose centres of power were located elsewhere. The inhabitants of the densely populated coastal plains were overwhelmingly Muslims and Hindus.

---

6 For more recent riverside offerings, see Van Schendel, Mey, and Dewan 2000, 158.

7 The cosmological significance of the Karnaphuli for various cultural groups has not been studied much. Some indications of Mizo beliefs about river-dwelling spirits can be found in Tonson 2021. For work on another river in the region, see Varte 2017.

8 The term "waterscape" (or "hydrosocial territory") considers the spatial dimensions of interactions between natural and human processes in, on, and around a river. Just like the term "landscape," it derives from Dutch. In this language, *waterschap* has a very specific meaning, in existence from the thirteenth century to the present: a regional governing body for the management of surface water in the environment. "Waterscape" is sometimes used as an equivalent to "riverscape," but this may be confusing because riverscape can also refer more narrowly to the ecology of a freshwater river habitat. Flaminio, Rouillé-Kielo, and Le Visage 2008; Haslam 2008; Dunham et al. 2018.

9 Figure 2.3. For a portrayal of a much larger river system linking "Southeast Asia" and "South Asia," see Ludden 2019.

And those who lived in the urban settlement at the mouth of the river formed an ever-changing amalgam of ethnic and religious groups living off trade, shipbuilding, administration, and fishing (Berthet 2020).[10] Apart from Chittagonian Bengalis, the town was, at different times, settled by Arabs, Persians, Afghans, and Arakanese; and it was visited by Chinese traders.[11] When Europeans became active in the region, the town became an important settlement for the Portuguese and later the British (Figure 2.4).[12]

FIGURE 2.4 Chittagong and its river harbour in 1837 CE
Note: Water colour with pen and ink by Jane Blagrave, 1837, signed "J.M.A. Blagrave. Calcutta /37. Tempest Hill. Fairy Hill. South View." (shelfmark: WD1704; item number: 1704, British Library, London, UK).

10 The religious and artistic networks of this zone can be traced for many centuries by way of archaeological finds such as a large hoard of 9th- to 12th-century bronzes discovered on the east bank of the Karnaphuli near Chittagong city, or a 12th–13th-century stone sculpture from its west bank. Ray 2021, 176; Bautze-Picron 2016.

11 "The port attracted foreign capital and empire builders, in addition to the Arakanese and Mughals, including the Portuguese, Afghans, Pathans, and eventually the British. Christian slaves and fugitives from Goa, Ceylon, Cochin, and Malacca settled in Chittagong and were gifted land grants by the king of Arakan … Bengali, Tamil, Telugu, Malay, Arab, Persian, and Dutch merchants also traded in its harbours and on its streets." Jhala 2019, xx.

12 Europeans also came to rank these four cultural zones, from the "wildest" in the uplands to the least "wild" on the coast. Sen 2018, 178.

As the Karnaphuli flowed between these distinct cultural worlds, it connected them in many ways and created a variegated hydrosocial territory.[13] These connections were not without tensions, as a Mizo legend, written down in 1912, indicates.[14]

The Karnaphuli was an important trade route between the ocean and the uplands, but not a major connection to regions further east, because it is navigable only up to the western edge of the uplands. As a result, it played only a

---

13 There was considerable borrowing, exchange, and fusion across these cultural worlds, making the region hard to read for administrators and outsiders. Jhala 2019; Pachuau 2022.

14 According to J. Shakespear, the Mizos explained the origin of the Tuichawng river, which flows into the Karnaphuli, as follows:

Nine miles from Demagri, on the Lungleh road, the traveller has to cross the Tui-chong river, one of the largest tributaries of the Kurnaphuli, on which Chittagong stands. This river, according to the Lushais, owes its origin to the self-denial of a girl called Tui-chongi, who, with her little sister Nuengi, was walking on the hills whence the river rises. It was April, and the sun blazed down on them. Nuengi began to cry for water. "How can I get you water on the top of a hill? Don't you know that all the springs are dry, for are not the jhums [fields] ready to be burnt?" "Water, water, or I shall die," wailed Nuengi. "Would you rather have water than me?" asked Tui-chongi. "If I don't get water, I shall die, and then of what use would you be to me?" replied the spoilt child. So Tui-chongi, to satisfy her youngest sister's thirst, changed herself into a river, and Nuengi drank and was satisfied. But the water flowed down among the hills and burst its way into the country of the Bengalis. The king of the Bengalis was astonished to see so mighty a river flowing past his palace, and sent some of his people to find out whence it came. They journeyed many days, till at length they reached the source of the stream, and there sat Nuengi, who, now that her thirst was satisfied, would gladly have had her sister back again to show her the way home. The explorers were astonished to find so beautiful a maiden sitting thus in the middle of the jungle, and decided that it would be wise to take her back to their master, who liked pretty girls. So Nuengi was added to the harem of the king of Chittagong, and in time became the mother of a most lovely boy. The king's chief wife, on seeing the child, thought to herself, "If my lord sees this jungle woman's brat, he will assuredly love her more than me who am childless." So she had the child thrown into the river, which flowed under the palace windows, and frightened Nuengi into keeping silence on the matter. Tui-chongi, however, in spite of the change in her circumstances, remembered her little sister, and cherished the child so that he grew and throve. In the same way six more children were born and thrown into Tui-chongi's fostering arms. When they were grownup Tui-chongi told them the circumstances of their birth, and sent them to dance on the roof of their father's palace, who, hearing the noise, came out to see the cause of the disturbance. When he saw seven handsome young men he was much astonished, and asked them who they were. "We are your sons," they replied. "Why do you lie to me?" said the king; "liars have short lives in my kingdom." "Nay, O king, we lie not; we are Nuengi's sons"; and they told him their story. So the king smote off the head of the bad queen, and installed Nuengi in her place (Shakespear 1895, 184–185; Shakespear 1912, 95–96; Lalthangliana 2005, 385–386).

minor role in long-distance trade to Burma and China, and it does not figure in imaginaries of the land-based "Southern Silk Road" (Pachuau 2022).

Humans used the river as a resource for gaining power, and over time this power shifted markedly between the cultural zones. For centuries, the uplands had the upper hand. The people of the lowlands were often controlled by larger coastal states but those of the foothills and uplands were not. In the dry season, armed men from the uplands would travel on foot through the riverbeds of the Karnaphuli and its tributaries to raid communities living in the foothills and lowlands, creating animosities and suspicions between "uplanders," "foothillers," and "lowlanders" that are still noticeable today (Hamilton 1825; Guite 2014).

Such attacks continued in the late eighteenth century when the lowlands became firmly integrated into an expansionist state—British India—with eyes on revenue collection beyond the lowlands.[15] Military confrontations followed. From the 1770s British raids into the foothills were met with organized resistance, palisaded villages, and counter attacks coordinated from Rangunia, a settlement on the Karnaphuli that was home to the Chakma chief (Serajuddin and Buller 1984; Van Schendel 1992, 106–07; Qanungo 1998; Chowdhury 2016).[16] In 1860 these imperial raiders annexed the foothills and gave them the name "Chittagong Hill Tracts" (now in Bangladesh). By the late 1860s, they had explored the Karnaphuli up to Tlabung/Demagiri, the boundary between the foothills and the uplands, and they stationed armed troops at this eastern border to ward off upland warriors (Lewin 1869, 2; Figure 2.5).[17] The seat of the Chakma chief, now a colonial subject, moved from Rangunia to Rangamati, the new district headquarters of the Chittagong Hill Tracts, where it is still located. A visitor described the lifestyle of the new rulers in Rangamati as follows:

> Here as all the world over, the English live after the English fashion, the old homely ways clinging to them like the very soles of their feet. They eat more curry, more rice, more poultry than in England; but in other respects they eat, drink and play lawn tennis, just as if residing within a few hours of London. (Riebeck 1885, 3)

15 There had been military confrontations with the precolonial state (which did not claim jurisdiction over the foothills but demanded tribute in the form of cotton in exchange for trade privileges) when tribute was not forthcoming. Serajuddin and Buller 1984, 93.

16 In the 1770s the cartographer James Rennell's knowledge of the Karnaphuli reached up to Rangamati, which he depicted as beyond the eastern borders of British control. The Karnaphuli and its tributary Chengri were shown to form short stretches of the border between the territory claimed by the British and the lands to the east, which he named "Roshaan" (Arakan). Rennell 1776.

17 For a map of Tlabung/Demagiri in the 1870s, see Lewin 1912, 305.

FIGURE 2.5 Colonial border troops on the Karnaphuli riverbank in Tlabung/Demagiri, the boundary between the British-controlled foothills and the independent uplands, in 1882. These Gorkha soldiers from present-day Nepal are pointing their guns eastwards (Riebeck 1885, 6; also Van Schendel, Mey and Dewan 2000, 55–57).

The Karnaphuli became a thoroughfare for multiple British raids—"punitive outings" (Shakespear 1929, 65)—into the eastern uplands. The historiography of these forays along the Karnaphuli relies heavily on British accounts. A military chronicler described one such raid in 1861 as:

> a military expedition, under Major Raban, of 230 rifles from the Native Regiment at Chittagong with 450 carriers. These, proceeding in small country boats, assembled at Burkul on the Kornaphuli river, whence they marched to Ruttunpoia's village—the offending chief, who submitted at once. No opposition was offered, only great difficulty is recorded in having practically to cut their way for five days through thick forest. (Shakespear 1929, 65–66)

In 1871–1872, British military columns and their suppliers used the Karnaphuli in a major attempt to annex the mountains east of Tlabung/Demagiri. They lost this confrontation and for years mutual attacks continued (Elly 1893, 15–17; Lewin 1912). In 1889–90 they tried again, using the Karnaphuli as an indispensable supply route as well as a conduit to take back hundreds of sick and injured troops and followers (Grant 1890; Figure 2.6). Once more the inhabitants of the uplands proved to be resistant to the imperial onslaught and they were hard to subdue (Shakespear 1929; Reid 1942, 1–44; Lalthlengliana 2013; Jackson 2023, 46–103). But by the late 1890s, the British had turned the Karnaphuli and its tributaries into an entirely British-dominated waterscape (Pachuau and Van Schendel 2015, 109–17). They named their new territory "Lushai Hills" (now: Mizoram in India).

FIGURE 2.6 Boats on the Karnaphuli that were used for the British attack on the Lushai Hills (Mizoram) in 1889–1890 (Shakespear 1929, facing 76).

2.2 *A Sinew of Empire*

British rule meant that military, economic and cultural power over the river shifted away from the uplands to the lowlands, and from locals to foreign administrators. The Karnaphuli lost its military importance, and British administrators began to penetrate the uplands, entering via the river. As motorboats became available for part of the journey, the travelling time from Chittagong to Tlabung shrunk to about five days. Beyond Tlabung, travel east continued on foot, or by pony on bridle paths.

The Karnaphuli became a route of imperial extraction. The export of timber from the Lushai forests to Chittagong was valuable but transport on the upper reaches of the Karnaphuli was hampered by rapids and narrows. So the British manipulated the river: "Till recently [Tlabung] was inaccessible even to small boats; but since then various narrows have been blown up with gunpowder and thrown open to river traffic" (Riebeck 1885, 5). Even so, transporting timber required navigational skills that were rare in the uplands: "Only experienced traders and cutters from the Chittagong hill tracts could navigate [the Karnaphuli and] the Mizo did not excel at boatmanship" (Tianghlima 2022, 96; also Pemberton 1835, 5–6; Lewin 1912, 143, 194). Bamboo was another prime resource. It was floated down the Karnaphuli in huge rafts that could also carry other commodities, such as cotton, thatching grass, wax, and rubber. Salt, dried fish, spices, and metal goods went in the opposite direction and were sold at three markets along the river.[18]

The Karnaphuli turned into a sinew of empire in another sense as well. It became a thoroughfare for Christian missionaries on a civilizing mission.[19] They travelled up the Karnaphuli to set up centres for spreading the gospel in the foothills and uplands, providing health services, and creating schools (Van Schendel, Mey and Dewan 2000, 166–79; Pachuau and Van Schendel 2015, 59–108). The Christian cosmology literally entered the Karnaphuli when missionaries used its waters to baptize converts by publicly submerging them, a scene that they often photographed (Van Schendel, Mey and Dewan 2000, 170–71). In 1907 they established a riverside hospital in Chandraghona that attracted many patients who arrived by boat (Hutchinson 1909, 64; Van Schendel, Mey and Dewan 2000, 180–89).

Missionaries travelling to the uplands still had to overcome the hurdle of the dangerous rapids between Barkal and Tlabung, but a trolley track, constructed during the invasion of the uplands in 1889–90, helped them to bypass these (Hutchinson 1906, 8). The trolleys were pushed by labourers imported for the purpose (Figures 2.7 and 2.8).

18 In Kasalong, Rangamati and Chandraghona.

19 A parallel civilizing mission was carried out by Buddhists in the lowlands who invited monks from Ceylon to establish Pali learning centres on either side of the Karnaphuli. Barua 2022.

FIGURES 2.7 AND 2.8 Missionaries and their servants abandoning boats at Barkal, and Santal labourers standing by, ready to push their trolley across the rapids, 1904 (Van Schendel, Mey and Dewan 2000, 222).

The river remained an all-important connection because roads were very poor beyond the coastal plains. As one observer noticed in the 1860s: "There are at present no roads in the district ... The favorite path throughout the district is the sandy bed of a stream" (Lewin 1869, 7; also Barbe 1845, 387). By the early twentieth century, there were still only bridle paths (Hutchinson 1906, 14–17), and twentieth-century maps of the region frequently mentioned "path follows [river] bed" (Figure 2.9).

FIGURE 2.9
Travellers in the Chittagong Hill Tracts, 1940s.
PHOTO BY MURIEL STARKE

There were no bridges across the Karnaphuli (and only bamboo foot bridges on some of its tributaries), so crossings had to be done by raft or boat (Figures 2.10 and 2.11). It was not until 1930 that a railway bridge was constructed in Chittagong city. By 1962 this Kalurghat Bridge was upgraded to a rail-and-road bridge. In 1989 a second road bridge was added; and it was upgraded in 2010.[20] Upstream, but still in the lowlands, there is only one other road bridge across the Karnaphuli, in Rangunia.

FIGURE 2.10 A bamboo raft used as a ferry to cross the Karnaphuli (here known as the Khawthlangtuipui) near the Tuichawng tributary in Mizoram, 1930s (Pachuau and Van Schendel 2015, 149).

FIGURE 2.11 A bus crossing the Karnaphuli, 1950s.
PHOTO BY ALAN TAYLOR

20 The Karnaphuli Bridge—a second-hand road bridge donated by the Netherlands—was later renamed the Shah Amanat Bridge. A new Shah Amanat Bridge, constructed by a Chinese company and with Kuwaiti support, was opened in 2010. Currently, a tunnel under the Karnaphuli riverbed is being constructed in Chittagong city.

2.3 *A State Border*

British rule introduced the idea of fixed administrative borders and the Karnaphuli river system was used for this purpose. Between the 1860s and the 1890s, stretches of its tributaries (Kawrpui and Sazuk/Tuilianpui) acted as the easternmost border of British India. When British troops pushed further east and subjugated upland populations in the 1890s, these tributaries became provincial borders (between Bengal and Assam), and they remained so up to 1947 (Figure 2.12).

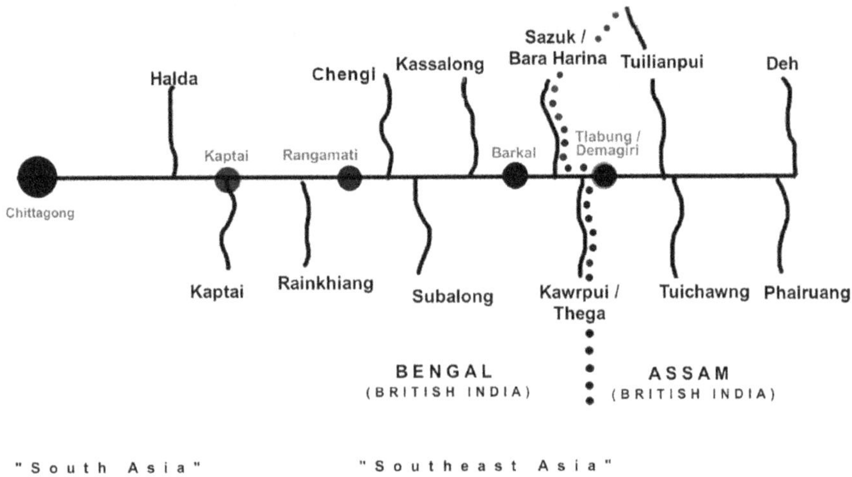

FIGURE 2.12 The Karnaphuli river system, 1890–1947.

In that year British India collapsed, and in the final days of imperial rule the territory was divided between two states: Pakistan and India. This "Partition" truncated the Karnaphuli. Its upper reaches became Indian territory, but the foothills and lowlands were allocated to Pakistan. This controversial decision in 1947 suddenly gave the Karnaphuli geopolitical prominence and turned the tributaries into border markers between two postcolonial states (Figure 2.13).[21]

21 The logic of this river truncation flew in the face of two considerations. First, the foothills were not a Muslim-majority region (the rationale for allocating regions to Pakistan). Second, the colonial habit of considering a watershed to be the border (which would have brought the Karnaphuli river system under a single state authority) was ignored. It has been suggested that the foothills (Chittagong Hill Tracts) were given to Pakistan to allow that country to use the Karnaphuli's hydroelectric potential. The uplands had been administered as part of the province of Assam, rather than the partitioned province of Bengal, and they were left out of the territorial division. Gardner 2019.

Border incidents began to occur, for example in 1963, when Pakistan border guards crossed to the Indian side of the Karnaphuli and seized a bamboo raft transporting hundreds of kilogrammes of paddy and rice.[22]

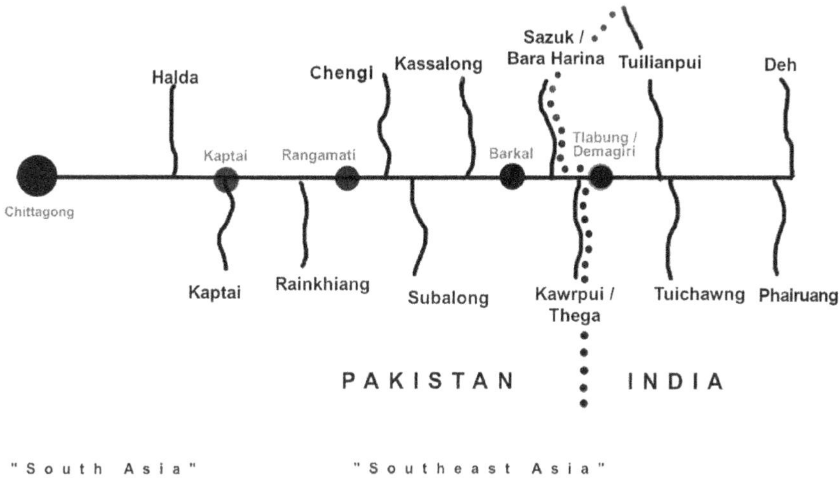

FIGURE 2.13 The Karnaphuli river system, 1947–1960.

This situation continued up to 1971, when war broke out and Pakistan disintegrated into Bangladesh and present-day Pakistan. Since 1971, the two tributaries have marked the border between Bangladesh and India (Figure 2.14).

FIGURE 2.14
Boatman on the Kawrpui (Thega) river: Bangladeshi territory to the left, Indian territory to the right, 2000.
PHOTO BY WILLEM VAN SCHENDEL

22 "Border Incident-Assam-East Pakistan Border. Trespass by some E.R.P. personnel into Indian territory, village Binasora in Mizo District, and kidnapping of Indian National namely Phoreya and Boki Chakma on 24-12-63." External Affairs—Pakistan Proceedings, Nos. 108(19)-PAKI, 1964; National Archives of India, New Delhi.

These permutations had far-reaching effects on communication along the Karnaphuli because relations between these postcolonial states fluctuated between tense and hostile (Van Schendel 2005). Since 1947 the border has been officially closed, and in recent years the Indian authorities have been fencing it. The river continues to flow freely across the border but there is not a single approved border crossing for goods or people, only surreptitious movements. State officials and politicians subordinated the interests of locals, traders, and travellers to their own. Despite these restrictions, the Karnaphuli river system continued to figure in trans-border movements and geopolitical games. It was at the centre of uprisings against the states of India and Bangladesh—notably the Mizo war of independence (1966–1968) and the Chittagong Hill Tracts war (1975–1992)—and it became embroiled in border-crossing warfare (Van Schendel 2016; 2020; Pachuau and Van Schendel 2015).

Over a period of some 200 years, the dynamics of human power around the Karnaphuli shifted markedly. Not only did new groups establish their dominance over the river, but local powerholders also strikingly lost power to outsiders. Violent processes of colonial annexation brought new rulers, traders, and cultural brokers, and so did postcolonial rule. The hydrosocial cycle of the Karnaphuli was often characterized by turbulence, most notably during the early colonial period and between the 1960s and 2000s, when two attempts to recoup local power by armed insurrection failed to redress the balance. Today the middle course of the Karnaphuli flows through a zone of heavy-handed state power: the militarized and restive Chittagong Hill Tracts.

3 Shifting Nonhuman Power

The power games around the Karnaphuli also brought a protracted shift in the power of humans over nonhuman organisms and the ecosystems on which they depended. Humans have long disturbed the Karnaphuli. Fishing with nets, weirs, harpoons, and poison was a brutal intervention in its aquatic ecosystem. Cutting riverside bamboos and trees caused soil erosion and landslides and led to topsoil muddying the river water. Washing clothes and bodies with soap and oil polluted the river, and so did waste from households and agriculture. Increasingly, the Karnaphuli became a focus of stepped-up exploitation and resource extraction, with far more devastating consequences for the river itself as well as the humans and nonhumans that relied on it.

3.1 *A Development Site*

Humans began seriously tampering with the natural flow of the Karnaphuli in the nineteenth century, when they used gunpowder to broaden its upstream narrows. Since the mid-twentieth century, however, human self-confidence increased and manipulation of the river intensified, especially in the downstream section administered by Pakistan. Under the heading of "development," the Pakistan state entertained "the notion of the river as a 'slice' of nature that can be engineered, packed, and delivered for the 'nation'" (Iqbal 2022, 105). The Karnaphuli was a resource to be manipulated. Two examples may demonstrate this. Both denied local people any say, so these were prime examples of non-participatory development (Jhala 2023). In 1953 the Karnaphuli Paper Mill came into production in Chandraghona, a village marking the boundary between the foothills and the plains. The mill sourced its raw material from upstream bamboo forests and sold its paper across Pakistan (Van Schendel, Mey and Dewan 2000, 190–196). Its effluents began to pollute the river (Figure 2.15).

FIGURE 2.15 Scum from the Paper Mill at Chandraghona, 1962.
PHOTO BY EDNA SKIRROW

Far more damaging was the idea that the river was primarily a flow of energy to be utilized for national development. The Pakistan state elite appropriated the Karnaphuli, hitherto understood as river commons, and commodified it to construct a dam to provide hydroelectricity and prevent annual floods downstream. This barrage was originally conceived at Barkal, during colonial times, but the political reality of Partition forced the Pakistan authorities to adapt the plan, partly because a dam at Barkal would create a lake extending into Indian territory. By 1962 a hydroelectric power station was completed downstream—at the village of Kaptai—made possible by heavy foreign investment and technical expertise. At the time, the project was widely promoted as a human triumph over nature. Local people experienced it as a nightmare (Chakma, Chakma, Dewan and Ullah 1995). The barrage created the vast Lake Kaptai, which flooded many villages, displaced about 100,000 people, and changed the lives of many more (Jhala 2023; Figures 2.16 and 2.17). It also had a devastating environmental effect: it submerged extensive forests and destroyed the riverine ecology—plants perished, and animals drowned or had to flee (Mountfort 1969, 91–130). The lake distorted the river's drainage system and its sedimentation pattern, and it hampered water transport (Van Schendel, Mey and Dewan 2000, 196–206; Figures 2.18 and 2.19). Producing electricity for Chittagong city and the regional capital, Dhaka, but not for local rural settlements, the ecologically devastating Kaptai project was of vital importance to the industrialization of energy-starved Pakistan.

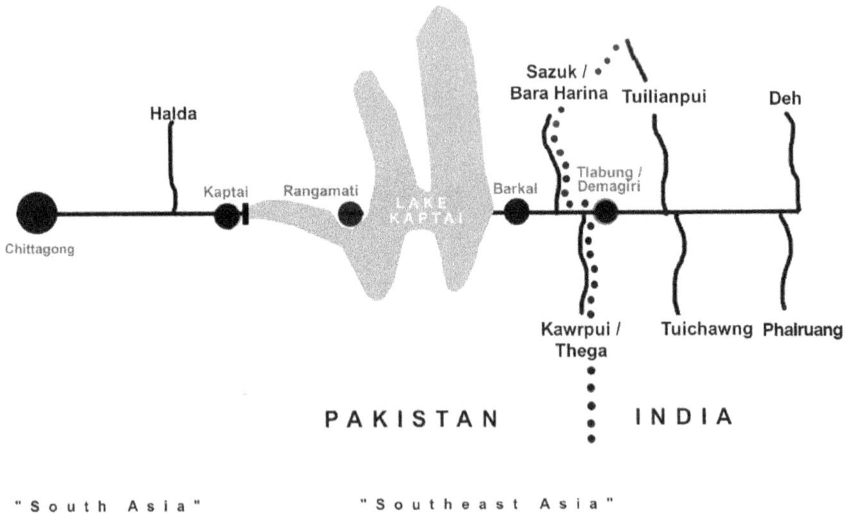

FIGURE 2.16 The Karnaphuli river system, 1960–1971.

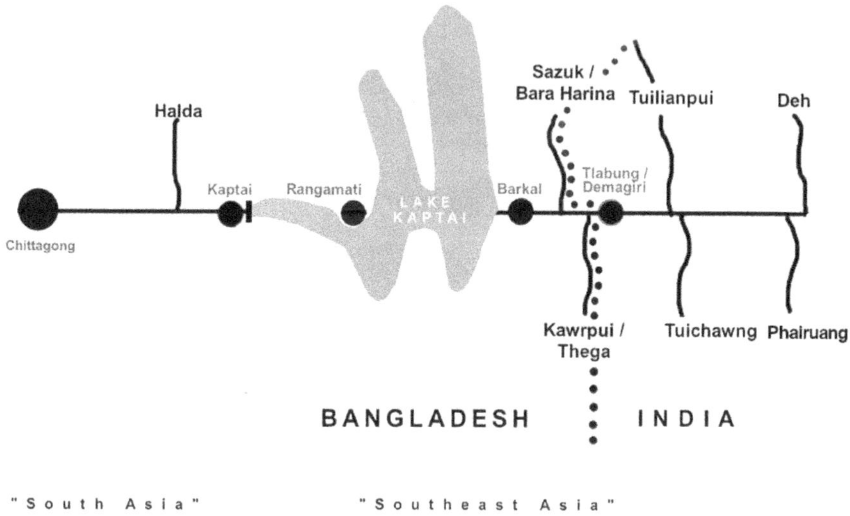

FIGURE 2.17 The Karnaphuli river system, 1971–present.

FIGURE 2.18 The Kaptai dam, 1964–65 (Van Schendel, Mey and Dewan 2000, 199).
PHOTO BY KEITH SANDERCOCK

FIGURE 2.19 Boats and bamboo rafts waiting to be transported across the Kaptai dam, 1970s.

The Kaptai barrage presented a stark confrontation between opposing "hydrosocial imaginations" (Hommes, Hoogesteger and Boelens 2022). The destructive force of the water, the loss of livelihoods, and the wilful silencing of local protests since the 1960s contributed to cultural alienation and a revolt that ignited the Chittagong Hill Tracts War. After 1975 the Bangladesh government responded with outright militarization and state-sponsored immigration of Bengali-speaking plains people, victimizing many "foothillers" (Figure 2.20).

FIGURE 2.20
Bangladesh army personnel and Bengali settlers terrorizing local inhabitants on the bank of Lake Kaptai, 1993 (Guhathakurta and Van Schendel 2013, 408).
Note: This image appeared on the cover of a locally published booklet, *Chitkar* (The Scream), which commemorated the 1993 Naniarchar massacre. The artist was not acknowledged.

3.2 *Ecological Invasion*

The Karnaphuli river system is not only a cultural layer cake, but also an ecological layer cake. The uplands and the Chittagong Hill Tracts form the westernmost edge of one of the world's biodiversity hotspots ("Indo-Burma"; Sloan et al. 2014). They provide an ecological contrast with the coastal plains. Humans interfered in the ecology of the uplands and foothills in many ways, some intended, others unanticipated. They introduced commercially useful plants, such as teak (from Myanmar) and Hevea rubber (from Brazil). New crops took the river route to the uplands. For example, "In 1909, an enormous first shipment of seed potatoes arrived by boat in Demagiri" (Jackson 2023, 122). Deadly diseases, such as cholera and smallpox, travelled upstream with human newcomers from the lowlands (Barbe 1845, 381; Jackson 2023, 88–89). Hunting, habitat loss, and the erection of a border fence between India and Bangladesh decimated many animal species, contributing to the regional extirpation of the tiger, three species of rhinoceros, sloth bear, banteng, wild buffalo, marsh crocodile, peafowl, white-winged duck, sarus crane, king vulture, and many more. The waterscape was also vulnerable to human-assisted—but unintentional—invasions of exotic species. Water hyacinth, a floating plant from the Amazon basin, proliferated in Lake Kaptai (Iqbal 2009; Figure 2.19). Bitter vine (*Mikania micrantha*), a fast-spreading creeper from the Americas, invaded the riverside forests. Carp and tilapia escaped from fish farms; and various carp species from East Asia and Europe are now found in the Karnaphuli river system (Bashar, Uddin and Basak 2018; Alam, Hossain, Monwar and Hoque 2013). Awareness of the ecological and economic costs of these changes increased apace.

3.3 *Hazardous Water*

As the water of the Karnaphuli was increasingly compromised, concerns about environmental degradation grew. Riverside human communities used the river for drinking water, bathing, washing, and irrigation. They also polluted it with their excrements and agricultural and industrial runoffs (Nahar 2022). The Karnaphuli received "numerous kinds of chemicals, gasoline and other petroleum products, agricultural fertilizer, insecticides, herbicides, fungicides, construction debris, paints and solvent, changed car oil, sediments, plastic bags, fishing line, hypodermic needles, industrial pollutants" (Sikder, Rahman, Sarker and Hossen 2012, 56–57; also Hossain et al. 2022). Despite environmental policies such as the creation of a fish sanctuary, fish populations suffered from polluted and turbid water as well as from overfishing (Nahar 2022; Arshad-Ul-Alam and Azadi 2015, 275–76; Arshad-Ul-Alam 2020; Das 2022).

Environmental degradation was more evident in the heavily industrialized lower section of the river than in the middle and upper sections (Islam et al. 2022; Hossen, Rafiq, Kabir, and Morshed 2019). That was particularly bad news for the Karnaphuli's largest aquatic mammal, the endangered freshwater Ganges river dolphin (*Platanista gangetica*). The Kaptai dam had blocked its access to the upper course of the river (apparently there are no dolphins in Lake Kaptai or beyond) but a small population can still be found in the lower Karnaphuli and its tributary, the Halda (Smith, Ahmed, Ali and Braulik 2001; Aziz 2019, 34–36; Khan 2019).

As the power of humans over nonhuman organisms shifted, humans increasingly destabilized the ecosystems on which both they and nonhumans depended. Manipulation of the Karnaphuli river system by means of dam building, introduced species, and pollution affected the natural drainage system and threatened biodiversity as well as human wellbeing. As the power of humans over nonhumans became more and more evident, the Karnaphuli river morphed into an apparently human-dominated waterscape.

3.4 *The Hydrosocial Cycle*

The Karnaphuli matters to many people. It is vital to those living near it as well as to many others living further away: city dwellers who depend daily on its hydroelectricity, displaced citizens who had to seek refuge abroad, and all those who use the port of Chittagong. It is a river that provides a good example of the hydrosocial cycle by which water and society make and remake each other. To begin with, over many generations the river's life-giving waters and aquatic organisms supported and shaped human cultures across four distinct ecozones. It enabled fisherfolk, boatmen, raftsmen, bamboo cutters, loggers, rapids labourers, agriculturists, and many others to make a living (Figure 2.21). Its monsoon rhythms moulded human existence. And its flash floods and storm surges occasionally took human lives.[23]

Second, humans used the river to establish their power over other humans. The Karnaphuli has long been a prized pathway for warfare, rule, and trade. Over the past few centuries power shifted away from the uplands to the lowlands,

23 For example, in 1902 a Chittagong newspaper urged the authorities to straighten the banks of the Halda, a tributary of the Karnaphuli, because its annual floods did "immense harm to life and property." *Jyoti* (24 July 1902). *Report* 1902, 465.

FIGURE 2.21 Fishermen on the Karnaphuli near Tlabung (India) in 2000. Here the river is known as Khawthlangtuipui.
PHOTO BY WILLEM VAN SCHENDEL

and from local powerholders to more distant ones who understood the river purely as a resource to be exploited. They remade the river and, with it, local communities and cultures. In their pursuit of hydropower and the production of commodities, they initiated projects that permanently distorted the drainage system and messed up the water quality—to the detriment of hundreds of thousands of local inhabitants.[24] Riverside communities were broken up, human health was jeopardized, and large numbers of developmental refugees fled across the state border to be resettled in faraway places.[25] The politics of water had turned against those whose daily lives depended on the river.

Third, as the power of humans over nonhumans grew, the ecosystems of the Karnaphuli suffered accelerating degradation. Human power remade water. Hydroelectric desire caused the flooding of vast agricultural lands and

24 The Kaptai hydroelectric project was directly linked to Cold War power politics. Funded by the USA, it was a tool to ensure the Pakistan state elite's loyalty. Jhala 2023.

25 Tens of thousands of victims of the Kaptai Hydroelectric Project are now living in the states of Arunachal Pradesh, Tripura, and Mizoram in India, as well as further afield.

forests—and rising water destroyed uncounted animals and plants in what had been the valleys of the Karnaphuli and its tributaries. Human activities caused increasing deterioration of the Karnaphuli's aquatic habitat along its full length, but especially in its lower reaches where water pollution was most pronounced (Figure 2.22). By the early twenty-first century, environmental concerns began to influence policies and practices affecting water to some extent, but the negative trend was not reversed. The life-giving qualities of the Karnaphuli river were compromised, and its water posed a direct threat to the health and lives of the many humans who used it for drinking, washing, and bathing—and who consumed its polluted fish.

FIGURE 2.22 Ships on the lower Karnaphuli near Chittagong. In the foreground is the Shah Amanat bridge and in the distance the Bay of Bengal.
PHOTO BY ABDUL MOMIN.

The hydrosocial pendulum has started swinging in the opposite direction, with the Karnaphuli remaking human society in unfamiliar ways as it becomes a growing threat to those who depend on its increasingly toxic liquid contents. The destructive impact of the river on human society is expected to increase further because global warming is widely projected to intensify rainfall in this region. This will lead to more water in the river, changes in agricultural production, and more devastating flash floods and storm surges. In the very long run, rising sea levels could shorten the Karnaphuli because sea water would permanently submerge the lowlands and wipe out everything that lives there. Such scenarios were unthinkable half a century ago, but they are seriously contemplated today.

4 Conclusion

Histories of individual rivers are, by nature, case studies. They explore the dynamic relationship between abiotic substances (water, silt, and rock) and a range of organisms (microbiota, fungi, plants, and nonhuman and human animals) in a particular setting. Social histories of water frame rivers as ever-changing entanglements between a precious resource, flowing water, and evolving human societies. Even smaller rivers are rich sources of cultural and scientific knowledge; and the future of societies depends on the sustainable use of these two forms of knowledge.

In this chapter we have sketched the social history of the Karnaphuli river over two centuries. It is a case study that demonstrates both general patterns and idiosyncrasies. The most recent episode resembles that of many other rivers today. Hydroelectric barrages, flooded valleys, and water pollution occur in numerous streams around the world. Some observers find it helpful to interpret these manipulations as examples of the Anthropocene, a new epoch of geological history that is marked by the emergence of humankind as a dominant factor shaping the evolution of our planet. In the case of the Karnaphuli, human manipulation of the river is age-old, but human dominance can be said to start roughly from the mid-twentieth century.[26] And this dominance was the domain of certain humans only, with others pondering: "whose river is it anyway?" (Baruah 2012).

This calls into question the usefulness of applying the comprehensive concept of the Anthropocene to studies such as these. A very influential notion, it remains quite unstable. Scholars strongly disagree about its onset, its planetary reach, and its geological and social characteristics.[27] The idea of an earth-

26 It is important not to frame river histories in what Greenough has called the standard environmental narrative: "a transition … from a condition of environmental harmony, distributive justice, and material abundance to one of ecological disruption, massive social inequity, and widespread misery." Greenough 2001, 141.

27 Currently, four starting dates are being debated: the Neolithic farming revolution (over 12,000 years ago—"the early Anthropocene"), European colonization (starting some 500 years ago—"the Kleptocene"), the "Industrial Revolution" (starting some 200 years ago),

spanning epoch has recently been challenged by a more flexible approach that views the Anthropocene as a process that has been gathering strength over time (Gibbard et al. 2022). In addition, the concept has spawned several offshoots: the Capitalocene, the Plantationocene, the Chthulucene, and the Age of Destruction, pulling theoretical discussions into conflicting directions. As a result, the Anthropocene does not lend itself easily to meticulous communication.

Even so, the notion of the Anthropocene has been effective in enhancing awareness of the human-induced nature of environmental change. This insight is crucial in studying waterscapes; it helps us to link approaches to river histories rooted in the two separate traditions of the natural sciences and the social sciences.[28] In considering the social history of rivers, geological epochs and planetary scales can highlight some aspects. Planet-wide patterns of human-induced change are clearly at work in the case of the Karnaphuli: global warming, rising sea levels, and crashing biodiversity are changing the scenario.

Significantly, the Karnaphuli is a child of the capricious South Asian monsoon, the future behaviour of which "has implications for the whole world. Arguably no other part of the global climate system affects more people, more directly" (Amrith 2018, chapter 1). In the Karnaphuli region, the South Asian monsoon is predicted to intensify, bringing more extreme rainfall and more frequent cyclones. In addition, rising sea levels threaten the future of all life forms inhabiting the Chittagong plains. The outlook is most uncertain, but the waterscape is clearly entering a new, uncharted phase. These insights have eroded the confidence that humans used to have that they were the dominant party in their dealings with water.

Because river histories are place-specific, we need locally adaptable concepts to explain many features (Schönach 2017). The "hydrosocial cycle," with its emphasis on how water and society make and remake each other, can be helpful. It grants a measure of agency to water in history: "Humans may try to alter rivers when our needs and the reality of a river's nature clash; but the river in the long run will always have its say" (Chakraborty 2018, 1). Rivers shape not only physical landscapes, but also, crucially, the political, cultural,

---

and "the Great Acceleration" of the 1950s. See Keeler 2022. For a distinction between "global" and "planetary," see Chakrabarty 2021.

28 For two studies exemplifying these approaches, see Coulthard and Van De Wiel 2012, and Dobbs 2003. For a study combining them, see Haslam 2008.

and economic histories of human societies. We need a water optic in writing these histories.

Perpetual "making and remaking" occurs all over our "rivered earth," wherever fresh flowing water and humans meet. But place and time are important, so detailed empirical case studies are key to understanding hydrosocial cycles. Comparing these cases allows us to explore "translocal" patterns[29] – similarities that may exist between the hydrosocial dynamics of, say, all monsoon rivers and their human societies. Such insights are indispensable if we are to suggest pathways towards a sustainable future for rivers and all the human and nonhuman organisms that depend on them.

References

Account, "An – of a Journey undertaken by Order of the Board of Trade through the Provinces of Chittagong and Tiperah, in order to look out for the places most proper for the cultivation of Spices," by Francis Buchanan M.D. (British Library, Add MS 19286: Mar 1798–May 1798).

Adnan, Mohammed Sarfaraz Gani, Ashraf Dewan, Khatun E. Zannat and Abu Yousuf Md Abdullah. 2019. "The Use of Watershed Geomorphic Data in Flash Flood Susceptibility Zoning: A Case Study of the Karnaphuli and Sangu River Basins of Bangladesh." *Natural Hazards* 99: 425–48. https://doi.org/10.1007/s11069-019-03749-3.

Alam, Mohammed Shahidul, M. Shahadat Hossain, Md. Mostafa Monwar and Md. Enamul Hoque. 2013. "Assessment of Fish Distribution and Biodiversity Status in Upper Halda River, Chittagong, Bangladesh." *International Journal of Biodiversity and Conservation* 5 (6): 34957. https://doi.org/10.5897/IJBC2013.0555.

Amrith, Sunil. 2018. *Unruly Waters: How Rains, Rivers, Coasts, and Seas Have Shaped Asia's History*. Basic Books.

Archer, Kenna Lang. 2015. *Unruly Waters: A Social and Environmental History of the Brazos River*. University of New Mexico Press.

Arshad-ul-Alam, Mohammad. 2020. "Monograph on Threatened Biodiversity in the Interconnected Rivers of Chattogram with Reference to New Records of Horned Carp and Mahseer Carp." *Journal of Chittagong College* 1: 19–31.

29 For the concept of translocality ("an intermediary concept which helps to better understand and conceptualise connections beyond the local which are, however, neither necessarily global in scale nor necessarily connected to global moments"), see Freitag and Von Oppen 2010, 3.

Arshad-ul-Alam, Mohammad, and Mohammad Ali Azadi. 2015. "Efficiency of Fishing Gears in the river Halda, Chittagong, Bangladesh." *Journal of Fisheries* 3 (3): 275–84. https://doi.org/10.17017/j.fish.104.

Aziz, Mohammad Abdul. 2019. *Atlas on Ganges River Dolphin and Irrawaddy Dolphin of Bangladesh*. Bangladesh Forest Department.

Barbe, M. 1845. "Some Account of the Hill Tribes in the Interior of Chittagong, in a Letter to the Secretary of the Asiatic Society." *Journal of the Asiatic Society of Bengal* 14 (1): 380–91.

Barua, D. Mitra. 2022. "Pāli Studies in Colonial Bengal: Bengali-Speaking Buddhists' Strategy to Distinguish themselves from Hindus." *The Journal of Hindu Studies*: 1–28. https://doi.org/10.1093/jhs/hiac003.

Baruah, Sanjib. 2012. "Whose River is it Anyway? Political Economy of Hydropower in the Eastern Himalayas." *Economic & Political Weekly* 47 (29): 41–52.

Bashar, Md. Abul, Kazi Belal Uddin and Sanjib Basak. 2018. *Sub-Project Completion Report on Production Enhancement of Carps and Tilapia in Creeks of Chittagong Hill Districts*. Bangladesh Agricultural Research Council.

Bautze-Picron, Claudine. 2016. "Images of Devotion and Power in South and Southeast Bengal." In *Esoteric Buddhism in Mediaeval Maritime Asia: Networks of Masters, Texts, Icons*, edited by Andrea Acri, 163–90. ISEAS Publishing.

Berthet, Samuel. 2020. "Circulations in Shadow Corridors: Connectivity in the Northern Bay of Bengal." In: *Shadow Exchanges along the New Silk Roads*, edited by Eva P.W. Hung and Tak-Wing Ngo, 97–123. Amsterdam University Press.

Bhattacharyya, Debjani. 2021. "A River Is Not a Pendulum: Sediments of Science in the World of Tides." *Isis* 112 (1): 141–49.

Bythell, W.J. 1891. "Report on the Survey Operations with the Chittagong Column, Chin-Lushai Expeditionary Force." In: *General Report on the Operations of the Survey of India Department Administered under the Government of India during 1889–90*, edited by H.R. Thuillier, LIII–LXI. Office of the Superintendent of Government Printing, India.

Chakma, Harikishore, Tapash Chakma, Preyasi Dewan and Mahfuz Ullah. 1995. *Boro Porong: The Tale of the Developmental Refugees of the Chittagong Hill Tracts*. Centre for Sustainable Development.

Chakrabarty, Dipesh. 2021. *The Climate of History in a Planetary Age*. University of Chicago Press.

Chakraborty, Abhik. 2018. "Introduction: The Rivers and Society Debate Revisited." In: *Rivers and Society: Landscapes, Governance and Livelihoods*, edited by Malcolm Cooper, Abhik Chakraborty and Shamik Chakraborty, 1–8. Routledge.

Chowdhury, Tamina M. 2016. "Raids, Annexation and Plough: Transformation through Territorialisation in Nineteenth-Century Chittagong Hill Tracts." *The Indian*

Economic and Social History Review 53 (2): 183–224. https://doi.org/10.1177/0019464616634874.

Cioc, Mark. 2002. *The Rhine: An Eco-biography, 1815–2000*. University of Washington Press.

Coulthard, T.J., and M.J. Van De Wiel. 2012. "Modelling River History and Evolution." *Philosophical Transactions of the Royal Society* A 370: 2123–42. https://doi.org/10.1098/rsta.2011.0597.

Das, Palash Chandra. 2022. "Socio-Economic and Livelihood Status of the Fishers' Community of Karnaphuli River in Chattogram District and an Overview of Fish Species Distribution in their Fishing Areas." MS thesis. Chattogram Veterinary and Animal Sciences University.

Dobbs, Stephen. 2003. *The Singapore River: A Social History, 1819–2002*. National University of Singapore Press.

Dunham, Jason B., Paul L. Angermeier, Shelley D. Crausbay, Amanda E. Cravens, Hannah Gosnell, Jamie McEvoy, Max A. Moritz, Nejem Raheem and Todd Sanford. 2018. "Rivers are Social–ecological Systems: Time to Integrate Human Dimensions into Riverscape Ecology and Management." *WIREs Water* 5, e1291: 1–10. https://doi.org/10.1002/wat2.1291.

Elly, E.B. 1893. *Military Report on the Chin-Lushai Country*. Government Central Printing Office.

Flaminio, Silvia, Gaële Rouillé-Kielo and Selin Le Visage. 2022. "Waterscapes and Hydrosocial Territories: Thinking Space in Political Ecologies of Water." *Progress in Environmental Geography* 1 (1–4), 33–57. 4. https://doi.org/10.1177/27539687221106796.

Freitag, Ulrike, and Achim von Oppen, 2010. "Introduction: 'Translocality': An Approach to Connection and Transfer in Area Studies." In *Translocality: The Study of Globalising Processes from a Southern Perspective*, edited by Ulrike Freitag and Achim von Oppen, 1–21. Brill.

Gardner, Kyle. 2019. "Moving Watersheds, Borderless Maps, and Imperial Geography in India's Northwestern Himalaya." *The Historical Journal* 62 (1): 149–70. https://doi.org/10.1017/S0018246X18000146.

Gibbard, Philip, Michael Walker, Andrew Bauer, Matthew Edgeworth, Lucy Edwards, Erle Ellis, Stanley Finney, Jacquelyn L. Gill, Mark Maslin, Dorothy Merritts et al. 2022. "The Anthropocene as an Event, not an Epoch." *Journal of Quaternary Science* 37 (3): 395–99. https://doi.org/10.1002/jqs.3416.

"Grant of extra allowance to Babu B.K. Saha, ass. surgeon in medical charges of civil station Chittagong Hill tracts and his subordinate for work done by them in connection with the Chin Lushai Expedition of 1889–90" (Government of India, Foreign Department, Proceedings, December 1890, Files 183–208). National Archives of India.

Greenough, Paul. 2001. "Naturae Ferae: Wild Animals in South Asia and the Standard Environmental Narrative." In: *Agrarian Studies: Synthetic Work at the Cutting Edge*, edited by James C. Scott and Nina Bhatt, 141–85. Yale University Press.

Guhathakurta, Meghna, and Willem van Schendel, eds. 2013. *The Bangladesh Reader: History, Culture, Politics*. Duke University Press.

Guite, Jangkhomang. 2014. "Colonialism and Its Unruly? The Colonial State and Kuki Raids in Nineteenth Century Northeast India." *Modern Asian Studies* 48 (5): 1188–1232. https://doi.org/10.1017/S0026749X12000674.

Hamilton, Francis. 1825. "An Account of the Frontier between Part of Bengal and the Kingdom of Ava." *Edinburgh Journal of Science* 2: 48–58.

Haslam, S.M. 2008. *The Riverscape and the River*. Cambridge University Press.

Hommes, Lena, Jaime Hoogesteger and Rutgerd Boelens. 2022. "(Re)making Hydrosocial Territories: Materializing and Contesting Imaginaries and Subjectivities through Hydraulic Infrastructure." *Political Geography* 97 (102698): 1–10. https://doi.org/10.1016/j.polgeo.2022.102698.

Hossain, Md. Jaker, Sheikh AftabUddin, Farjana Akhter, Nabila Nusrat, Atikur Rahaman, Mohammad Nurul Azim Sikder, Md. Mostafa Monwar, Mohammed Shah Nawaz Chowdhury, Shan Jiang, Huahong Shi, et al. 2022. "Surface Water, Sediment, and Biota: The First Multi-compartment Analysis of Microplastics in the Karnafully River, Bangladesh." *Marine Pollution Bulletin* 180, 113820: 1–15. https://doi.org/10.1016/j.marpolbul.2022.113820.

Hossen, M.A., F. Rafiq, M.A. Kabir and M.G. Morshed. 2019. "Assessment of Water Quality Scenario of Karnaphuli River in Terms of Water Quality Index, South-Eastern Bangladesh." *American Journal of Water Resources* 7 (3): 106–110. https://doi.org/10.12691/ajwr-7-3-3.

Hutchinson, R.H. Sneyd. 1906. *An Account of the Chittagong Hill Tracts*. Bengal Secretariat Book Depôt.

Hutchinson, R.H. Sneyd. 1909. *Eastern Bengal and Assam District Gazetteers: Chittagong Hill Tracts*. Pioneer Press.

Iqbal, Iftekhar. 2009. "Fighting with a Weed: Water Hyacinth and the State in Colonial Bengal, c. 1910–1947." *Environment and History* 15 (1): 35–59. https://doi.org/10.3197/096734009X40465.

Iqbal, Iftekhar. 2020. "From Zomia to Holon: Rivers and Transregional Flows in Mainland Southeastern Asia, 1840–1950." *Suvannabhumi* 12 (2): 141–55. https://doi.org/10.22801/svn.2020.12.2.141.

Iqbal, Iftekhar. 2022. "Rivers of Mobility: Multi-ethnic Societies and Ecological Commons in a Fluvial Asia." In: *Yunnan–Burma–Bengal Corridor Geographies: Protean Edging of Habitats and Empires*, edited by Dan Smyer Yü and Karin Dean, 105–121. Routledge.

Islam, Md. Sirajul, Natun Basu Chakma, Prayasi Chakma, Md. Humayun Kabir, Rifat Shahid Shammi, Mir Md. Mozammal Hoque and Kazi Farhed Iqubal. 2022. “Seasonal Variation of Water Qualities in River, Lake and Waterfall in Rangamati and Khagrachhari Hill Tracts, Bangladesh.” *Grassroots Journal of Natural Resources* 5 (2): 59–79. https://doi.org/10.33002/nr2581.6853.050205.

Jackson, Kyle. 2023. *The Mizo Discovery of the British Raj: Empire and Religion in Northeast India, 1890–1920*. Cambridge University Press.

Jhala, Angma Dey. 2019. *An Endangered History: Indigeneity, Religion, and Politics on the Borders of India, Burma, and Bangladesh*. Oxford University Press.

Jhala, Angma D. 2023. “Dam(n)ing the Hills: Indigeneity, American Aid, and Cold War Politics in the Kaptai Dam, East Pakistan, 1957–1964.” *Modern Asian Studies* 57 (1): 70–99. https://doi.org/10.1017/S0026749X21000561.

Jordan, Terry G. 1978. “Perceptual Regions in Texas.” *Geographical Review* 68 (3): 293–307. https://doi.org/10.2307/215048.

Keeler, Kyle. 2022. “Before Colonization (BC) and after Decolonization (AD): The Early Anthropocene, the Biblical Fall, and Relational Pasts, Presents, and Futures.” *Environment and Planning E: Nature and Space* 5 (3): 1341–60. https://doi.org/10.1177/25148486211033087.

Khan, M. Monirul H. 2019. *Management Plan for the Ganges River Dolphin in Halda River of Bangladesh*. UNDP/Bangladesh Forest Department.

Kühn, Julius. 1884. “Zoological – Remarks on the Gayal Skull Brought by Dr. Emil Riebeck from Chittagong.” In Emil Riebeck. 1885. *The Chittagong Hill-Tribes: Results of a Journey Made in the Year 1882*. Asher & Co: 1–3.

Lalthangliana, B. 2005. *Culture and Folklore of Mizoram*. Ministry of Information and Broadcasting, Government of India.

Lalthlengliana, C. 2013. *The Lushai Hills: Annexation, Resistance and Pacification 1886–1898*. Akansha Publishing House.

Lewin, T.H. 1869. *The Hill Tracts of Chittagong and the Dwellers Therein, with Comparative Vocabularies of the Hill Dialects*. Bengal Printing Company.

Lewin, Thomas H. 1912. *A Fly on the Wheel, or How I Helped Govern India*. Constable & Company.

Linton, Jamie, and Jessica Budds. 2014. “The Hydrosocial Cycle: Defining and Mobilizing a Relational-dialectical Approach to Water.” *Geoforum* 57: 170–80. http://dx.doi.org/10.1016/j.geoforum.2013.10.008.

Ludden, David. 2019. “India’s Spatial History in the Brahmaputra-Meghna River Basin.” In: *Landscape, Culture, and Belonging: Writing the History of Northeast India*, edited by Neeladri Bhattacharya and Joy L.K. Pachuau, 23–37. Cambridge University Press.

Mountfort, Guy. 1969. *The Vanishing Jungle: The Story of the World Wildlife Fund Expeditions to Pakistan*. Collins.

Nahar, Ainoon. 2022. "Analyzing Requirements for Environmental Protection of the Halda River in Bangladesh." *Global Mainstream Journal* 1 (1): 1–6.

Pachuau, Joy L.K. 2022. "'Circulations' along the Indo-Burma Borderlands Networks of Trade, Religion, and Identity." In *Flows and Frictions in Trans-Himalayan Spaces Book: Histories of Networking and Border Crossing*, edited by Gunnel Cederlöf and Willem van Schendel, 105–36. Amsterdam University Press.

Pachuau, Joy L.K., and Willem van Schendel. 2015. *The Camera as Witness: A Social History of Mizoram, Northeast India*. Cambridge University Press.

Pemberton, R. Boileau. 1835. *Report on the Eastern Frontier of British India, with an Appendix and Maps*. Baptist Mission Press.

Plaisted, Bartholomew. 1764. "Islamabad or Chittagong river: surveyed in 1764." In *Dalrymple's 52 Charts of the Bay of Bengal, 1772–1802*. Alexander Dalrymple.

Qanungo, Suniti Bhushan. 1998. *Chakma Resistance to British Domination, 1772–1798*. Shanti Press.

Ray, Himanshu Prabha. 2021. *Coastal Shrines and Transnational Maritime Networks across India and Southeast Asia*. Routledge.

Reid, R.N. 1942. *History of the Frontier Areas Bordering on Assam, from 1883–1941*. Assam Government Press, 1942.

Rennell, James. 1776. "A map of Bengal, Bahar, Oude & Allahabad: with part of Agra and Delhi, exhibiting the course of the Ganges from Hurdwar to the sea by James Rennell, F.R.S., late Surveyor Genl. in Bengal." Wm. Faden 1786.

Report on Native Papers in Bengal for the Week ending August 2, 1902. 1902. Bengal Secretariat Press.

Riebeck, Emil. 1885. *The Chittagong Hill-Tribes: Results of a Journey Made in the Year 1882*. Asher & Co.

Saikia, Arupjyoti. 2019. *The Unquiet River: A Biography of the Brahmaputra*. Oxford University Press.

Schönach, Paula. 2017. "River Histories: A Thematic Review." *Water History* 9: 233–57. https://doi.org/10.1007/s12685-016-0188-4.

Sen, Anandaroop. 2018. "Early Years of East India Company Rule in Chittagong: Violence, Waste and Settlement *c.* 1760–1790." *The Indian Economic and Social History Review*, 55 (2): 147–181.

Sen, Dineschandra. 1928. *Eastern Bengal Ballads, Vol 3*. University of Calcutta.

Sen, Dineschandra. 1932. *Eastern Bengal Ballads, Vol 4*. University of Calcutta.

Serajuddin, A.M., and John Buller. 1984. "The Chakma Tribe of the Chittagong Hill Tracts in the 18th Century." *The Journal of the Royal Asiatic Society of Great Britain and Ireland* 116 (1): 90–98.

Seth, Vikram. 2011. *The Rivered Earth*. New Delhi: Hamish Hamilton/Penguin.

Shakespear, J. 1895. "The Lushais and the Land They Live In." *Journal of the Society of Arts* 43: 167–85.

Shakespear, J. 1912. *The Lushei Kuki Clans*. Macmillan and Co.

Shakespear, L.W. 1929. *History of the Assam Rifles*. Macmillan and Co.

Sikder, Md. Sowayib, Md. Mustainur Rahman, Md. Monzer Hossain Sarker and Tamanna Hossen. 2012. "A Comparative Study on Water Quality Criteria between Industrial and Non-Industrial Zones of Karnaphuli River of Bangladesh." *Bangladesh Journal of Public Administration* 21 (1): 51–61.

Sloan, Sean, Clinton N. Jenkins, Lucas N. Jappa, David L.A. Gaveau and William F. Laurance. 2014. "Remaining Natural Vegetation in the Global Biodiversity Hotspots." *Biological Conservation* 177: 12–24. https://doi.org/10.1016/j.biocon.2014.05.027.

Smith, Brian D., Benazir Ahmed, Muhammad Edrise Ali and Gill Braulik. 2001. "Status of the Ganges River Dolphin or Shushuk Platanista gangetica in Kaptai Lake and the Southern Rivers of Bangladesh." *Oryx* 35 (1): 61–72.

Tianghlima, R.V.L. 2022. "British Forest Policy and Resource Management in the Lushai Hills." PhD thesis, Mizoram University.

Tonson, T. Vanlal Remruata. 2021. "Sacrificial Chants of Pre-Christian Mizo Belief System." *Historical Journal Mizoram* 21: 79–93.

Tvedt, Terje. 2021a. *The Nile: History's Greatest River*. I.B. Tauris.

Tvedt, Terje. 2021b. *Water and Society: Changing Perceptions of Societal and Historical Development*. I.B. Tauris.

Tvedt, Terje, and Terje Oestigaard. 2006. "Introduction." In: *A History of Water*, edited by Terje Tvedt and Eva Jakobsson, Series 1, Vol. 3, IX–XXII. I.B. Taurus.

Van Schendel, Willem (ed.). 1992. *Francis Buchanan in Southeast Bengal: His Journey to Chittagong, the Chittagong Hill Tracts, Noakhali and Comilla*. University Press Ltd.

Van Schendel, Willem. 2005. *The Bengal Borderland: Beyond State and Nation in South Asia*. Anthem Press.

Van Schendel, Willem. 2015. "Spatial Moments: Chittagong in Four Scenes." In: *Asia Inside Out: Connected Places*, edited by Eric Tagliacozzo, Helen F. Siu, and Peter C. Perdue, 98–127. Harvard University Press.

Van Schendel, Willem. 2016. "A War within a War: Mizo Rebels and the Bangladesh Liberation Struggle." *Modern Asian Studies* 50 (1): 75–117. https://doi.org/10.1017/S0026749X15000104.

Van Schendel, Willem. 2020. *A History of Bangladesh*. Cambridge University Press.

Van Schendel, Willem, Wolfgang Mey, and Aditya Kumar Dewan. 2000. *The Chittagong Hill Tracts: Living in a Borderland*. White Lotus Press.

Varte, Immanuel Zarzosang. 2017. "In Search of the 'Holy' Confluence: A Journey to the Barak River." In *Doing Autoethnography*, edited by Queenbala Marak, 268–85. Serials Publications.

Wesselink, Anna, Michelle Kooy and Jeroen Warner. 2017. "Socio-hydrology and Hydrosocial Analysis: Toward Dialogues Across Disciplines.' *WIREs Water* 4, e1196. https://doi.org/10.1002/wat2.1196.

Wilson, Roderick I. 2021. *Turbulent Streams: An Environmental History of Japan's Rivers, 1600–1930*. Brill.

CHAPTER 3

River Territory: Spatial History, Power and Precarity

David Ludden

Rivers have names, mythologies, spirits, secrets, bodies, histories, and personalities. Their social life is deeply woven into civilization: throughout human history, they animate aquatic ecosystems; they are nature's vital engine of productive mobility, flowing down from mountains into above and below ground in streams made from vapour in clouds, from ice in glaciers, and from saltwater seas, providing humans the fresh water that is literally life itself.

Water forms 60 per cent of our body mass. Humans can live for only about three days without fresh water. A 60-year-old person drinking a healthy minimum of three litres per day would have consumed about seventy thousand litres (18.5 thousand gallons) in their lifetime. We also need lots of fresh water for cleaning, cooking, and producing food. Moisture is everywhere; fresh water is not. Oceans hold all but 3 per cent of Earth's water; ice traps 70 per cent of all fresh water; leaving less than 1 per cent of the Earth's water to support all living things (including tiny creatures that kill a million humans annually and sicken millions more with water-borne disease) (WHO 2023). Clean fresh water that humans need must spend part of its life flowing.[1]

The spatial and social distribution of flowing fresh water frames human history. Most people live near natural fresh water sources; aqueducts, pumps, and pipes expand its proximity (Kummu et al. 2011). More than half the world's people live along Asian rivers, whose flooding has sustained farming for millennia and also caused damage that becomes more severe as populations grow and heavy rains and rising seas accompany climate change (Floodlist; Chen et al. 2012). River transportation and productivity channel human migration and capital investments into an ever-increasing exploitation of freshwater ecosystems, rendering them ever-more essential for the political economy of empires and nations and also increasingly vulnerable to poisonous pollution. Meanwhile, the demands for clean fresh water keep growing, with two billion more people in the last twenty years, and urbanization, farming, food

1 For Earth water statistics, see US Geological Survey https://www.usgs.gov/. Also the Hinrich foundation, Visual Capitalist, "Deep Dive: The rising contest over Asia's water", https://www.hinrichfoundation.com/research/wp/sustainable/the-rising-contest-over-asia-water/.

 | HTTPS://DOI.ORG/10.1163/9789004758100-005

processing, and energy production further increasing demand. Results today include increasing conflict over rivers (Pacific Institute; Visual Capitalist) in a world where 2.2 billion people have no safe water (UN Water), aggravating hunger and sickness (UNICEF), and where water control has become a weapon of subjugation (Sharma 2018; Amnesty International 2017).

1 River Territory

In that troubling contemporary context, scholars are paying more attention to rivers (Wagner and Jacka 2018), and public attention to water problems is increasing. In 1993, the UN declared 22 March as World Water Day, "to tackle the global water crisis" (UNICEF). On World Water Day, in 2012, the Asia Foundation declared Asia "the world's most water-stressed continent" (The Asia Foundation). The online open access resource journal, *Water Alternatives*, began in 2008; the journal *Water History* began in 2009.[2]

Rivers have emerged as a distinct form of aquatic terrain, as bounded territories of contested, contingent control over flowing water's ubiquitous transformative unbounded mobility, which takes a great many forms, including inside the cells of living things. In that context, we can think of water as essentially mobile. Aquatic ecosystems flow inside organisms and travel all around the planet. Rivers take shape under forces variously amenable to human cognition: they are natural and also human, constructed physically and conceptually, and densely woven into social worlds (Cunha 2018).

Weaving flowing water into social life is a naturally human project, fundamental for cultures and politics, as we can see in ancient Tamil bardic poems that praise kings by saying,

> Those who give food give life to all living beings,
> who cannot live without water.
> Food is first for all bodies made of food,
> and food is but soil and water mixed.
> So those who unite water and soil
> create living bodies and life in this world.
> Even kings with vast domains
> strive in vain when the land is dry,

2 https://www.water-alternatives.org/, and https://link.springer.com/journal/12685.

when fields sown with seed
must look to the sky for rain.

So Pandyan king who makes dreadful war,
do not disdain my words:
expand quickly those watery places
built where water moves across the land!
For those who harness water
bring rewards for themselves,
and those who fail cannot endure. (*Purananuru* No.18, 15–30)

Those "watery places" were very small spaces of river drainage, dammed with dirt and stone to train moving water into paddy fields to grow rice to feed people who praised kings who deployed labour to water the land. Various renditions of that dynamic of interwoven interactive space of rain, water, soils, plants, food, culture, and power formed a vast diversity of river territories in Monsoon Asia, where the social life of rivers flows with natural and human input at expanding levels of scale over the centuries.

2 Spatial History

Rivers are linear forms of watery placemaking living in Asian monsoon spaces that connect planetary and earthly spatial scales of mobile aquatic ecosystems. Monsoon airborne spaces of flying wind and of falling rain travel through nature's topography into flowing spaces where people make places to work with water on land. In those spaces, the Anthropocene is a constructed historical time, inside the geological idea of the Holocene (Emmett and Lekan 2016; Carrington 2024), where climate warming is causing extreme water scarcity (aridity) and water excess (flooding) (World Meteorological Association) at a global scale (Handwerk 2019). The physical impact of climate change on aquatic ecosystems is nevertheless still regionally and locally specific, as it has been throughout geological time. River territory spatial histories continue to be shaped at various levels of scale by nature's topography and by human construction, and, in addition, by social power relations that today pile the worst climate suffering onto people least responsible for the pollution which is aggravating Anthropocene extremes (Ludden 2012).

River territory spatial histories participate in the Anthropocene acceleration of rural–urban migration and sprawling urban construction during industrial capitalist globalization, which is increasing the spatial and social inequity and making Asian megacities ever-more gigantic.[3] Earlier epochs of climate warming have also produced major spatial transformations of watery places. Around the turn of the Common Era, four warm ancient centuries of increasing rain fed farms and increased the population and imperial territory in Rome, Han China, and India. Cooler centuries that followed (*circa* 400–800) hosted massive migrations of Turkic and Arabic peoples who galloped and marched across arid lands in west, central, and east Asia; Buddhism and Islam travelled with them among river oases in the heyday of the Silk Road. A second set of warm centuries (800 to 1200) (Quinn 1992) encouraged a broad southern shift of Asia's growing population: China moved south; the Yangtze became China's engine of growth, and its delta led a long upward trend in population, trade, productivity, and living standards (Shaffer 1994; Pomeranz 2000). Tibeto-Burman peoples migrated south in river valleys to conquer inland Southeast Asia (Lieberman 2009). In India, the Chola Empire conquered and expanded irrigated territory across the South, as royal investments in productive labour control fed the rise of Brahmanism, decline of Buddhism, and expansion of productive territorial social inequity in the name of caste, all across South Asia (Ludden 2018).

Nature's water spaces and human territorial placemaking have always travelled together among rivers in Monsoon Asia: these shifting mobile spaces formed the Asia territories that we see on maps today, where rivers host over half of the total human population on planet Earth, and probably much more of its wealth. River valleys have always been major arteries for migration and lifelines for merchants, arts, religion, and rulers. River banks are prime locations for settlers and investors who build the land-and-river-facing central places that anchor networks of mobility travelling along rivers and overland across river catchments among river valleys, creating spatial frames for social, economic, and political territory. Major cities and towns multiply and grow along rivers; they fill river-connected watery places with construction material, economic opportunity, roads, railways, airports, and pollution. Rivers organize expansive inland and overseas mobility: routes along river valleys form the warp, and routes crossing rivers form the woof, of the mobilities that weave together imperial territories in Monsoon Asia, from ancient to modern times (Ludden 2003).

3 Seven of the top ten most populous cities in the world are in Asia. All are on rivers and/or seaports: Tokyo (37,435,191), Delhi (29,399,141), Shanghai (26,317,104), Dhaka (20,283,552), Mumbai (20,185,064), Beijing (20,035,455), Osaka (19,222,665).

3 Imperial Modernity

River territories have changed dramatically over the centuries. There is a broad tendency to view modern history in terms of major breaks and turning points, when the new replaces the old, like "the great transformation" of the nineteenth century (Polanyi 2001; Pomeranz 2000; Rosling 2009) and now the Anthropocene, which "the Anthropocene Working Group agreed … began in the year 1950, when the Great Acceleration, a dramatic increase in human activity affecting the planet, took off" (National Geographic Education). Historians tend to narrate the impact of European conquest in that way, as a dramatic break; but the picture looks more complex in Asia (Ludden 2002; Gommans and Zurndorfer 2008), when viewed in the long term and focused on everyday local agency.[4] So it is with the social life of Asia's many hundreds of rivers.

My own perspective is anchored in a tightly knit family of rivers in the Tirunelveli region of Tamil Nadu, at the tip of India, where the ancient bards praised their Pandyan king. This small but complex interconnected multi-river system drains the Western Ghat mountains and flows east fifty miles toward the Gulf of Mannar between India and Sri Lanka. It has one famous named river, the Tambraparni (Wikimedia Foundation 2025). It is technically perennial (a rarity in dry south India), but none of its water reaches the sea because man-made channels feeding hundreds of low embankments to hold water feed crops for months at a time in a dry region where rain is scarce and water is precious.

Water is precious because it mingles with soil to feed life and because "those who harness water bring rewards for themselves", as the ancient bards say. Local, regional, and imperial families of wealth and power followed the bard's advice. Early inscriptions indicate that river diversion to feed rice was underway upstream early in the Common Era; it continued in place after place downstream, over two millennia, on all water routes converging on the Tambraparni.

As state territorial power expanded over the centuries, imperial systems of resource control became more elaborate and multi-layered, fed by productive river systems. The social life of rivers and the agency of river communities energized imperial modernity with agrarian productivity and human capital invested in the wealth of water control. Profits and revenues from rivers fed imperial regimes from medieval Cholas to modern India. The size of dams, expanse of irrigated acreage, and scale and diversity of farm production

4 Two early examples are Doumani, Beshara. 1995. *Rediscovering Palestine: merchants and peasants in Jabal Nablus, 1700–1900*. University of California Press and Ludden, David. 1985. *Peasant History in South India*. Princeton University Press.

increased to feed the accumulation of wealth and power in larger and larger political territories with capitals farther and farther north, with bigger and bigger armies, needing more tax revenues, and with more personnel and financial assets to invest in building irrigation dams, channels, and paddy fields to enrich local elites and feed their workers and imperial elites.

The Tirunelveli family of rivers attracted conquerors, investors, and migrant settlers who added complexity and diversity to its social life. Migrant settlers also turned surrounding dry grass and scrub lands into cotton fields that fed textile manufacturing in irrigated localities where intensive cultivation sustained increasingly complex commercial life and growing populations, providing revenues for rulers, profits for inland and overseas merchants, and attractions for conquerors, eventually, the Portuguese, Dutch, French, and English (Ludden 2004; 1966).

The nineteenth century launched a “great transformation” that accelerated long trends in imperial expansion and the intensification of water control. Industrial technologies, engineering, and the rapacious financial hunger of the British Empire and then the Republic of India took ever-more intense control of rivers in their grasp. The imperial political economy concentrated more wealth, power, and population in higher-ranking urban centres, where the ruling elites focused investments on large projects centrally managed by professional bureaucracies. Imperial institutions—and eventually global development financiers, like the World Bank—ranked the potential of river projects by the scale of their financial benefits. Large dams to irrigate large tracts and the supply of massive amounts of urban electricity became the order of the day (Paranjpye 1990).

The Punjab Canal Colonies were the epitome of imperial river system control in British India: they turned a vast arid pasture landscape into the most productive farmland in the world, reshaping whole societies in the process (Gilmartin 2015). Large imperial project imperatives are domineering to this day, exemplified by Narmada dams built to enrich Gujarat and destroy tribal homelands in the hills, by the analogous Kaptai Dam in the Chittagong Hill Tracts, and by the ongoing surge of Anthropocene hydroelectric schemes (Baviskar 1995; Nayak 2019; Vagholikar and Das 2010).

A large dam in the lower Tambrarparni stopped water from reaching the sea, and a high dam at the origin of the river took control of the flow of the whole river. Domineering large project imperatives led to the decimation of mountain ecologies and to the decline of small-scale drainage irrigation and local community control over river investments in the plains, where a vast programme promoting private well-digging would feed the Green Revolution (Ludden 1979). Long before 1947, the social life of water flowing in South Asia was locked into a techno-imperial bureaucracy under the legal authority of a modern state based in New Delhi under economic and financial controls exerted by the operational imperatives of a capitalist development regime with headquarters first in London and then Washington (Ludden 1992; 2006).

The social life of Asian rivers today thus inhabits very long histories of imperial globalization. Tirunelveli provides one example of how large-scale trends translate in river localities. For two thousand years, rulers and local elites invested in controlling Tirunelveli river water by patronizing Brahmans, Hindu temples, and a caste social order to discipline labour for intensive irrigated rice production. River territory became *dharma* territory (Ludden 1999; 1985).[5] Migrant settlers increased and diversified the population; Brahmans came from all over India; and non-Brahman farmers settled in dry cotton-growing farm localities; the result was a dynamic spatial economy of commercial manufacturing weaving the Tirunelveli region around the elite imperial authority led by Brahmans who dominated the regional government, business, and culture in urbane segregated hometowns on the banks of the Tambraparni (Ludden 1996; 1988).

During the twentieth-century imperial transition from British to independent India, Brahman social mobility led them to high-status educational and employment opportunities in the higher imperial ranks of central places: in India's capital cities, there was a general Brahmanization of India's national elite. That triggered an upsurge of non-Brahman politics in various forms, which spawned the Partition of British India, and the eventual capture of politics by competing majority non-Brahman parties in South India (Washbrook 1976; Subramanian 1999; Ludden 2002).

In the process, Tambraparni social life lost its elite Brahman allure, and political and economic power moved up the imperial ranks to more important places, as river problems of increasing urbanization set in along the Tambraparni. As the Anthropocene spatial inequities of capitalist economic development increased, favouring urban over rural areas, more urban migration, construction, water pollution, drainage obstruction, and river damage ensued. Raw sewage and paper mill and industrial effluents were dumped into the river. Though sand mining was banned, it continued along with encroachments onto riverbanks that altered the river flow. Urban construction blocks drainage, increases flooding, and aggravates flood damage.

All the issues in Tirunelveli's river history are relevant for Asian rivers generally. Rivers typically host early, continuous, and accelerating agricultural

5 For details, see Ludden, *Peasant History in South India*. Ludden, *An Agrarian History in South Asia*.

development, imperial state-building, commercialism, manufacturing, urbanism, and elite formation (Mostern and Ryan 2021; Storozum et al. 2008; Greer 1979). Their imperial value increases when imperial mobility and territorial authority based on river resources expand, as they do dramatically in modern times (Ludden 2008; 2017; Romanos 2024). Their ecosystems sustain a wide range of lifeways and many claims to river resources, which become subject to documentation, study, control, and adjudication by state authorities (Baqee 1998; Chaphekar and Mhatre 1986; Swain 1998). The larger they are—and Asia has the largest array of the world's largest rivers, mostly descending from the Tibetan plateau—the bigger the problems that they pose for state politics (Crow 1995; Saikia 2019; Chaphekar and Mhatre 1986; Meenakshisundaram et al. 2010), economic development, public health and wellbeing (Baqee 1998), and imperial order (Hill 1997).[6] The more distant a river territory is from centres of state power, the more marginal it is for imperial power, and the more likely that it will host lifeways that resist incorporation into imperial territory (Ludden 2011) and thus into modern political economies (Khan 2023). The more limited their water supplies—and aridity is increasing in much of Asia—the more vulnerable that rivers are to destruction as rivers: small seasonal rivers that once ran through Chennai are now drains with no outlet except city streets; the Tambraparni is in effect a series of channels ending in paddy fields; and the Syr Darya and Amu Darya have been dammed and pumped dry to feed cotton fields before they reach what was once, but is no longer, a thriving Aral Sea (Edelstein et al. 2012).

The obstruction and pollution of rivers which have piled up during centuries of industrial capitalist global expansion and atmosphere pollution, which gave us the Anthropocene, now aggravate our climate traumas. In addition to the concrete sprawl of cities and roads, over 35 thousand obstructions like dams and locks damage ecologies and throw spanners into the work of water management (Kornei 2020). In the Sundarbans located in the Brahmaputra-Ganga delta, and other coastal areas, low-level obstructions to the outflow of fresh water create saline polders for shrimp farming that are praised in high places as a climate change adaptation for investors facing the inevitability of rising seas, but in fact, shrimp farms displace vibrant fresh water paddy farms and kill thriving river ecosystems, poisoning the land, and driving poor farmers into city slums (Paprocki 2021). This is another example of how capitalist innovation and creative destruction in the name of profit increase the current

6 In 1820, when the East India Company sought to integrate Assam into British Bengal, they found the old city of Dibrugarh submerged in the Brahmaputra. Deepali Barua, *Urban History of India (A Case Study)*. New Delhi: Mittal Publications, 1994, 4.

inequity, impoverishing poor people, and worsening the urban sprawl of poor neighbourhoods most vulnerable to flooding. As cities expand into rivers, they block drainage, causing floods; they create more flood victims, and more damage and recovery costs (Dasgupta et al. 2015; Halder and Majed 2023; Hosseinzadeh 2005). The uncontrolled urban sprawl is also massively polluting rivers with untreated sewage and all kinds of garbage.

This is where I will end, having travelled historically from ancient farming villages in southernmost India though centuries of slow population growth and farming frontier expansion around Asia into the modern age of high industrial capitalist imperial inequity where cities are dumping millions of tonnes of poisonous junk into the sea (Ludden 1999). It is sadly appropriate to end with The World Economic Forum report on research indicating that eight of the ten rivers that deliver almost all the plastic are in Asia: the Yangtze, Indus, Huang He, Hai, Ganga, Pearl, Amur, and Mekong. (Two are in Africa: the Nile and the Niger.) If they continue to dump eight million tonnes of plastic annually, there could be more plastic than fish in the sea in thirty years (Gray 2018).

References

Amnesty International. 2017. "The Occupation of Water." https://www.amnesty.org/en/latest/campaigns/2017/11/the-occupation-of-water/.

Baqee, Abdul. 1998. *Peopling the Land of Allah Jaane: Power, Peopling, and Environment: The Case of Char-Lands of Bangladesh*. University Press Limited.

Barua, Deepali. 1994. *Urban History of India (A Case Study)*. Mittal Publications.

Baviskar, Amita. 1995. *In the Belly of the River: Tribal Conflicts Over Development in the Narmada Valley*. Oxford University Press.

Carrington, Damian. 2024. "Geologists reject declaration of Anthropocene." *The Guardian*, March 22. https://www.theguardian.com/profile/damiancarrington.

Chaphekar, S.B. and G.N. Mhatre. 1986. *Human Impact on Ganga River Ecosystems; An Assessment*. Concept.

Chen, Yunzhen, James P.M. Syvitski, Shu Gao, Irina Overeem, and Albert J. Kettner. 2012. "Socio-Economic Impacts on Flooding: A 4000-year history of the Yellow River, China." *Ambio* 41: 682–698.

Crow, Ben. 1995. *Sharing the Ganges: The Politics and Technology of River Development*. Sage.

Cunha, Dilip Da. 2018. *The Invention of Rivers: Alexander's Eye and Ganga's Descent*. University of Pennsylvania Press.

Dasgupta, Susmita et al. 2015. *Urban Flooding of Greater Dhaka in a Changing Climate: Building Local Resilience to Disaster Risk*. World Bank Publications.

Doumani, Beshara. 1995. *Rediscovering Palestine: Merchants and Peasants in Jabal Nablus, 1700–1900*. University of California Press.

Edelstein, Michael R., Astrid Cerny, and Abror Gadaev, eds. 2012. *Disaster by design: The Aral Sea and its lessons for sustainability*. Emerald Group Publishing Limited.

Emmett, Robert and Thomas Lekan, eds. 2016. *Whose Anthropocene? Revisiting Dipesh Chakrabarty's 'Four Theses*. RCC Perspectives: Transformations in Environment and Society 2.

Floodlist. "Asia." https://floodlist.com/asia.

Gilmartin, David. 2015. *Blood and Water: The Indus River Basin in Modern History*. University of California Press.

Gommans, Jos, and Harriet Zurndorfer, eds. 2008. *Roots and Routes of Development in China and India: Highlights of Fifty Years of the Journal of the Economic and Social History of the Orient* (1957–2007). BRILL.

Gray, Alex. 2018. "Around 90% of all river-borne plastic that ends up in the ocean comes from just 10 rivers." In *World Economic Forum*. Accessed April 1, 2024. https://www.weforum.org/agenda/2018/06/90-of-plastic-polluting-our-oceans-comes-from-just-10-rivers/.

Greer, Charles. 1979. *Water Management in the Yellow River Basin of China*. University of Texas Press.

Halder, Anupom, and Nehreen Majed. 2023. "The effects of unplanned land use and heavy seasonal rainfall on the storm-water drainage in Dhaka metropolitan city of Bangladesh." *Urban Water Journal* 20 (6): 707–722. DOI: 10.1080/1573062X.2023.2209065.

Handwerk, Biran. 2019. "Modern Climate Change Is the Only Worldwide Warming Event of the Past 2,000 Years." *Smithsonian Magazine*, July 24. https://www.smithsonianmag.com/science-nature/modern-climate-change-only-worldwide-warming-event-past-2000-years-180972719/.

Hill, Christopher V. 1997. *River of Sorrow: Environment and Social Control in Riparian North India, 1770–1994*. Association for Asian Studies, Inc.

Hinrich foundation and Visual Capitalist. "Deep Dive: The rising contest over Asia's water." https://www.hinrichfoundation.com/research/wp/sustainable/the-rising-contest-over-asia-water/.

Hosseinzadeh, Seyed Reza. 2005. "The Effects of Urbanization on the Natural Drainage Patterns and the Increase of Urban Floods: Case study Metropolis of Mashhad-Iran." In *Sustainable Development and Planning* II, vol. 1: 423–32 (WIT Transactions on Ecology and the Environment 84), edited by A.G. Kungolos, C.A. Brebbia, and E. Beriatos. Wit Press.

Khan, Naveeda. 2023. *River Life and the Upspring of Nature*. Duke University Press.

Kornei, Katherine. 2020. "Europe's Rivers Are the Most Obstructed on Earth." *EOS*. Accessed March 25, 2024.https://eos.org/articles/europes-rivers-are-the-most-obstructed-on-earth.

Kummu, Matti, Hans De Moel, Philip J. Ward, and Olli Varis. 2011. "How close do we live to water? A global analysis of population distance to freshwater bodies." *PloS one* 6 (6): e20578.

Li, Lillian. 1991. "Life and Death in a Chinese Famine: Infanticide as a Demographic Consequence of the 1935 Yellow River Flood." *Comparative Studies in Society and History* 33 (3): 466–510.

Lieberman, Victor. 2009. *Strange Parallels.* Vol. II: Southeast Asia in Global Context, c. 800–1830. Mainland Mirrors Europe, Japan, China, South Asia, and the Islands. Cambridge University Press.

Ludden, David. 1966. "Tirunelveli in the Economic History of Early-Modern India." In *History and Society: Essays in Honour of Professor S. Kadhirvel*, edited by K.A. Manikumar. Madras.

Ludden, David. 1979. "Patronage and Irrigation in Tamil Nadu: A Long-Term View", *Indian Economic and Social History Review* 16 (3): 347–65, reprint in *Environmental History of Early India: A Reader*, edited by Nandini Sinha Kapur, Oxford University Press. 2011.

Ludden, David. 1985. *Peasant History in South India.* Princeton University Press.

Ludden, David. 1988. "Agrarian Commercialism in Eighteenth Century South India: Evidence from the 1823 Tirunelveli Census." *Indian Economic and Social History Review*, 25 (4): 493–519, reprinted in *Merchants, Markets and the State in Early Modern India.* Edited by Sanjay Subrahmanyam, Oxford University Press, 1990.

Ludden, David. 1992. "India's Development Regime." In *Colonialism and Culture,* edited by Nicholas Dirks. University of Michigan Press.

Ludden, David. 1996. "Caste society and units of production in early modern South India." In *Institutions and Economic Change in South Asia: Historical and Contemporary Perspectives.*, edited by Burton Stein and Sanjay Subrahmanyam. Oxford University Press.

Ludden, David. 1999. *An Agrarian History of South Asia.* The New Cambridge History of India, vol. iv. Cambridge University Press.

Ludden, David. 2002a. "Spectres of agrarian territory in southern India." *The Indian Economic & Social History Review* 39 (2–3): 233–257. Reprinted in *Land, Politics, and Trade in South Asia*, edited by Sanjay Subrahmanyam. Oxford University Press, 2004.

Ludden, David. 2002b. "Modern Inequality and Early Modernity: A Comment for the AHR on articles by R. Bin Wong and Kenneth Pomeranz." *The American Historical Review* 107 (2): 470–480.

Ludden, David. 2003. "Presidential address: Maps in the mind and the mobility of Asia." *The Journal of Asian Studies* 62 (4): 1057–1078.

Ludden, David. 2004. "The Formation of Modern Agrarian Economies in South India." In *The History of Indian Science, Philosophy and Culture* (*Economic History of India, 18th-20th Centuries*), vol. VII, edited by Binay Bhushan Chaudhuri. Oxford University Press.

Ludden, David. 2005. "Development Regimes in South Asia: History and the Development Conundrum." *Economic and Political Weekly* 40 (37): 4042–51.

Ludden, David. 2008. "Investing in nature around Sylhet: An Excursion into Geographical History." In *Contested Grounds: Essays on Nature, Culture and Power*, edited by Amita Baviskar. Oxford University Press.

Ludden, David. 2011. "The process of empire: frontiers and borderlands", in *Tributary Empires in Global History*, edited by Peter Fibiger Bang and C.A. Bayly. Palgrave Macmillan.

Ludden, David. 2012. "Imperial Modernity: History and Global Inequity in Rising Asia." *Third World Quarterly* 33 (4): 581–602, republished in *Identity, Inequity and Inequality in India and China: Governing Difference*, edited by Ravinder Kaur and Ayo Wahlberg, Routledge, 2014.

Ludden, David. 2017. "Country Politics and Agrarian Systems: Land grab on Bengal frontiers, 1750–1800." *Modern Asian Studies* 51 (2): 319–349.

Ludden, David. 2018. "The Decline of Buddhism, Revisited: The Rise of Hindu Territorial Hegemony, *c*.950–1250." In *Clio and Her Descendants: Essays for Kesavan Veluthat*, edited by Manu V. Devadevan. Primus Books.

Meenakshisundaram, S.S. et al. 2010. *The Cauvery Conflict: Backgrounders on Conflict*. National Institute of Advanced Studies.

Mostern, Ruth, and Ryan M. Horne, eds. 2021. *The Yellow River: A Natural and Unnatural History*. Yale University Press.

National Geographic Education. "Anthropocene." Accessed 31 March 2024. https://education.nationalgeographic.org/resource/anthropocene/.

Nayak, Arun Kumar. 2019. "Involuntary Displacement and Human Security: A Study of the Kaptai Dam in Bangladesh." *Jadavpur Journal of International Relations* 23 (2): 199–231. https://doi.org/10.1177/0973598419863234.

Pacific Institute. "Water Conflict Chronology." https://www.worldwater.org/conflict/list/.

Paprocki, Kasia. 2021. *Threatening Dystopias: The Global Politics of Climate Change Adaptation in Bangladesh*. Cornell University Press.

Paranjpye, Vijay. 1990. *High Dams on the Narmada; A Holistic Analysis of the River Valley Projects*. INTACH.

Polanyi, Karl. 2001. *Great Transformation: The Political and Economic Origins of Our Time*. Beacon Press.

Pomeranz, Kenneth. 2000. *The Great Divergence: China, Europe, and the Making of the Modern World Economy*. Princeton University Press.

Purananuru No.18, 15–30.

Quinn, William A. 1992. "A Study of Southern Oscillation-Related Climatic Activity for AD 622–1900 Incorporating Nile River Flood Data." In *El Niño: Historical and Palaeoclimatic Aspects of the Southern Oscillation*, edited by Henry F. Diaz and Vera Markgraf. Cambridge University Press.

Romanos, Chris. 2024. "Liquid Territories: The Configuration of Geographic Space in the Cartography of the Mekong River's Catchment Areas." *Water History*: 1–16.

Rosling, Hans. 2009. "200 years that changed the world." Gapminder Foundation. Video, 4 min., 37 sec. https://www.gapminder.org/videos/200-years-that-changed-the-world/.

Saikia, Arupjyoti. 2019. *The Unruly River: A Biography of the Brahmaputra.* Oxford University Press.

Shaffer, Lynda. 1994. "Southernization." *Journal of World History* 5 (1): 1–21.

Sharma, Mukul. 2018. *Caste and Nature: Dalits and Indian Environmental Politics.* Oxford University Press.

Storozum, Michael J., Qin Zhen, Ren Xiaolin, Li Haiming, Cui Yifu, Fu Kui, and Liu Haiwang. 2018. "The collapse of the North Song Dynasty and the AD 1048–1128 Yellow River Floods: Geoarchaeological Evidence from Northern Henan Province, China." *The Holocene* 28 (11): 1759–1770.

Subramanian, Narendra. 1999. *Ethnicity and Populist Mobilization: Political Parties, Citizens, and Democracy in South India.* Oxford University Press.

Swain, Ashok. 1998. "Fight for the Last Drop: Inter-State River Disputes in India", *Contemporary South Asia* 7 (2): 167–80.

The Asia Foundation. "Asia: The World's Most Water-Stressed Continent." https://asiafoundation.org/2012/03/21/asia-the-worlds-driest-continent/.

UN Water. "World Water Day." https://www.unwater.org/our-work/world-water-day.

UNICEF. "Water Scarcity." https://www.unicef.org/wash/water-scarcity.

US Geological Survey https://www.usgs.gov/.

Vagholikar, Neerah and Partha J. Das. 2010. "Damming Northeast India: Juggernaut of hydropower projects threatens social and environmental security of the region." Kalpavriksh, Aaranyak and ActionAid India. Pune/Guwahati/ New Delhi.

Visual Capitalist. "Deep Dive: The rising contest over Asia's water." https://www.hinrichfoundation.com/research/wp/sustainable/the-rising-contest-over-asia-water/.

Wagner, John R. and Jerry K. Jacka, eds. 2018. *Island Rivers: Fresh Water and Place in Oceania.* Australian National University Press.

Washbrook, David A. 1976. *The Emergence of Provincial Politics: The Madras Presidency, 1870–1920.* Cambridge University Press.

WHO. 2023. "Drinking Water." https://www.who.int/news-room/fact-sheets/detail/drinking-water.

Wikimedia Foundation. 2025. "Thamirabarani River." Last modified March 3, at 13:40 (UTC). https://en.wikipedia.org/wiki/Thamirabarani_River.

World Meteorological Association. "Climate change impacts increase in Asia." Accessed March 22, 2024. https://wmo.int/news/media-centre/climate-change-impacts-increase-asia

CHAPTER 4

Monsters and Mysteries: Rivers in the Contemporary Bruneian Imaginary

Kathrina Mohd Daud

The river features prominently in the mythopoetics of Brunei Darussalam. In classical Bruneian Malay literary conceptions of the birth of the Sultanate, such as in the epic poem *Syair Awang Semaun*, for example, the original settlements that became Negeri Brunei were chosen for their strategic proximity to what is now the Brunei river (Haji Tassim 2023). In more prosaic modern conceptions of the nation, such as Prince Mohamed Bolkiah's memoir, *Time and the River*, Bruneian identity is entwined in the mythic language of Bruneians having "come from the river". This relationship between the Bruneian people and the river is most commonly and iconically encapsulated through the portrayal of Kampong Ayer (translated as the "Water Village"), a site of enormous historical and cultural significance, consisting of stilted houses, schools, mosques and communities on the Brunei river. Kampung Ayer is one of the main tourist attractions of Brunei (Izni and Yong 2022), even though the actual population of inhabitants of the village has decreased from year to year (Yong 2021) and it has subsequently "devolved into an unexceptional settlement complex [...] which is bereft of economy and industry" (Yong et al 2026). Despite this, Kampong Ayer continues to be one of Brunei's most recognizable and iconic landmarks, with efforts made in recent decades to register this "cultural hearth" (Yong 2021, 83) as a UNESCO World Heritage Site.

The relationship between the Bruneian people and the river dates back to Brunei's development into an "important thalassocracy (a seaborne empire)" (Druce et al. 2023, 102) which is considered by historians to have occurred over the fifteenth and sixteenth centuries. Its decline in the late sixteenth and seventeenth centuries was followed by a "perilous" (Druce et al. 2023, 104) nineteenth century, during which Brunei's territories continued to decrease in continuous concessions to then governor of Sarawak, James Brooke. During the period of British Residency in Brunei, beginning in 1906, "inhabitants of Kampong Ayer were encouraged to leave and settle on land in order to boost the economy, although very few did" (Druce et al. 2023, 108). According to Haji Tassim, the primary reason for refusing to migrate, despite the British Residents Order in 1910 "prohibiting the residents from building new houses"

 | HTTPS://DOI.ORG/10.1163/9789004758100-006

in Kampong Ayer (146), was "ancestral ties" and that "they could not envisage living on the mainland" (148). Cultural practices relating to living on the water, including the physical houses that were shaped by family units and practices, were reportedly as significant a barrier to the migration to land scheme as a reliance on the "maritime-based economic activities" (148) that formed the basis of the Kampong Ayer livelihoods. As Yong notes, "Kampong Ayer was the main city of the Brunei Sultanate before modernization and transition to a nation state on land. The process has relegated the historic water-based settlement to be just one of many residential areas in the country" (2021, 83). Despite the development of land areas in parallel with the decrease in life on the water, however, the river continues to be of historical and cultural significance in the sultanate.

Today the Bruneian national identity is most closely associated with the national philosophy of *Melayu Islam Beraja* (MIB), which can be translated as Malay Islamic Monarchy. This philosophy was proclaimed as central to Brunei's nationhood during Sultan Hassanal Bolkiah's decree upon Brunei achieving independence in 1984 (Ooi 2022, 35). MIB weaves together as "mutually interrelated" and complementary (ibid) the three facets of "firstly, Malay ethnicity that comprises the Malay language, culture, customs and traditions; secondly, Islam as the principal religious faith; and thirdly, the sultanate as the monarchical form of governance" (ibid). The philosophy governs and shapes almost all aspects of Bruneian daily life, despite a not-insignificant number of the population falling outside the "classificatory umbrella" (Ooi 2022, 41) of Malay Muslim identity.

This chapter aims to articulate the relationship between this national philosophy and governing ideology, and the physical landscape of Brunei, with particular attention to rivers, as imagined by contemporary Bruneian writers. How does the national philosophy shape the way the physical landscape is perceived and conceived? How does the popular imaginary imbue the physical landscape with symbolism about national identity? And what can snapshots of this imaginary at different points in Brunei's history, tell us about the shifting perception of the landscape in the age of the Anthropocene?

These questions are shaped by Brian S. Osborne's work on the interaction between artists, national consciousness and the representation of landscapes in visual art (1992). In this work, Osborne asserted that the "landscape is part of national identity" (230). In considering "the transition of societies from *Gemeinschaft* (community) to *Gesellschaft* (society) [...] an important dimension of the relationship between the continua of community/society and nation state is the varying degree to which they are grounded in a sense of place" (230).

Osborne (1992) asserts that one of the differences between *gemeinschaft* (communities) and *gesellschaft* (societies) is the extent to which physical, tangible materialities are supplemented or replaced by symbolic value. In the former, Osborne (1992) contends that the landscape is "a fundamental concept to a people's sense of community, heritage and nationhood" because "it is the setting for the day-to-day economic, social and ideological activities that serve to unite a distinctive people" (230). In the latter, because "localisms are replaced by new centralizing structures of government, communications and bureaucracy" (230), states must work harder to "consciously nurture shared symbolic constructs that reinforce the sense of membership to the group and the development of a sense of distinctiveness" (231). Although the physical landscape no longer plays the same immediate, everyday role in uniting the group, it can be transformed into a symbolic construct or myth that can represent "collective memories" or "imagined communities" (231). Indeed, the influence of print culture on the development of Benedict Anderson's conceptualization of the nation as an "imagined community" (2006) underpins the way that I read the relationship between national ideology and the physical landscape as produced by writers and artists in this chapter.

Osborne's (1992) elucidation of the role that artists play in the construction of these symbolic myths underscores this reading.

> Some have argued that artists have had a powerful role in this process as interpreters of the essence of the experience of nationhood. Often, artistic imaginations are rooted in their lived-in worlds and their creative responses may informally contribute new and insightful dimensions to their culture-group's identification with their locus. Other artists, however, have gone further. They have attempted to nurture a sense of nationalism by intervening formally in the creation of a discrete cultural identity and national consciousness. They have marshalled their imagery in a campaign to create a national imagination (232).

Other scholars such as Alan R.H. Baker have also traced the relationship between ideology and landscapes (be they natural or altered/developed), arriving at the understanding that "a landscape is a social construction, both intentionally and unintentionally" (Baker 1992, 2). It follows that all representations of landscapes are likewise entwined with ideology. Artistic depictions of landscapes, whether in narrative or visual form, are interpretations as much as they are reproductions.

Osborne noted that central to the relationship between the artist and national consciousness is "the question of access" (242), that is, the extent to which artistic representations are part of the communal consciousness. For this reason, Osborne argues that since "the challenge of new nationalisms has been to nurture the process of developing new identities […] and indeed, to accelerate the cultivation of a distinctive national consciousness" (231), visual art was historically the most inclusive mode of communication, more so than, say, the written word. In the Bruneian context, storytelling as a mode of communication and cultural transmission certainly has a more developed history and form than other mediums, and although still nascent, literature is one of the most dominant narrative forms in the country, its history intertwined with the nation-building project.

Keeping in mind the centrality of the river to Bruneian culture, and the pervasiveness of the national ideology in considering Bruneian identity, in this chapter, I propose that the portrayal of rivers in contemporary Bruneian stories reflects the ways in which the physical landscape has become imbued with properties that problematize national values and official narratives about nationhood and citizenship as expressed in the national philosophy of *Melayu Islam Beraja*. I suggest that these narratives have become additionally fraught in the age of the Anthropocene. To examine this issue, this chapter will consider the representation of the river in three fictional Bruneian texts which explicitly engage with environmental change and national identity: Muslim Burmat's Malay novel *Puncak Pertama* (1988), Aammton Alias' Anglophone novel *The Last Bastion of Ingei* (2016) and the Anglophone superhero comic series *Tebuanman* (2019–2021).

I propose that critical reading offers an insight into how local fictions engage with the concept of national identity as articulated by the national philosophy of *Melayu Islam Beraja*; the dissimilarities and spaces of fissure between this national philosophy and how local fictions construct Bruneian identity offer insights into how the relationship between the national philosophy and lived realities is perceived. I offer in the next section a brief explanation as to how these fictive spaces operate distinctly from other spaces that shape and govern the Bruneian imagination when it comes to narratives of the environment.

1 The Environment in Bruneian "stories" in the Age of the Anthropocene

In a study published in 2022 and conducted over 6 months in 2018, S.N.H. Alkaff et al. (2022) reviewed environmental stories in the most prominent Bruneian newspaper. The study revealed that the majority of stories on the environment consisted of hard news generally presented without commentary, with the remaining stories consisting of press releases (10%), and advisories or government notices (10%). Five out of the six opinion pieces that were published were letters to the editor related to waste (illegal and careless dumping). Alkaff et al. (2022) concluded that this coverage of environmental news, which focused on the efforts undertaken by the relevant agencies to tackle environmental issues, including climate change, "reflects the values of MIB, which promotes highly idealized principles of good conduct, decorum, and respect towards authority figures while discouraging aspects that are considered as threatening to the social harmony" (220).

The ecosensibilities[1] of the local media can thus be characterized as solution-oriented, while avoiding being critical of the "damages caused to the environment and the causes of environmental hazards" (2022, 220). Alkaff et al. (2022) suggest that this avoidance may be due to the local media landscape being "tightly controlled" (211) by the government, and "dominated by the country's national philosophy, *Melayu Islam* Beraja" (220). Thus, the local media's avoidance of engaging with more open dialogue regarding national environmental policies or climate change and sticking to straight news commentary is reflective of the "overt and subtle sociopolitical restrictions operating in the country" (211).

In contrast, environmental issues can be said to be an increasingly explicit and diverse concern of contemporary Bruneian fiction, and in turn of literary scholars in Brunei. This difference between the government-controlled press and literature produced outside of the immediate scrutiny of the relevant authorities (for example, ebooks produced and available exclusively in digital form, or books published outside of Brunei but which take Brunei as their setting, both of which form a significant portion of contemporary Anglophone Bruneian fiction) has been noted by literary scholars (Kathrina et al. 2016; Kathrina 2019). Contemporary Bruneian fiction runs along diverse ideological lines in relation to Bruneian identity construction,[2] including but not limited to those which align with and reaffirm the MIB philosophy.

Indeed, as per the work of literary scholar, Rinni Haji Amran, Bruneian fiction showcases or has the potential to showcase a more diverse and alternative range of perspectives on environmental issues, by being more willing to engage

1 In this chapter I use Rinni Haji Amran's definition of ecosensibilities, as "one's awareness, perceptions of, and attitudes towards one's environment. While similar to the geographic concept of place identity or place-based identity, ecosensibility brings to the fore ecocritical connotations of the relationship and interactions between individual and place" (2021).

2 This diversity is influenced by various aspects of the literary ecosystem, including the means of production, language policies and choices, and modes of dissemination.

with identities which lie outside the dominant Malay-Muslim form. In her consideration of how local ecosensibilities are portrayed in Bruneian novels, Rinni concludes that the sheer range of attitudes and engagements with the environment depicted suggest that "various discourses, ideologies, materials and bodies [co-exist] within the same space" (2021). She also posits that "the presence of several possible narratives in the national environment consciousness contributes to fractured, incomplete, incoherent self-identities on the one hand, as well as more fluid, adaptable, hybrid, transcultural Bruneian identities on the other" (2021).

There are a handful of other studies which explore the relationship between Bruneian identity and the representation of the physical landscape. Of particular interest has been Aammton Alias' sci-fi supernatural novel, *The Last Bastion of Ingei* (2016). In "The Violence of Othering and (non-) Indigenous Revival", Hannah Ming Yit Ho argues that the novel "asserts a revival of indigenous (non-white, non-Western) culture" (2020, 57) through a reading of the physical setting of the novel, mainly "rural areas [subsumed] within interior forests" (64). In another exploration of the rainforest setting, "Into the Haunted Forest: Reading Anti-Extractivism in Aammton Alias' *The Last Bastion of Ingei*" (2016), Rinni Haji Amran explores how the portrayal of the rainforest resists resource extractivism at the ideological as well as at the physical level, and suggests that critical readings of the novel require a shedding of "human-centred perspectives [...] to allow nonhuman agencies to emerge and intervene in this age of the Anthropocene" (77).

In considering the novel and its sequel, both of these studies foreground readings of cultural spaces, approaches and ethnic diversities that "go beyond Malay-, Islam-, and monarchy-centered narratives" to reflect Bruneian identity, "indicative of the enduring Indigenous-inflected presence of forests in the contemporary cultural imaginary" (80). In this way, the rainforest becomes a site to speak to and of minority and marginalized identities within Brunei, with its changing nature in the age of the Anthropocene gesturing concurrently towards a "constant negotiation of identity" (Ho and Rinni 2021, 173), a negotiation which orbits, resists and expands MIB as an articulation of Bruneian-ness.

These studies have been primarily terracentric (Ryan 2021, 2), echoing Boomgard's contention that "in the study of Southeast Asia, there has always been strong emphasis on everything terrestrial with a concomitant neglect of aquatic aspects" (1). Using postcolonial, ecoGothic and other lenses, these scholars have examined how the representation of Bruneian rainforests and its inhabitants, human and non-human and supernatural, have offered alternative ways of thinking about nation, territory, and the environment in the age of the Anthropocene. In some ways, this terra-centrism has been a corrective to and an expansion of the contemporary conventional or primary mythopoetics of the nation, in which a symbolic affinity with the water has minimized those indigenous communities whose livelihoods and origins are land-based and/or agriculturally focused (in contrast to fishing).

This chapter will intervene in this discourse by offering a reading of the river in order to engage with and develop these analyses of the portrayals of the rainforest. If the rainforest represents a space in which official discourse of national identity can be expanded through the representation of "multiethnic" characters (Ho 2020, 56), how do the portrayals of the river supplement these readings? What would a critical reading of these multiple ecological spaces in the nation reveal about constructions of Bruneian identity?

As mentioned previously, I will consider three different texts in this chapter—two novels, Muslim Burmat's *Puncak Pertama* (1988) and Aammton Alias' *The Last Bastion of Ingei* (2016), and the comic series *Tebuanman* written by Khai Anwar and illustrated by Adi Farhan (2019–2021). These texts have been chosen for their explicit engagement with environmental change and national identity, both within the texts and in their paratextual and material production. While no selection of texts can be considered fully representative of cultural beliefs and practices (and indeed this reading is not meant to be comprehensive), I propose that considering these texts offers an insight into the evolving ways that the physical landscape has become intertwined with the national ideology of MIB, including how different landscapes establish dialogues about Bruneian identity. The temporal scope of these texts—from Burmat's novel, published some four years after independence in 1988, and Khai and Adi's comic series, published in the middle of the COVID-19 pandemic—offer different snapshots of Brunei at significant points in its national history.

In the following section, I offer a brief overview of the positionality of the novel within the socio-cultural and ideological landscape of Brunei, which is relevant to how the physical landscapes are represented and can be interpreted within their scope.

2 Modes of Storytelling: Novels and Other Mediums

The Bruneian novel has a relatively young history. The first Bruneian novel, which was written in the Malay language, *Mahkota Yang Berdarah*, by Yura Halim, was published in 1951. A second novel was published in 1952, followed by a period of almost complete silence until the 1980s (the exception being a novel published in the 1960s), which saw a short-lived but definite flourishing of literary production.

The first Bruneian novel in English—Selamat Munap's *The Wild Men of the East*—was published some fifty-eight years after the first Malay novel, in 2009. Since then, there has been increased but still minimal production, with approximately 15 Bruneian novels in English and a corresponding number in Malay.

In a comparative analysis of fiction in the two language mediums, Kathrina, Chin, and Maslin (2016) argued that despite the majority of Malay language novels being published in print by the Dewan Bahasa dan Pustaka (National Language and Literature Bureau), and the majority of English language novels being self-published (either digitally or in print), or published outside of the country and therefore not subject to the same internal scrutiny,[3] both genres were "ideologically invested in perpetuating a vision of Brunei in which MIB is central" (246).

Subsequent studies such as those by Kathrina (2017, 2020, 2021), Rinni (2022) and Ho (2020) have interrogated the parameters and niceties of this vision. The findings suggest that while Bruneian literature repeatedly engages, implicitly and explicitly, the concept of MIB in its depictions of Bruneian identity, affirming the concept's inescapability in the discourse around nationhood and identity, there is an increasing sense that, as Ho suggests, "the contemporary Bruneian Anglophone literary scene [...] offers an avenue to explore alternative representations of self that are often overlooked by official configurations of national identity" (2020, 59). I would propose in fact that even Malay language texts such as Muslim Burmat's work and the *Tebuanman* series, which are endorsed by the National Language and Literature Bureau, and therefore can be considered to be in alignment with these "official configurations of national identity", can be read in a way that shows that they have also found a space for these alternative representations.

Another related aspect of the discourse to note is that there is a historical separation between Malay language and Anglophone Bruneian novels (Kathrina, Chin, and Maslin 2016), and indeed the Malay language and Anglophone literary communities in Brunei, which has its roots in different literary influences, and the different treatment of the Malay and English languages in national language policies and education. Additionally, Anglophone literature's emergence in the twenty-first century has been made possible in response to very different conditions in the local, regional, and global publishing industry than were present in the inception of Malay literature in the 1980s. This

3 The National Language and Literature Bureau has a stated purpose of upholding and making known the national identity and Bruneian identity, and so their publications are aligned with this purpose and the national ideology.

historical separation has led to the development of distinct bodies of literature which have been observed to have very few, if any, interactions between each other. Despite this, recent scholarship by Rinni Haji Amran (2021) has considered Malay and Anglophone texts together, as part of a common Bruneian imaginary. This chapter works in the same vein, proposing that although the two bodies of literature are materially, aesthetically and discursively separate, investigation into a common theme, that is, how the physical landscape that Bruneians inhabit is represented, offers a way to begin to understand a common or transcendent Bruneian imaginary and consciousness.

3 The River as Origin

> The start of my life, like that of my country, Brunei Darussalam, can be described very simply. In the beginning was the river. (Bolkiah 2000)

The sentiment of the mythic language deployed by Prince Mohamed Bolkiah in his memoir is echoed in the Bruneian imaginary across the texts that this chapter will be discussing. The strength of this sentiment is such that the river is presented not just as tied to the origins of the kingdom, but also a saviour and sanctuary from contemporary social and environmental ills.

In the Malay novel *Puncak Pertama,* usually translated as the *First Summit,* the river is the site of the central protagonist, Ahmad's lost home. The 500-page novel, published in 1988, is a richly detailed, immersive historical fiction that takes as its setting Brunei in the 1950s to 1970s. The novel opens with the protagonist, Ahmad, being relocated from his *kampung* in Kampung Ayer (the Water Village) on the Brunei river. Ahmad's *kampung* is being demolished as part of a national project to build a grand mosque, which in the novel was to be officially opened to the public on the day of declaration of Brunei's constitution. Yet Ahmad, whose small home is being sacrificed in the name of national progress, "could not picture the proposed magnificence of the holy building. All he could see was an old building on the banks of the Brunei river, surrounded by water and homes" (11–12).

The loss of this first home is never recovered. Over the course of the novel, Ahmad and his family are forced to move across the developing landscape of the country in search of a new and permanent home. Ahmad and his wife, Urai's most enduring wish, to have a home of their own, sees them initially moving far inland to Seria, a newly bustling oil town in the Belait district; then back to the Brunei Muara district to live with relatives, then to a shanty house that they build themselves on the edges of a friend's property. The development of this property into a commercial area sees Ahmad and his wife displaced once again.

At every turn, Ahmad's quest for a home on land is stymied by national development and its demands on citizens. Ahmad misses out on the government offer to be awarded his own piece of land because he is worried about losing his job in Seria within the fast developing oil economy. At the close of the novel, Ahmad is forced to give up his quest for a home on Bruneian terra firma. He reflects that his family's "history on land had to be completely forgotten, because on land he had never known the meaning of happiness, peace, and success" (546). His return to the Brunei river to live out the rest of his days suggests that it is in fact the water which offers these. The promise of land, and what it symbolizes about inclusion in Brunei's national development, has never been fulfilled. Ahmad's return to a life of familiar precarity, dependent on fishing for sustenance, and a space in the home of relatives, proposes that national progress has been achieved at the expense of particular (indigenous) communities. The fact that fishing, despite its historical and cultural importance, continues to be a precarious livelihood in the novel,[4] existing as it does outside the safety net of national development, suggests a fissure between a cultural emphasis on traditional values, and the reality of the price of national progress (and indeed, exposes the linearity of the idea of progress).

In *The Last Bastion of Ingei* (2016) (henceforth *Ingei*), similarly, the waters of Brunei, and in particular the rivers of Brunei are presented as spaces which protect citizens because of this originary relationship. As one character says, "You were born from water, and it is water that will save you" (288). In *Ingei*, billed as a sci-fi supernatural thriller, a narrative of conservationists seeking to protect endangered species in the rainforest runs parallel with the supernatural world of the *Bunian*[5] entities, brought together by a portal located in the Ingei site.

The novel's action takes the characters on an adventure through the natural landscape. Within this mythology, the river stands in for a traditional, lost world. "My house, a stilted house built on the river, stands undisturbed for the past hundreds of years, which my ancestors built with no machinery" (118), one character reflects as they traverse the landscape. In other instances in the novel, characters escape danger by running towards and through the river (170).

The tying of the mythic origins of Brunei to the river, and then presenting the river as a potential source of solutions for the threats faced by Bruneians,

4 And will become increasingly precarious and endangered in an age of climate change.

5 In local folklore, the *bunian* are invisible people of the forest.

imbues Brunei's history with a sense of transcendental, primordial, almost divine wisdom. It suggests that Bruneians must look to their traditions and histories in order to survive in the present; in fact, that the uniqueness of Bruneian culture offers a global audience an alternative basis for survival in an age of upheaval.[6] It echoes the sentiment of a well-known Malay idiom, *Seperti kacang lupakan kulit*,[7] a cautionary idiom against forgetting where you come from and where your roots are. In *Ingei*, the river symbolically represents how Bruneian identity is rooted in its history and traditions.

This sensibility is also pronounced in *The Chronicles of Tebuanman* (henceforth *Tebuanman*), the first Bruneian superhero comic series (2019–2021). The series spans four comic books, introducing the character Azri Ahmad as an academic researcher who is deployed to a research centre in the heart of the Temburong rainforest to try to discover what is causing unpredictable climate changes that are threatening the region with floods and other natural disasters. The reason why Brunei has survived thus far, in contrast to other countries in the region, is because "our history started in Kampong Ayer. The water isn't foreign to us" (2020, 7). Like *Puncak Pertama* and *Ingei,* the series holds cautionary hints about the ambiguous price of progress and development—it is suggested that the "untouched" (2019, 9) nature of Brunei's rainforests and rivers is central to Brunei's comparative although threatened safety. "While across the country, riverbanks are flowing with increased sea levels, a national landslide crisis and people going missing near bodies of water, Temburong stands unaffected", the researchers note. "Almost like a time capsule, undisturbed by the chaos around it. Could that be it? Did the rest of the world develop itself to doom?" (2019, 9). They agree that "the secret to the situation in Bandar and across the world, lies here" (2019, 9), referring to the Temburong rainforest and river.

All three texts refer to the mythological centrality of the river to Bruneian history and identity, and insist on bringing that centrality into the contemporary.

4 River as Threatened and Threatening

And yet even as these texts affirm the mythopoetic significance of the river as originary and transcendental, they do not assert its status as protected or

6 This is a sentiment identified and echoed across other themes in Bruneian literature, including a novel in which Brunei's MIB values ensure her survival in an apocalyptic future (Kathrina 2019).

7 An idiom that can be roughly translated as "like a peanut forgetting its shell", i.e., forgetting one's roots.

inviolate, and in fact highlight how threatened the river (and thus symbolically, Bruneian history) is. For example, in *Puncak Pertama,* Ahmad's return to life on the Brunei river at the close of the novel throws into stark relief the difference between the values of those on land and those of the water culture. Throughout the novel, Ahmad strives for stability, independence and permanence, encapsulated literally and symbolically through land ownership. Not only does he fail to achieve this, Ahmad remains till the end of the novel dependent on the kindness of others, and living in other people's homes. His final home, shared with old acquaintances, is indeed a return to his beginnings, but it is also framed as a return to a life of precarity and unpredictability, with his daily sustenance dependent on the tides. Ahmad clearly views this as a failure on his part, although it is possible to read this ending not as an individual failure but as a critique of how national development has failed to provide for its own citizens.

Given that the novel poses the river as an alternative site of provision for Bruneians who are crowded off the land, it becomes clear that national development is tied to terrestrial development, and that the river is envisioned as a site of deprivation, precarity and dependence. As mentioned in the previous section, the novel does not offer a simple resolution to this dichotomy; it does not villainize national development or valorize yearning for the traditional ways. Instead it offers a clear-eyed perspective on the price of progress. A home on the Brunei river saves Ahmad from a continued life of struggle and potential homelessness, with some degree of independence; however, this way of life is also increasingly diminished and reduced in the face of national progress which (at the time) urged migration to land and reduced the infrastructural support provided to the inhabitants of Kampong Ayer. This can be interpreted as a perceived dismissal of the "ancestral" values and traditions so cherished by the inhabitants of Kampong Ayer, or indeed the price of progress.

The Last Bastion of Ingei offers a more explicitly ecological, and as Rinni (2022) suggests, anti-extractive perspective on the river as a threatened space. In the novel, it is suggested that "we have given up our lands to Man, we have given up our mountains to Man, we have given up our hills to Man, and we have even given up our rivers to Man" (61). If Man is viewed in the novel as a symbol of national progress, lines such as "this beautiful and otherwise, gentle river, made choppy by the acts of Man" (118) take on additional symbolic significance. Thus, the natural world in general, and the river in particular, are envisioned as a mutually exclusive contrast to urban expansion and national progress.

This fraught relationship between the development of the land and the concurrent neglect of the water lies at the heart of the *Tebuanman* series. Indeed, the ecosensibility of this series mythically links contemporary climate change to a historical battle between land and river. In *Tebuanman*, the narrative impetus for the story is how the rivers have begun to behave in unpredictable, erratic ways, requiring Azri and his team to begin their research with samples from the river, introducing the idea of the river not just as threatened, but as an agentic and threatening force. The researchers see the water as the source of answers, as well as a threat to the citizens—as Boomgaard puts it, "oscillating between a life-giving force and a deadly one" (2007, 3). The river is a threat in an age of unpredictable climate change, especially for a coastal sultanate, and Azri observes that "everyone's afraid. As they should be. People have been swept away by rising water levels nationwide. When almost half the country lives near a body of water, the risk becomes scary" (10).

Thus in *Tebuanman,* the sense of the river as a mythopoetic source of culture and identity does not render it a neutral or even wholly benevolent space for Bruneians. Indeed, the series sets the space of the river in conflict with the land. This is made explicit when during his research, Azri says, "We can't risk losing more lives ... more land" (2020, 7), confirming that the river is a threat not just to individuals but to the land itself, signalling a symbolic fracture within the nation state. This antagonistic relationship manifests in *Tebuanman* in the figure of the villain, Bintang, a sea monster and avatar of the river, who seeks revenge on the land dwellers for the murder of his family.

This portrayal of water–land relations must be read in the context of the imaginary developed by *Puncak Pertama* and *Ingei. Puncak Pertama* saw land development as necessary for national progress, to the extent of sacrificing homes in Kampung Ayer for the sake of the project of land expansion; *Tebuanman* reverses this dynamic, suggesting that the river's potent, latent power can challenge that of the land—symbolically—and that failing to heed and value Brunei's history and traditions will have dire consequences for national development.

Ingei posited that the natural landscapes of both land and water are in danger of being sacrificed in the name of modernization. In *Tebuanman*, the natural landscape itself is fractured, with the terra-communities and water-communities possessing different agendas, alliances, and tribal identities, both in the past and present. This fracture evolves the discourse beyond a consideration of natural landscape in contrast with urban built environment, towards an acknowledgement that the natural landscape is diverse, complex, and resists simplistic binaries of progress and tradition.

Thus, it is significant that the final confrontation between Bintang and Azri happens in one of Brunei's central retail areas. The sorrowful defeat of Bintang as the river avatar in one of the most modernized areas in Brunei highlights the need to address the symbolic and physical presence of the past in the present.

It is only by acknowledging the relevance and the crucial importance of the physical, localized past, that Azri and his team are able to resolve the climate crisis of the present.

Thus, the three texts indicate, in their different ways, the perceived negotiations that the fictive nation and its characters have to make between progress and the preservation of values. While in reality these choices may not be quite so binary in nature, it is clear that in the popular imaginary, writers are articulating a tension, with the river functioning as a symbol for traditional ways of life and values, and the land, including both the urban built environment and the rainforest, often standing in for modernization and progress.

5 Conclusion

In these three texts, the opacity of the river is emphasized. In *Ingei,* the Brunei river is "light brown and murky" (118), a sentiment echoed by the visual representation of the various rivers in *Tebuanman*. Peter Boomgaard notes in his introduction to *A World of Water: Rain, rivers and seas in Southeast Asian histories* that people living in coastal areas "have to be constantly alive to these two faces of the sea, switching effortlessly from an open, welcoming approach towards it to a defensive one when necessary" (2). As mentioned previously, water, in Boomgard's work, is "always in flux, oscillating between a life-giving force and a deadly one" (3). In this regard, the representation of the river in Bruneian texts echoes what Boomgard identifies as a regional water culture—a sense that the water in general and the Brunei river in particular, are uncontrollable, unknowable spaces that offer the capacity for resistance to singular, rigid notions of Bruneian identity. In fact, these texts insist on the capacity of the river to hold both mythological value for nation-building and national identity, but also resist the concept of immutability and homogeneity in reference to national identity. The river thus behaves as an agent that holds the efforts of national development and progress accountable to a kind of cultural truth or integrity anchored by ideas of tradition and history.

In this way, representations of the physical landscape, river and of the rainforest are offered as sites for alternatives to and expansions of the Bruneian identity in fictive spaces, especially in contrast to other narrative spaces. This raises a meaningful and significant question of what is to be understood as the space between the water and the rainforest, between the fickle unknowability of the sea, the mysterious governing of the jungle and the possibility and price of urban built environments. The physical landscape offers a rich site for ideological inquiry, that, as demonstrated by this reading of the river, resists a static interpretation that reflects the evolving relationship of citizen, nation and landscape in Brunei Darussalam in the age of the Anthropocene.

References

Aammton, Alias. 2016. *The Last Bastion of Ingei.* M Content Creations.

Alkaff, Sharifah Nurul Huda, Khairunnisa Ibrahim, and Najib Noorashid. 2022. "The Reporting of Environmental News in an English Language Newspaper in Brunei Darussalam." In *Environment, Media and Popular Culture,* edited by Jason Paolo Telles, John Charles Ryan, and Jeconiah Louis Dreisbach. Asia in Transition, Vol 17. Springer.

Anwar, Khai and Adi Farhan. 2019–2021. *The Chronicles of Tebuanman.* AD Comics.

Baker, A. R. H. 1992. "Introduction." In *Ideology and landscapes in historical perspective: Essays on the meaning of some places in the past,* edited by Alan R. H. Baker and Gideon Biger. Cambridge University Press.

Boomgaard, P. 2007. "In a state of flux: Water as a deadly and a life-giving force in Southeast Asia." In *A World of Water: Rain, Rivers and Seas in Southeast Asian Histories,* edited by P. Boomgaard. KITLV Press.

Druce, Stephen, and Nani Suryani Haji Abu Bakar. 2023. "Contextualising the History of Brunei." In *Routledge Handbook of Contemporary Brunei,* edited by Ooi Keat Gin and Victor T. King. Routledge.

Haji Tassim, Haji Abu Bakar. 2023. "Migration from Water to Land: A Study of Brunei's Kampong Ayer Residents, Circa 1906 to 1941." In *Routledge Handbook of Contemporary Brunei,* edited by Ooi Keat Gin and Victor T. King. Routledge.

Ho, H. M. Y. 2020. "The Violence of Othering and (non-)Indigenous Revival: Aammton Alias." in *The Last Bastion of Ingei: Imminent* as Postcolonial Speculative Fiction of Brunei Darussalam. *SARE* 57 (1): 55–79.

Ho, H. M. Y., and Rinni Haji Amran. 2021. "Constructs of the Modern Home: Negotiating Identity in Anglophone Bruneian Literature." In *Engaging Modern Brunei: Research on language, literature, and culture,* edited by Hannah Ming Yit Ho and David Deterding. Springer.

Kathrina, Mohd Daud, Chin, G. V. S., and Maslin Jukim. 2016. "Contemporary English and Malay literature in Brunei: A comparison." In *The use and status of language in Brunei Darussalam: A kingdom of unexpected linguistic diversity,* edited by Noor Azam Haji-Othman, J. McLellan, and D. Deterding. Springer.

Kathrina, Mohd Daud. 2017. "Articulating Female Citizenship in Norsiah Gapar's Pengabdian." In *The Southeast Asian Woman Writes Back: Gender, Nation and Identity in the Literatures of Brunei Darussalam, Malaysia, Singapore, Indonesia and the Philippines,* edited by Grace V. S. Chin and Kathrina Mohd Daud. Springer.

Kathrina, Mohd Daud. 2019. "Packaging MIB: Representations of Islam in Anglophone Bruneian Fiction." In *Muslim Piety as Economy: Markets, Meaning and Morality in Southeast Asia*, edited by Johan Fischer and Jeremy Jammes. Routledge.

Kathrina, Mohd Daud. 2021. "Translating Islam: Conversion and Love in Bruneian Fiction." In *Translational Politics in Southeast Asian Literatures: Contesting Race, Gender, and Sexuality*, edited by Grace V. S. Chin. Routledge.

Muslim, Burmat. 1988. *Puncak Pertama*. Dewan Bahasa dan Pustaka.

Pannell, S. 2007. "Of Gods and Monsters: Indigenous Sea Cosmologies, Promiscuous Geographies and the Depths of Local Sovereignty." In *A World of Water: Rain, Rivers and Seas in Southeast Asian Histories*, edited by P. Boomgaard. KITLV Press.

Osborne, B. S. 1992. "Interpreting a nation's identity: artists as creators of national consciousness." In *Ideology and Landscape in Historical Perspective*, edited by Alan R.H.Baker and Gideon Biger. Cambridge University Press.

Rinni, Haji Amran. 2021. Examining the Imagined Environments in Contemporary Bruneian Fiction: Developing Southeast Asian Ecocriticism. In *Engaging Modern Brunei: Research on language, literature and culture*, edited by Hannah Ming Yit Ho and David Deterding. Singapore.

Rinni, Haji Amran. 2022. Into the Haunted Forest: Reading Anti-Extractivism in Aammton Alias's *The Last Bastion of Ingei*. *SARE*, Vol 59 (1): 70–87, 2022.

Ryan, J. C. 2021. Hydropoetics: The Rewor(l)ding of Rivers. *River Research and Applications*: 1–8. https://doi.org/10/1002/rra.3844.

Yong, Gabriel Yit Vui Yong. 2021. Identifying Cultural Traits of the Historic Kampong Ayer of Brunei Darussalam using Biomimetic Analysis. *International Journal of Environment, Architecture and Societies*. Vol 1, No 2, pp. 83–94.

Yong, Gabriel, Izni Azrein bin Noor Azalie, Noor Hasharina Hassan. 2026. "Riverine Lifelines of the Brunei Sultanate". In *Fluvial Phenomena: The River and Anthropocene Life-World in the Asia-Pacific*, edited by Iftekhar Iqbal, Noor Hasharina Hassan and Asiyah Kumpoh. Brill.

CHAPTER 5

Indigenous Hydropoetics: Voicing Rivers through Aboriginal Australian Poetry

John Charles Ryan

I am a child of the Dreamtime People
Part of this Land, like the gnarled gumtree
I am the river, softly singing
Chanting our songs on my way to the sea.

HYLLUS MARIS (1988, 60, ll. 1–4)

∵

1 Introduction

Regarded from a technocratic standpoint as channels for agriculture, industry, trade, transportation and urban development, rivers are the most biodiverse yet vulnerable ecosystems in the world (Wu et al. 2021). The factors imperiling river basins globally include erosion from land-use changes, sedimentation from dam construction and the intensifying effects of climate disturbance (Dethier, Renshaw, and Magilligan 2022). Australia's longest river, the Murray, is severely threatened by escalating human demands on the environment and the overconsumption of water for irrigation, both exacerbated by continuing ecological degradation (Hart et al. 2021, XXII). In conjunction with practical conservation strategies, creative perspectives on human-water interrelations are vital to reimagining the spiritual, social and cultural significance of riverscapes. Through the framework of the transdisciplinary environmental humanities (Hubbell and Ryan 2022), the idea of *Indigenous hydropoetics* will be delineated in this chapter as an approach to rivers that incorporates the sustainable hydrological values of Indigenous people with an emphasis on the Aboriginal people of Australia. Cultural groups across Australia hold in common a custodial ethics of rivers shaped by ancestral connections to water. Together with land and sky, water forms part of Country, an Aboriginal term signifying "a nourishing, responsive, relational, living terrain imbued with

 | HTTPS://DOI.ORG/10.1163/9789004758100-007

spirit that is at once the recipient and provider of love and care" (Wooltorton, Poelina and Collard 2022, 401, n9). As Country, land is an integrated natural, cultural and spiritual whole.[1]

The sacredness of freshwater bodies—rivers, streams, creeks, wetlands, billabongs, soaks, springs and aquifers—continues to mold Indigenous values, identities and livelihoods in Australia (Moggridge and Thompson 2021). Rivers are encompassed with kincentric ecologies that seamlessly integrate ecological, social and family relations (Wooltorton, Poelina, and Collard 2022, 394). For Queensland poet Samuel Wagan Watson (2020), "From the southern banks of the Brisbane river my Mununjali blood flows south. My Birri-Gubba blood is far in the north, on the Dawson river, another river, another world away" (47). Indigenous approaches to river management are thus grounded in "reciprocal relationships" between people and the more-than-human world (Martuwarra RiverOfLife, Taylor, and Poelina 2021, 40). Accordingly, "speaking with and as River on the basis of cultural knowledge and authority, observation and recognition, is normal" (Martuwarra RiverOfLife et al. 2022, 432, capitalization original). Regarded as sacred entities, rivers moreover are central to Aboriginal creation narratives or Dreaming stories.[2] As a case in point, among the Nyikina people, the Martuwarra, or Fitzroy river, of Western Australia emerged during the Bookarrarra, the dawn of time, through the actions of the ancestral beings Woonyoomboo, the night heron, and Yoongoorrookoo, the rainbow snake (Martuwarra RiverOfLife et al. 2022, 425). Woonyoomboo planted the freshwater mangrove (majala, or *Barringtonia acutangula*) along the river and conferred First Law to the Nyikina (Martuwarra RiverOfLife et al. 2022, 425).

Recognizing the agencies of water, an Indigenous hydropoetics is further evident in contemporary Aboriginal Australian poetry that narrates human-river interdependencies. In the context of biospheric change, two aspects of an Indigenous hydropoetics—corporeality and interrelationality—prove especially integral to articulating the concept's potential as a response to Anthropocene urgencies. In Yorta Yorta writer Hyllus Maris' poem "Spiritual Song of the Aborigine" (1988), for instance, the speaker identifies ancestrally, empathetically and sonically with rivers as kin, "chanting our songs on my way to the sea" (60, l. 4). Similarly, the hydropoetic verse of Aboriginal Australian writers Jack Davis, Samuel Wagan Watson and Jeanine Leane reveals embodied relations to—and relational epistemologies of—rivers and their habitats

1 Throughout this chapter, Country is capitalized to reflect the Indigenous Australian understanding of land as an animate presence.

2 In keeping with Aboriginal cultural conventions, Dreaming is capitalized here and elsewhere as a proper noun signifying living ancestral narratives.

through attention to Derbal Yerrigan (the Swan river) of Western Australia, the Murrumbidgee river of New South Wales and Maiwar (the Brisbane river) of Queensland, respectively. Rather than giving voice to rivers—and thus reinforcing an anthropocentric stance on water—the hydropoetics of Davis, Watson and Leane acknowledges the intrinsic languages of rivers.

2 Towards Indigenous Hydropoetics: Embracing Potamocentrism

The aim of hydropoetics is to counter terracentrism—the privileging of the terrestrial over the aquatic—and to embrace hydrocentricism or, more precisely, potamocentricism oriented towards new perspectives on rivers beyond their normative social construction as industrial conduits or exploitable commodities (Ryan 2022). Stemming etymologically from *potamos* for river, potamocentrism denotes an ecological worldview reflecting physical, emotional and spiritual identification with rivers as sentient entities. Potamocentrism disrupts an ontology predicated on the perceived constancy of terra firma and "demands a more dynamic approach to organizing life on Earth" (Bennett 2020, 178). In contrast, terracentricism reflects the evolutionary constitution of *Homo sapiens* as a ground-dwelling species. The Western inclination to background the aquatic consequently arises from an incapacity to perceive and cognize the poietic transformation of rivers and other waterbodies over time. Unlike a relatively perdurable phenomenon—a flower, tree, boulder, mountain or other ostensibly static object in the environment—a river is a process, a perpetual in-becoming whose subjectivity lies distributed across inhuman spatial and temporal scales. As such, the idea of *poiesis* is valuable to elucidating the lives of rivers and human-water interdependencies. Broadly speaking, poiesis brings attention to the transformational capacities inhering in that which exists. The term denotes the idea of "bringing forth"—the lively potential of beings, things and elements—including rivers and their habitats—to change, evolve, coalesce, dissociate, intensify and dissipate.

The idea of poiesis is integral to delineating hydropoetics as a framework for rethinking the cultural significance of rivers, articulating the complexities of human-water interactions and confronting river conservation urgencies in the Anthropocene. Cultural theorist Warwick Mules (2014) maintains that an object is indistinguishable from the transformation it manifests and, moreover, that poiesis "identifies the being of things in their becoming other: in their creative, shaped and connected possibilities" (22). Correspondingly, philosopher Arkady Plotnitsky (2013) asserts that "all poiesis, all creative making (the original ancient Greek meaning of poiesis) is material and performative" (275).

The conceptualization of poiesis as transformation is prominent in Humberto Maturana and Francisco Varela's (1980) influential notion of autopoiesis as "what takes place in the dynamics of the autonomy proper to living systems" (XVII). For Maturana and Varela (1980), autopoiesis underscores the interdependent nature of entities as "composite unities realized through neighborhood relations" (XXIII). In this sense, beings are constituted by their ecological interchanges within living systems. Characterized by their relations, rivers and other fluvial phenomena emerge from exchanges between environmental agents (Maturana and Varela 1980, 118). A living system, therefore, is always intrinsically autopoietic insofar as the diminishment of its autopoiesis leads to biological disintegration, decline and death (Maturana and Varela 1980, 112).

Over the last two decades, the concept of poiesis has been vital to the emergence of environmental poetry—or ecopoetics—characterized as poetry or poetic thought that expresses ecological values, critiques environmental degradation and instills in audiences an ethics of nature (Bryson 2005; 2002; Hume and Osborne 2018; Killingsworth 2004; Knickerbocker 2012; Morrison 2015; Rigby 2004). Catherine Rigby (2004), for example, defines the concept of ecopoiesis in terms of human interactions with the more-than-human that enhance ecological vitality. In contrast, Scott Knickerbocker (2012) proposes the idea of *sensuous poiesis* as "the process of rematerializing language specifically as a response to nonhuman nature" (2). Broadened beyond its literary provenance, however, the term *ecopoetics* also encompasses diverse biocultural practices concerned with conservation, sustainability and more-than-human justice (Hume and Osborne 2018). As the intensive study of environmental poetry, the scholarly field of ecopoetics nonetheless has exhibited a prominent terracentric partiality throughout its short history. Seminal publications such as J. Scott Bryson's *The West Side of Any Mountain* (2005) allude only briefly to rivers and other waterbodies as symbolic figures in American environmental poetry. Lynn Keller's *Recomposing Ecopoetics* (2017), furthermore, foregrounds the role of terrestrial organisms—beetles, butterflies and birds—in recent environmental poetry. While ecopoetic theory asserts planetary values, scholars such as Keller place minimal emphasis on the biosphere's hydrological systems including its precious and precarious freshwater ecologies. Among ecocritics and environmental humanists, the problem of terracentrism persists, notwithstanding the inclusion of river-themed poems in ecopoetic anthologies (see, for example, Fisher-Wirth and Street 2013).

The terracentric focus of ecopoetics as a critical practice signifies the wider social, cultural and political privileging of the terrestrial over the aquatic. In response, Indigenous hydropoetics aims to blur the rigid Western distinction between the grounded human and inundated non-human—between terra firma and aqua pellicientes—by cherishing, materially and figuratively, "the freedom of the open water" (Bennett 2020, 173–74). Energized by poetry yet not constrained to creative writing genres, Indigenous hydropoetics is an outlook that advances "a river-centric view of the world," acknowledging rivers as "sentient entities that [or who] are allowed to express their own voice [or voices]" (Brierley 2020, VII, VIII). A hydropoetic perspective regards rivers as "place-beings" bearing moral status and with whom humankind is interconnected in myriad ways (Brierley 2020, VIII). As interdisciplinary geographer Gary J. Brierley (2020) elaborates, "if we are part of the river and the river is part of us, it is subject to sensations, it experiences responses, it has a form of consciousness. In other words, it is sentient" (17). Conceptualizing rivers in this mode, Indigenous hydropoetics counters an ethos of Western dominance over rivers and instead accepts the responsibility of kinship, a process also invoked by Maya Dania in this volume in relation to kin-making along the Mekong river (Chapter 8). From this standpoint, rivers are fellow beings—percipient interlocutors—who command dialogue and garner respect: "When working with a river, the first place to start is to talk to it, finding out who it is and what it needs to flow" (Brierley 2020, 21). Correspondingly, as hydrologist Luna Leopold (1977) contended more than four decades ago, "economic views are too insensitive to be the only criteria for judging the health of the river organism. What is needed is [...] reverence for rivers" (430). With these ideas in mind, the discussion now turns to Derbal Yerrigan (the Swan river) of Western Australia and the hydropoetics of Noongar writer Jack Davis.

3 Derbal Yerrigan and Wagyl: Jack Davis' Swan River Hydropoetics

Comprising fourteen language groups and approximately 30,000 individuals, the Noongar are the Aboriginal people of Southwest Australia including the metropolitan areas of Geraldton, Perth, Bunbury and Albany. Derbal Yerrigan (the Swan river) flows from the Avon river in the Wheatbelt region of the Southwest 175 kilometers west to its mouth at the port of Fremantle on the Indian Ocean. The name Derbal Yerrigan denotes "mixing and flowing" where *derbal* refers to mixing because, at locations such as Burswood peninsula near Perth, "you have the sweet water and the salt water coming together. That is where you have got your tidal movement. Water flows under there too" (Nannup 2008, 109). The Noongar recognize the Wagyl (spelled alternately as Waakarl and Waakal) as a major creation being who protects Derbal Yerrigan and other freshwater sources (Wooltorton, Poelina, and Collard 2022, 398).

The Wagyl represents the interdependent relationship between all freshwater bodies—seas, estuaries, rivers, wetlands and groundwater—and safeguards *kayepa dordok* or "living water" (Bracknell et al. 2022, 404, 405). The twists and turns of Derbal Yerrigan from source to sea represent the path journeyed by Wagyl as the rainbow serpent's movements across the ancient landscape produced the region's hydrological features (Bracknell et al. 2022, 406). Noongar Elder Noel Nannup speaks of Wagyl and the emergence of Derbal Yerrigan in his story "The Carers of Everything" (2003).

In Nannup's account of Noongar Dreaming, while the spiritual forebears gathered to debate who should become carers, Wagyl physically opened a space between the sky and earth. As the plant, tree, bird, mammal and reptile ancestors watched:

> they saw the great serpent use all its muscles together and, with all its strength and energy, it lifted the sky, and in total defiance move across the land creating a smooth trail beckoning all the others to follow [...] At times this great serpent went under the ground and came up again forming the area where there would be lakes. (Nannup 2003, 1)

Later, the ancestral crocodile Yondock travelled down from the north, inciting natural disasters including the flooding of Derbal Yerrigan with salt water. As the custodian of freshwater, Wagyl journeyed the length of the river to intervene in the emergency. The inevitable clash between Yondock and Wagyl saw the rainbow serpent triumph as Yondock's bodily remains turned into Meandip (Garden Island), a narrow island located about five kilometers off the coast of Western Australia (Robertson et al. 2016, 49). Crucial to Wagyl's Dreaming relationship to Derbarl Yerrigan is the seahorse. Although the first Wagyl was male, a powerful female Wagyl later took up residence in the ocean. When she attempted to access the river, two male serpents prevented her, so she transformed into a seahorse: "That is why the seahorse is part of our story for the river. There she was, swimming along like a seahorse [...] the guards couldn't help staring at her. They forgot what they were meant to be doing" (Nannup 2008, 111). After the female Wagyl passed the distracted guards, she jettisoned her seahorse disguise in Djenalup ("place of the footprint" or Blackwall Reach), the deepest part of Derbarl Yerrigan.

Wagyl and the sacred hydrologies of the Southwest appear prominently in the writing of Jack Davis (1917–2000), a Noongar poet, playwright and activist. The publication of *The First-Born and Other Poems* (1970) distinguished Davis as the second Indigenous Australian writer, after Oodgeroo Noonuccal (née Kath Walker), to publish a poetry collection. Subsequently, Davis released the collections *Jagardoo: Poems from Aboriginal Australia* (1977) and *Black Life* (1992) as well as the plays *Kullark: The Dreamers* (1982) and *No Sugar* (1986), all of which confront the challenges of preserving Aboriginal identity. For instance, *Kullark* (1982) reinterprets colonial history, notably the murder of Yagan (1795–1833), the Noongar freedom fighter who resisted British settlement in Western Australia. Eventually apprehended then decapitated, his body was transported to England and exhibited as an antipodean curiosity at European museums until the 1960s. In the play, Yagan is described as "a member of the Swan river (Tjuaht) people," indicating his ancestral identification with Derbarl Yerrigan (Davis 1982, 5). *Kullark* depicts Wagyl as an embodiment of Aboriginal cultural vitality and a voice of the deep Indigenous history that vastly precedes European settlement. In Scene Two, through direct address, Yagan commemorates Wagyl's creation of lakes, swamps and "the wide clear river, / As you travelled onward to the sea" (Davis 1982, 12).

Davis' hydropoetic sensibility is further palpable in *No Sugar* (1986), a four-act play set in the 1920s and 1930s at Moore river Native Settlement and other locales in Western Australia. The play centres on an Indigenous family, the Millimurras, and their struggle to endure the imposition of assimilationist policies during the Great Depression. At the end of Act Two, Scene Six, Joe Millimurra, the family's eldest son, reveals to Mary Dargurru, a girl from the Kimberley region of Western Australia, his intent to escape the Native Settlement and return home to the town of Northam 175 kilometers away. Joe implores Mary, "Come on. I'm gunna show you my country. Got a big river, swans, beautiful white swans" (Davis 1986, 70). The "big river" to which Joe alludes is presumably Gogulgar Bilya (the Avon river), a tributary of Derbal Yerrigan (Ballardong NRM Working Group 2022, 16). Joe associates his Country with swans and the river, demonstrating the centrality of local hydroecologies to Indigenous identity. Waterbodies connect cultures to landscapes; consciousness of water catchments thus vitalizes culture (Nannup 2008, 107). As Noel Nannup observes, traditionally within each catchment, people "needed to know where the water came from because water is the giver of life and everything in that catchment is a part of you and you are a part of it" (Nannup 2008, 108). Also in Act Two, Scene Six, the character Jimmy Munday performs his grandfather's song in Noongar language, entreating crabs "to come up the river and for the fish to jump up high so he can catch them in the fish traps" (Davis 1986, 65–6). Davis' English translation of the chant announces, "crabs, crabs, crabs, crabs / In the river mouth, / They are coming in the river mouth, river mouth" (1986, 110). Jimmy's hopeful song of increase and praise narrates Aboriginal people's corporeal relationships to rivers and fluvial creatures through acts of subsistence.

Davis' poem "The Fight To Save Bennett Brook" (1992, 61) exhibits catchment consciousness while protesting the disturbance of the brook's sacred hydrology. Bennett Brook physically connects Derbarl Yerrigan to the Gnangara Mound, a major aquifer that supplies most of Perth's drinking water. A contextualizing note by Davis explains that "Bennett Brook is one of the few sacred places left to the Aboriginal people of Perth. A company intends to build houses on the land and the Aborigines are fighting to retain it" (Davis 1992, 61). In the poem, "The brook now bears / an alien name," a reference to the overwriting of Noongar place names by Anglo-European toponyms (Davis 1992, 61, ll. 5–6). The Whadjuck Noongar of the Perth area know Bennett Brook as Korndiny Karla Boodjar or "big fire and camping place," signifying the cultural importance of the brook to cooking, eating, sheltering and socialising. Despite the brook's sacred standing, "The sound of trucks / and whorls of dust" epitomize the ongoing fragmentation of the cultural ecologies of Korndiny Karla Boodjar (Davis 1992, 61, ll. 7–8). Human-water-plant intercorporeality, however, offers a mode of resistance through solidarity between people and other life forms:

> We are but swamp reeds
> standing in the sun
> but resilient
> bending with each blow. (Davis 1992, 61, ll. 14–17)

Speaking on behalf of Noongar people—as a "we"—the narrator identifies with the robust plant life of the brook's wetland habitats. Similarly, Noel Nannup (2008) draws attention to the correlation between human and fluvial bodies: "When the river is healthy, we are healthy" (114). Davis' "The Land at the Brewery" (1992, 22), furthermore, maintains Noongar peoples' right to access land and water yet laments, "All that is left / is this minute patch of sand" (ll. 1–2). In 1989, Indigenous protesters occupied Goonininup, the location of the Old Swan Brewery beside Derbarl Yerrigan, to voice opposition to the ruination of the site (Foley 1999). Situated between the river and Kaarta Garup (Mount Eliza)—and abundant in natural freshwater springs—Goonininup is an especially sacred place in Wagyl Dreaming. In poems and plays such as these, Davis develops a hydropoetics steeped in Noongar creation narratives and giving prominence to the embodied imbrications between humans, rivers and all that exists.

4 Maiwar and Moodagurra: Samuel Wagan Watson's Brisbane River Hydropoetics

As the longest river in Southeast Queensland, Maiwar (the Brisbane river) flows 344 kilometers from Mount Stanley to Moreton Bay on the Coral Sea, passing through the traditional lands of the Wakka Wakka, Yagera and Turrbal nations. The Turrbal are the Indigenous people of Mianjin (present-day Brisbane City) whose Country surrounds the river from Moreton Bay through Brisbane CBD to the town of Dinmore. Inland from Turrbal Country are the traditional lands of the Yagera (or Jagera) people encompassing Maiwar's route from Dinmore to Lake Wivenhoe, a reservoir formed by the Wivenhoe Dam. The rest of Maiwar runs through Wakka Wakka land from Lake Wivenhoe to the river's source at Mount Stanley. In the Southeast Queensland area, "countries were divided by the creeks, rivers and mountains" (Ruska in Wang, Ruska and Kerkhove 2018, min. 1:06–1:10). For Maiwar's Indigenous caretakers, the rainbow serpent, Moodagurra, created the river during Dreamtime. In the narrative, Moodagurra becomes stuck as she travels up a dry creek, so she appeals to Yara (rain) and Ngalan (clouds) for assistance. The two Ancestor Beings bring about thunderous rain that washes under Moodagurra's belly, enabling her to squirm free from the creekbed. As Moodagurra proceeds farther up the creek, her writhing motions produce Maiwar's curving path (Cook et al. 2022, 29–30). Yuggera, Noonuccal and Yugumbir cultural custodian Shannon Ruska clarifies that "Moodagurra—the rainbow serpent—she's the one that will bring the floods. She decides when the big rains come" (Ruska 2016). In a different Dreaming narrative by Wakka Wakka and Turrbal Elder Joe Kirk, Duelgum (the freshwater eel) creates the river to journey to Moreton Bay to lay her eggs in the seagrass (Kirk 2014).

Shannon Ruska (2021) further elaborates the Dreaming stories of Maiwar told to him by his Elders. In Birana—a flat treeless space without mountains, hills, wetlands and rivers—two ancestors, Biami and Birali, create the land, goannas, dolphins, eagles and kangaroos. The brothers, Yalgara (the goanna) and Gowanda (the dolphin), have a disagreement and declare war on one another. Yalgara rounds up the terrestrial animals while Gowanda gathers the salt water creatures. Mibin (the eagle) joins the struggle with an ancient spear from the south. On Mibin's arrival, Yalgara leaps up, seizes the weapon and lances Gowanda who then blows the spear out of his body. Gowanda's blood pours forward, scorching the land and congealing into the red rocks at Moreton Bay. Now an old man, Gowanda settles near the river, becomes wise and grows a mane of white hair that transforms into the fin of the dolphin. For the Indigenous people of Southeast Queensland, women are responsible for waterways and, during dances, invoke the tracks of Moodagurra who gave rise to Maiwar. Manifesting as subterranean and aboveground waterways, the paths Moodagurra made during the Dreaming link one nation to another. Maiwar's Indigenous nations, moreover, recognize goannas, dolphins and other creatures as totems connecting cultural groups from source to sea and delimiting custodial obligations. As the basis of ancestral law, each totem must be respected by neighboring tribes. Rather than a commodity that humans buy, sell and exploit, Maiwar Country is an ever present force commanding reciprocity (Ruska 2021, mins. 13:46–26:41).

As a material and spiritual omnipresence, Maiwar figures extensively in the writing of Samuel Wagan Watson, a poet of Irish, German, Dutch and Indigenous (Munaldjali and Birri Gubba) descent. Born in Brisbane in 1972, Watson spent his childhood in the Sunshine Coast region of Southeast Queensland but later returned to Brisbane to pursue a literary career. In addition to publishing his work in print format, Watson has inscribed poetry in Brisbane's built environment such as the Eleanor Schonell Bridge, or Green Bridge, that traverses Maiwar near the suburb of St Lucia (Heiss and Minter 2008). His poetry collections include *Of Muse, Meandering and Midnight* (2000), *Smoke Encrypted Whispers* (2004) and *Love Poems and Death Threats* (2014), all of which interrogate Australian settlerism and promote Indigenous sovereignty. In the short documentary *Bound in Bitumen*, Watson reads "Last Exit to Brisbane," a poem about Boundary Street in south Brisbane. During the nineteenth century, the thoroughfare imposed a perimeter that Indigenous people were not allowed to cross at night and on Sundays:

> Boundary St
> that forged black scratch
> a vein from Southbank to West End
> with a tail swallowed by the chocolate river. (Watson 2005, secs. 10–14)

Watson's image of Boundary Street's "tail swallowed by the chocolate river" invokes the Creation Serpent, Moodagurra, whose sinuous movements defy Brisbane's colonial-era gridlike structure. In his analysis of modes of resistance and accommodation in this volume, Andrew Alan Johnson highlights the emergence of fluid narratives in response to shifting power relations in the Mekong bioregion (Chapter 15). Comparable to the naga, the Mekong's great water serpents, Watson's Moodagurra is rooted in ancient Dreaming narratives that adaptively accommodate the colonialist reconfiguration of Country. Watson's short fiction collection, *New and Used Ghosts* (2023), moreover, celebrates Indigenous heritage through stories of the Brisbane environment including of Maiwar and the Dreaming Ancestors of the fluvial environment.

Imprinted on the reinforcing crossbars of Eleanor Schonell Bridge, Watson's "Poetry on the Green Bridge" evinces an Indigenous hydropoetics of Maiwar shaped by the Dreamings of Brisbane-area Aboriginal cultures. Evoking Maiwar's power, presence and poiesis, the poem comprises two short parts, "On the Transom of Ghosts" and "Dreaming river Triptych" (Watson 2014, 33). Inviting passersby to appreciate Maiwar flowing beneath them, the first part is "*an interactive poem for the commuter*" (Watson 2014, 33, l. 1, italics original). In the poem-imprint, the river is a songline or Dreaming track: "*On this transom, the river's dawning skin* ... stand / here [...] *and whisper upon night's canvas, whirlpool eyes, the / songlines of Kurilpa ghosts*" (Watson 2014, 33, ll. 2–3, 5–6, italics original). Meaning "place for water rats," the term *Kurilpa* is the Indigenous appellation for the land occupied by the suburbs of West End and South Brisbane. The poem's second part captures the perpetual transformation and renewal of Maiwar over immense stretches of time as Watson mediates the presence of Moodagurra through the figure of the serpent: "Never the same this river—archaic vein, snaking through / the land's dreaming cortex" (Watson 2014, 33, ll. 7–8). The river is eternal, timeless, beyond the normative linear construction of time as *chronos*, "the past, present and future—an / aquarelle triptych *cultured* on every tide" (Watson 2014, 33, ll. 10–11, emphasis added). Through the use of the rhetorical technique of anthimeria—where *culture* (noun) becomes *cultured* (adjective) and *culture* (verb)—Watson indicates how Wakka Wakka, Yagera and Turrbal nations care for Maiwar and, in reciprocal return, have been *cultured* by the river's all-encompassing, ever-changing nature.

The historical context of "Finn" is the Queensland flooding of 2010–11 during which the banks of the river broke, resulting in the evacuation of Brisbane CBD, West End and Fortitude Valley along with the inundation of 20,000 homes (Watson 2014, 34–35). The poem tracks the poietic transformation of Maiwar from an object of inspiration to a source of terror through the use of fluvial signifiers such as *recession, recidivism, flotsam, excess* and *wake*. In the narrative, "the river lurks in the shadows of our apocalyptic horizons [...] No one knows the limits of excess it may break" (Watson 2014, 34, ll. 2, 4). Unlike the evocation of Maiwar in "Poetry on the Green Bridge," the river in "Finn" is a volatile, destructive and bewitching entity:

> Naïve as we had been, this river is no longer a vent
> of childhood adventure.
> In a flash we have all relinquished association.
> It has been known as *Miramar* and *Brisbane* and Lord knows
> any other alias in these days of terror and suspicion.
>
> WATSON 2014, 34, ll. 15–19

The crisis of flooding discloses society's broader estrangement from Maiwar, an urban ghost in the heart of Brisbane. Anglo-European toponyms reinscribe the historical "terror and suspicion" of settler culture towards Maiwar. Nonetheless, as also evident in "Poetry on the Green Bridge," Maiwar is a changeable presence with the capacity for metamorphosis: "This is not the same river as yesterday" (Watson 2014, 34, l. 24). The catastrophe asserts a fluvial ontology defying perceived terrestrial permanence in "a floating world" in which nothing is "as anchored as it seems" (Watson 2014, 34, l. 29). Rather than a static phenomenon, Maiwar is a dynamic being who embodies Moodagurra and kindred Dreaming Ancestors: "Tomorrow the tide will shed another serpentine identity" (Watson 2014, 34, l. 34). More than a celebration of Maiwar from an Indigenous standpoint, "Finn" narrativizes the river's multidimensional agency.

At the same time, Watson's hydropoetic body of work foregrounds the industrialization of Maiwar. The satirical "Recipe for Metropolis Brisbane," for instance, decries the degradation of the river as "1 utopian landscape with a blue river" is combined with conservative politics, bitumen, steel, concrete and exhaust then stirred "until the blue river turns brown" (Watson 2008, 217, ll. 3, 15–16). Likewise, Maiwar's ecological decline features saliently in "On the river" as "deserted factories," "abandoned mechanical bits," "grey smoke billowing," "electric fences and weeds" engulf the riverbank, adding a dystopian backdrop to a story of nascent love between two writers (Watson 2000, 7, ll. 8, 9, 15, 16). The prose poem "Smoke Signals," additionally, opens with the speaker's memory of "construction cranes like herds of praying-mantis, high on the steamy Bjelke-Petersen plateau above a brown snake-coiled river" (Watson 2004, 147). The poem shifts from this littoral image to the narrator's memory of learning, as a child, to distinguish between the black smoke of Brisbane's working class, industrial neighbourhoods and white smoke of affluent, residential suburbs. In this way, Watson's hydropoetics incorporates Maiwar Dreaming and elicits the river's poiesis while drawing attention to the despoliation of aquatic ecologies.

5 Marrambidya Bila and Girawu: Jeanine Leane's Murrumbidgee River Hydropoetics

Stretching 1,485 kilometers through New South Wales and the Australian Capital Territory, Marrambidya Bila (the Murrumbidgee river) is Australia's third longest river and a major tributary of Dhungalla (the Murray river) (Australian Government 2021). In her contribution to the current volume, Heather Downey identifies the multidimensional meanings of the Murray river for two rural Australian river communities, concluding that many residents recognize their deep enmeshment in the river system (Chapter 9). Marrambidya Bila flows west from Peppercorn Hill in the Australian Alps of New South Wales through the capital region of Canberra to the river's convergence with Dhungalla near the town of Boundary Bend, Victoria. Along the way, Marrambidya Bila crosses the traditional lands of several Aboriginal cultural groups including the Barapa Barapa, Muthi Muthi, Nari Nari, Ngarigo, Ngunawal, Wamba Wamba and Wiradjuri nations. The mountainous eastern area from which Marrambidya Bila emerges belongs to the Ngunawal and Ngarigo whereas the westernmost reaches of the Lower Murrumbidgee fall within Muthi Muthi and Nari Nari Country. The Wiradjuri are the largest Indigenous nation living on the river's plains and slopes. Denoting "three river people," the name *Wiradjuri* refers to the Marrambidya (Murrumbidgee river), Galari (the Lachlan river) and Wambuul (the Macquarie river), all of which traverse Wiradjuri Country (Saddler 2020, mins. 1:32–1:44). In the Wiradjuri language, Marrambidya Bila signifies "big boss, plenty big water" (Saddler 2020, secs. 11–15). Wiradjuri cultural custodian Mark Saddler (2020) emphasizes that, in his people's language, "*bila* is river. Any *bila* or *darrun* (creek) that runs through any Aboriginal nation is extremely important [...] Marrambidya is one of our *murru* (walkways), and it's a special place we come back to constantly to reconnect to Country" (mins. 0:28–1:04). When rivers become sick, then humans, animals, plants, reptiles and Country as a whole also turn ill.

In a traditional story narrated by Lucinda Firebrace (2012), the Creation Being connected to Marrambidya Bila is Girawu, the goanna wife who unleashed water from the mountains, forming an extensive hydrological system. During the Dreaming, there is no river. Animals die of thirst on the dry land. Only the goannas are healthy. Other creatures begin to wonder if the goannas have a clandestine water supply. Growing impatient with their husbands, the goanna wives decide to search for the water cache. They agree that the best way to discover the water is to stay overnight on the mountain, now known as Peppercorn Hill. Frightened of malevolent spirits, though, many women refuse to go but a brave woman named Girawu volunteers to venture into the wilderness and save the dying animals from thirst. On arriving at the mountain, Girawu sits down but awakens suddenly to a band of little men. The group's elder reassures Girawu that they will lead her to the sacred water. Reaching the source, she drinks to her fill then asks the elder how to assist the desperate animals. The elder instructs Girawu to bring her sister goannas to the mountain so she departs and soon returns with them all. The little men tell Girawu to thrust her yam-digging stick into the mountain and run away immediately. On heeding the directions, a water torrent gushes from the earth, pours down the valley to Dhungalla (the Murray river) and obliterates the goanna men's camps. Rejecting their selfish husbands, the goanna wives turn into teal ducks and glide into the swamp where they live today, caring for Marrambidya Bila, their lifeline, blood and mother.

As a fluvial entity saturated with stories, Marrambidya Bila recurs throughout the work of Jeanine Leane, a Wiradjuri poet, novelist and researcher from south-west New South Wales. Leane was born in Wagga Wagga in 1961 and grew up in on a sheep farm outside Gundagai, a rural town situated beside the Murrumbidgee river. Made of iron and timber, the heritage-listed Prince Alfred Bridge spanning Marrambidya Bila in Gundagai appears frequently in her verse and prose to symbolise the intersection of settler and Indigenous cultures. Leane has published the poetry collections *Dark Secrets After Dreaming: AD 1887–1961* (2010) and *Walk Back Over* (2018) as well as the poetry anthology *Guwayu—For All Times: A Collection of First Nations Poems* (2020) and the short story collection *Purple Threads* (2010). For *The Disappearing*, a project by Red Room Poetry mapping poetry to place, she gained more familiarity with the Prince Alfred Bridge, Marrambidya Bila and the water systems of Gundagai Country (Leane 2019, mins. 7:39–8:31). The river is a palpable figure, as well, in *Purple Threads*, a "short story cycle," or collection of "independent yet interrelated short narratives" told from the perspective of Sunny, a Wiradjuri girl, and her sister, Star, growing up outside Gundagai (Kadmos 2014, 1–2). In the opening story, "Women and Dogs in a Working Man's Paradise," Sunny recollects Marrambidya Bila's devastating floods: "When the waters subsided, the carcasses of drowned stock hung high from the bridge poles and gum trees on the flat below" (Leane 2011, 3). Moreover, in the story "Lilies of the Field," during an exchange between Nan and her Anglo-Australian husband, Sunny's grandmother declares proudly, "I was born round here, on this country, bit farther back from the river 'tween two creeks, placed called Murrumburrah" (Leane 2011, 44). Nan's assertion demonstrates the intertwining of riverscapes, memory and identity for Wiradjuri people.

More specifically, Leane's hydropoetics probes the dialectical nature of memory. The idea of river memory developed in her work encompasses Aboriginal people's memories of Marrambidya Bila over deep cultural time, from the Dreaming through the turmoil of the settler era through to the present. Yet the river is also memory incarnate—a remembering subject and "a deep archive"—with the capacity to hold Country, Wiradjuri people and all beings in memory (Leane 2022a, n.p., l. 39). Published as part of *The Disappearing* project, "Tracks Wind Back" opens, "Gundagai is the bend, the curve, / the turn in the Murrumbidgee river" (Leane 2022b, n.p., ll. 1–2). In Wiradjuri, the word *Gundagai* means "to cut with a hand-axe behind the knee," a reference to the town's position on a major curve of Marrambidya Bila visually reminiscent of a bent knee (Ardha 2019). Eternally transforming yet maternally constant, the river is "ebbing and flowing—the ancient mother / of Wiradjuri children" (Leane 2022, n.p., ll. 3–4). Invoking the suffering inflicted on Indigenous societies during the colonial period, "Tracks Wind Back" is addressed to Marrambidya Bila as a listening presence. Altering river ecologies with axes and muskets, settlers were nevertheless "awestruck by your beauty, / your promise" (Leane 2022, n.p., ll. 5–6). In sharp contrast to the colonial instrumentalisation of the river as a resource, Marrambidya Bila remains, for Wiradjuri people, a memory-keeper who "remembers all her children, / holds all our stories—keeps our Dreaming" (Leane 2022, n.p., ll. 23–4).

The title of Leane's second poetry collection, *Walk Back Over* (2018), alludes to Gundagai's Prince Alfred Bridge but also suggests the critical importance of interrogating the past through memory. In "Archive" (Leane 2018, 8), memory is a transcorporeal phenomenon pulsing through the bodies of people, land, trees and rivers. Leane's most concertedly fluvial poem, however, is "Bridge Over the river Memory," appearing as "River Memory" in *Walk Back Over* (2018, 25). As similarly evident in "Tracks Wind Back," memory in the poem traverses human and the more-than-human domains. The river is both a *subject of memory* and a *remembering subject*. Multidimensional modes of memory mix together like currents as the poet recalls the force of Marrambidya Bila rising so high the river "swallowed / the bridge and the town" (Leane 2022a, n.p., ll. 47–8). The narrative's mnemonic focal point is the bridge, the "longest wooden bridge in the world" (Leane 2022a, l. 9), constructed of local trees such as river red gum, a species growing throughout Australia in floodplains and along the banks of watercourses. Like Marrambidya Bila, the trees afforded the raw materials of colonial expansion yet their histories have been elided:

> How many river gums were felled? What
> were their names before they were rearranged
> across the river—once their life blood.
> What was their history? (Leane 2022a, n.p., ll. 26–9)

For gum trees—as for Wiradjuri people—the river is "their life blood." Reasserting Indigenous memory, then, necessitates excavating fluvial and arboreal histories. The poem evokes the coursing of memory through place, people, trees and rivers in an inter-mnemonic flow:

> The water under the bridge ripples over
> my memory now. The bend of the
> Murrumbidgee—a deep archive—
> flows steady and slow. (Leane 2022a, n.p., ll. 37–40)

As the Dreaming of Girawu, the goanna wife, also demonstrates, Marrambidya Bila is "a deep archive" of Wiradjuri memory. Accordingly, Leane's hydropoetics narrativizes river memory as a corporeal and interrelational means of thinking through Australian history.

6 Conclusion: Indigenous Hydropoetics as River Poiesis

The hydropoetic writing of Davis, Watson and Leane incorporates Dreaming narratives and elicits river poiesis, in the form of poetry, while also confronting aquatic conservation urgencies in Australia. Taken as a whole, their work forwards an Indigenous hydropoetics immersed in Aboriginal creation narratives and emphasizing the embodied relations between humans, rivers and all life. Their poetry evokes the poiesis—the dynamic emergence or bringing forth—of sacred rivers whose origins lie in the Dreaming while also presenting a medium for reverent listening to rivers. Although representing different Indigenous nations, the three poets indeed share a concern for rivers as deep cultural heritage. Yet, while Leane's hydropoetics centres, in particular, on the mediating role of memory—both hers and the river's—Watson's poetry calls attention to fragmented river ecologies in the urban landscape of Brisbane. For Davis, additionally, bodily relationality between humans, rivers and plants constitutes a potent means of ecopolitical resistance through multispecies solidarity. Through the poietic transformation of language, Indigenous hydropoetic writing such as theirs energizes new imaginings of rivers and encourages human receptivity to their biocultural complexities. In an era of global river imperilment, hydropoetics as river poiesis calls attention to the vital importance of the sustainable hydrological values of Indigenous people in Australia and elsewhere.

References

Ardha, Tia. 2019. "Indigenous Meanings of Australian Town Names." *SBS Indonesian* website, 9 July. Accessed 7 December 2022. www.sbs.com.au/nitv/article/indigenous-meanings-of-australian-town-names/ttvhkw3sf.

Australian Government. 2021. "Murrumbidgee." Murray–Darling Basin Authority website. 27 October. Accessed 7 December 2022. www.mdba.gov.au/water-management/catchments/murrumbidgee.

Ballardong NRM Working Group. 2022. *Ballardong Noongar Budjar 'Healthy Country – Healthy People.'* Avon Catchment Council.

Bennett, Joshua. 2020. *Being Property Once Myself: Blackness and the End of Man.* Belknap Press of Harvard University Press.

Bracknell, Clint, Pierre Horwitz, Trevor Ryan and Jonathan W. Marshall. 2022. "Performing *Kayepa Dordok* Living Waters in Noongar Boodja, South-Western Australia." *River Research and Applications* 38 (3): 404–11. https://doi.org/10.1002/rra.3868.

Brierley, Gary J. 2020. *Finding the Voice of the River: Beyond Restoration and Management.* Palgrave Macmillan.

Bryson, J. Scott. 2002. *Ecopoetry: A Critical Introduction.* University of Utah Press.

Bryson, J. Scott. 2005. *The West Side of Any Mountain: Place, Space, and Ecopoetry.* University of Iowa Press.

Cook, Margaret, Lionel Frost, Andrea Gaynor, Jenny Gregory, Ruth A. Morgan, Martin Shanahan, and Peter Spearritt. 2022. *Cities in a Sunburnt Country: Water and the Making of Urban Australia.* Cambridge University Press.

Davis, Jack. 1970. *The First-Born and Other Poems.* Angus and Robertson.

Davis, Jack. 1977. *Jagardoo: Poems from Aboriginal Australia.* Methuen Australia.

Davis, Jack. 1982. *Kullark: The Dreamers.* Currency Press.

Davis, Jack. 1986. *No Sugar.* Currency Press.

Davis, Jack. 1992. *Black Life: Poems.* University of Queensland Press.

Dethier, Evan N., Carl E. Renshaw and Francis J. Magilligan. 2022. "Rapid Changes To Global River Suspended Sediment Flux By Humans." *Science* 376 (6600): 1447–52. https://doi.org/10.1126/science.abn7980.

Firebrace, Lucinda. 2012. "Dreamtime Stories – Girawu The Goanna." YouTube. 13 February. Accessed 7 December 2022. www.youtube.com/watch?v=tWvoTZxvEs8.

Fisher-Wirth, Ann and Laura-Gray Street. 2013. *The Ecopoetry Anthology.* Trinity University Press.

Foley, Gary. 1999. "The Battle for Goonininup." *Aboriginal Studies Occasional Paper*: 62–69.

Hart, Barry, Nick R. Bond, Neil Byron, Carmel A. Pollino and Michael J. Stewardson. 2021. "Preface." In *Murray-Darling Basin, Australia: Its Future Management*, edited by Barry T. Hart, Nick R. Bond, Neil Byron, Carmel A. Pollino and Michael J. Stewardson: xxi–xxviii. Elsevier.

Heiss, Anita and Peter Minter. 2008. "Samuel Wagan Watson." In *Anthology of Australian Aboriginal Literature*, edited by Anita Heiss and Peter Minter: 216–17. McGill-Queen's University Press.

Hubbell, J. Andrew and John C. Ryan. 2022. *Introduction to the Environmental Humanities.* Routledge.

Hume, Angela and Gillian Osborne. 2018. "Ecopoetics as Expanded Critical Practice: An Introduction." In *Ecopoetics: Essays in the Field*, edited by Angela Hume and Gillian Osborne: 1–16. University of Iowa Press.

Kadmos, Helena. 2014. "'Look What They Done to This Ground, Girl!': Country and Identity in Jeanine Leane's *Purple Threads*." *JASAL: Journal of the Association for the Study of Australian Literature* 14 (3): 1–11.

Keller, Lynn. 2017. *Recomposing Ecopoetics: North American Poetry of the Self-Conscious Anthropocene*. University of Virginia Press.

Killingsworth, M. Jimmie. 2004. *Walt Whitman and the Earth: A Study in Ecopoetics*. University of Iowa Press.

Kirk, Joe. 2014. *Duelgum: The Story of the Mother Eel*. Scholastic Australia.

Knickerbocker, Scott. 2012. *Ecopoetics: The Language of Nature, the Nature of Language*. University of Massachusetts Press.

Leane, Jeanine. 2010. *Dark Secrets After Dreaming: AD 1887–1961*. PressPress.

Leane, Jeanine. 2011. *Purple Threads*. University of Queensland Press.

Leane, Jeanine. 2018. *Walk Back Over*. Corditebooks.

Leane, Jeanine. 2019. "Jeanine Leane Speaks at Bundanon Trust – Interview with Kirli Saunders for Red Room Poetry." YouTube. Accessed 7 December 2022. www.youtube.com/watch?v=RFq0gxMzkGU.

Leane, Jeanine. 2020. *Guwayu – For All Times: A Collection of First Nations Poems*. Magabala Books.

Leane, Jeanine. 2022a. "Bridge over the River Memory." Red Room Poetry website. Accessed 7 December 2022. https://redroompoetry.org/poets/jeanine-leane/bridge-over-river-memory/.

Leane, Jeanine. 2022b. "Tracks Wind Back." Red Room Poetry website. Accessed 7 December 2022. https://redroompoetry.org/poets/jeanine-leane/tracks-wind-back/.

Leopold, Luna B. 1977. "A Reverence for Rivers." *Geology* 5: 429–30.

Maris, Hyllus. 1988. "Spiritual Song of the Aborigine." In *Inside Black Australia: An Anthology of Aboriginal Poetry*, edited by Kevin Gilbert: 60. Penguin Books.

Martuwarra RiverOfLife, Katherine S. Taylor and Anne Poelina. 2021. "Living Waters, Law First: Nyikina and Mangala Water Governance in the Kimberley, Western Australia." *Australasian Journal of Water Resources* 25 (1): 40–56. https://doi.org/10.1080/13241583.2021.1880538.

Martuwarra RiverOfLife, Unamen Shipu Romaine River, Anne Poelina, Sandra Wooltorton, Laurie Guimond and Guy Sioui Durand. 2022. "Hearing, Voicing and Healing: Rivers as Culturally Located and Connected." *River Research and Applications* 38 (3): 422–34. doi: 10.1002/rra.3843.

Maturana, Humberto and Francisco Varela. 1980. *Autopoiesis and Cognition: The Realization of the Living*. D. Reidel Publishing Company.

Moggridge, Bradley J. and Ross M. Thompson. 2021. "Cultural Value of Water and Western Water Management: An Australian Indigenous Perspective." *Australasian Journal of Water Resources* 25 (1): 4–14. https://doi.org/10.1080/13241583.2021.1897926.

Morrison, Susan Signe. 2015. *The Literature of Waste: Material Ecopoetics and Ethical Matter*. Palgrave Macmillan.

Mules, Warwick. 2014. *With Nature: Nature Philosophy as Poetics Through Schelling, Heidegger, Benjamin, and Nancy*. Intellect Press.

Nannup, Noel. 2003. *The Carers of Everything*. Swan Catchment Council.

Nannup, Noel. 2008. "Caring for Everything." In *Heartsick for Country: Stories of Love, Spirit and Creation*, edited by Sally Morgan, Tjalaminu Mia, and Blaze Kwaymullina: 102–14. Fremantle Press.

Plotnitsky, Arkady. 2013. "Afterword." In *Poiesis and Enchantment in Topological Matter*, edited by Xin Wei Sha: 271–91. The MIT Press.

Rigby, Catherine E. 2004. "Earth, World, Text: On the (Im)Possibility of Ecopoiesis." *New Literary History* 35 (3): 427–42.

Robertson, Francesca, Glen Stasiuk, Noel Nannup and Stephen D. Hopper. 2016. "*Ngalak Koora Koora Djinang* (Looking Back Together): A Nyoongar and Scientific Collaborative History of Ancient Nyoongar Boodja." *Australian Aboriginal Studies* 1: 40–54.

Ruska, Shannon. 2016. "A City Submerged: The 2011 Floods Ten Years On." Museum of Brisbane website. Accessed 7 December 2022. www.museumofbrisbane.com.au/a-city-submerged/.

Ruska, Shannon. 2021. "Welcome to Country – Shannon Ruska (Healthy Land & Water Showcase Conference 2021)." YouTube. Accessed 7 December 2022. www.youtube.com/watch?v=nsS6EUeNWoI.

Ryan, John Charles. 2022. "Hydropoetics: The Rewor(l)Ding of Rivers." *River Research and Applications* 38 (3): 486–93. https://doi.org/10.1002/rra.3844.

Saddler, Mark. 2020. "Marrambidya Bila – This Place." ABC Indigenous YouTube. Accessed 7 December 2022. www.youtube.com/watch?v=kpyLBNuytJg.

Wang, Beverley, Shannon Ruska and Ray Kerkhove. 2018. "Life in Irons: Stories of Brisbane's Aboriginal People and Convicts." ABC RN website. Accessed 7 December 2022. www.abc.net.au/radionational/programs/lifematters/brisbanes-convict-stories/9911584.

Watson, Samuel Wagan. 2000. *Of Muse, Meandering and Midnight*. University of Queensland Press.

Watson, Samuel Wagan. 2004. *Smoke Encrypted Whispers*. University of Queensland Press.

Watson, Samuel Wagan. 2005. *Bound in Bitumen*. Helen Kassila website. Accessed 7 December 2022. https://helenkassila.wixsite.com/portfolio/video.

Watson, Samuel Wagan. 2008. "Recipe for Metropolis Brisbane." In *Anthology of Australian Aboriginal Literature*, edited by Anita Heiss and Peter Minter: 217. McGill-Queen's University Press.

Watson, Samuel Wagan. 2014. *Love Poems and Death Threats*. University of Queensland Press.

Watson, Samuel Wagan. 2020. "The Grounding Sentence." In *Fire Front: First Nations Poetry and Power Today*, edited by Alison Whittaker: 46–48. University of Queensland Press.

Watson, Samuel Wagan. 2023. *New and Used Ghosts*. University of Queensland Press.

Wooltorton, Sandra, Anne Poelina and Len Collard. 2022. "River Relationships: For the Love of Rivers." *River Research and Applications* 38 (3): 393–403. https://doi.org/10.1002/rra.3854.

Wu, Jiang, Nan Xu, Yichu Wang, Wei Zhang, Alistair G. L. Borthwick and Jinren Ni. 2021. "Global Syndromes Induced By Changes in Solutes of the World's Large Rivers." *Nature Communications* 12 (1): unpaginated. https://doi.org/10.1038/s41467-021-26231-w.

CHAPTER 6

Ethico-Political and Analytic Ambivalence of a Posthumanist Approach to Environmental History

Rommel A. Curaming

Posthumanism is a set of ideas and practices that blurs the long-upheld distinction and presumed hierarchy between humans and non-humans. Among other things, it challenges the modernist conceptions of humans as rational, autonomous agents, superior over non-humans, and exceptional or privileged in the scheme of things (Hollinger 2020; Keeling and Lehman 2018). It envisions the de-centring or de-privileging of humans, giving proportionate attention to other sentient beings, as well as the physical environment. As an analytic approach, it has a fairly recent origin, developing gradually since the 1960s and gaining traction since the 1990s (Domanska 2018; Hollinger 2020). As an idea, however, it has roots in the antihumanism of the nineteenth century, and even deeper into the thousands of years of Asian, Western and indigenous cultural and philosophico-religious traditions. Such deep roots lend credence to the startling claim that humans have always been posthuman (Haraway 1991; Hayles 1999; Ferrando 2016). In its recent iteration, posthumanism is viewed as a reaction to the excesses of the European version of humanism, as exemplified for instance in the idea attributed to Protagoras that "Man is the measure of all things" (Braidotti 2019; Hollinger 2020). It is appreciably different from humanism, say, in the Confucian tradition, which is communally oriented rather than individualistic, and emphasizes harmony with nature and moral self-cultivation (Li 2009). The individualistic elements of European humanism encouraged humans, slowly at the beginning but exponentially in recent centuries and decades, to exceptionalize and privilege themselves, and to overcome and manipulate the environment to satisfy their needs. Their incessant pursuit of progress in various spheres of endeavour resulted in the "Great Acceleration" (Steffen, Crutzen and McNeill 2007; McNeill and Engelke 2014; Steffen et al. 2015) of the adverse, anthropogenic impact on the Earth system with possibly catastrophic consequences for humanity and the whole planet. Anthropocene is the buzzword that encapsulates this idea, without forgetting that as a concept and a phenomenon it remains disputed.

Posthumanism poses a serious dilemma to history, being one of the most humanistic and anthropocentric disciplines (Chakrabarty 2009; Domanska

 | HTTPS://DOI.ORG/10.1163/9789004758100-008

2018). The progressive envisioning of history, both liberal and Marxist, has endowed humans with the preeminent role as the vanguard and engine of historical transformation ostensibly for a continually better future. With humans as volitional agents, the question of accountability is central and unequivocal. Just as humans have the power to make and chart history, they ought to be held responsible, too, for the outcomes of their choices and decisions. To the extent that human collective actions are held accountable for the Great Acceleration, it is also their responsibility to undertake drastic measures to delay or forestall reaching the tipping point and, hopefully, reverse the situation. This idea underpins the global efforts, such as the Paris Agreement (2015), to reverse or slow down the apocalyptic endgame predicted of the Anthropocene.

The posthumanist aspiration to decentre humans in the scheme of things raises important questions: Does posthumanism imply a distributed or shared agency between humans and non-human actors, or among *actants*, in Latour's (1996) term? Does shared agency flag a diminished responsibility for humans? Is there a necessary contradictory relationship between the posthumanist stance and the demand for human accountability?

These questions are among those that figure in the debates on the Anthropocene (Chernilo 2017; Clark and Gunaratnam 2017; Gibbard et al. 2022; Haraway 2015; Lewis and Maslin 2015; Malhi 2017; Malm and Hornborg 2014). In this chapter, I shall revisit the questions of human accountability and analytic contradictions in posthumanism by engaging with two outstanding works on environmental history and philosophy. The first is Dipesh Chakrabarty's magisterial *The Climate of History in the Planetary* Age (Chakrabarty 2021), and the second is Astrida Neimanis's profoundly illuminating *Bodies of Water: Posthuman Feminist Phenomenology* (Neimanis 2017). This chapter mulls over whether the posthumanist approaches necessarily undermine efforts to address the challenges of the Anthropocene.

The choice of Chakrabarty's and Neimanis's books as a springboard for my reflections emanates from the profound insights they offer, as well as from the contrasting, but ultimately complementary, approaches to understanding and addressing the challenges of the Anthropocene. I argue that the contradictions often complained about concerning the posthumanist approaches (e.g. Herbrechter 2013; Wolfe 2010) operate mainly at the intellectual level, which Chakrabarty's approach exemplifies. If we draw from people's lived experience, as Neimanis's phenomenological approach does, contradiction ceases to be of central importance as the materiality of embodied experience trumps the intellectually imposed binaries such as self–others, body–mind, and nature–culture. Contradictions operate in the mind; bodily experience or performance takes precedent by synthesizing the seemingly contrasting binaries into a coherent whole.

1 The Intellectualist Approach and Its Contradictions

As the conventional timeline indicates, various communities and the first nation groups in the US, Ecuador and Bolivia pushed since 2006 for legalization of the rights of nature (Center for Democratic and Environmental Rights (CDER), n.d.). In 2006, certain local municipalities in Pennsylvania adopted the rights of nature in their ordinances. In 2008, Ecuador became the first country to enshrine in its national constitution the rights of nature, which legally acknowledges and protects the intrinsic value of the ecosystem and the sustainable relationship between human and environment. In 2010 and 2012, Bolivia passed the two Mother Earth laws, which were landmark pieces of legislation, though they are fraught with controversy as indigenous communities opposed them for being a watered-down version of what was initially agreed between the government and the indigenous communities (Munoz 2023). In another remarkable development, in 2017, the New Zealand government granted the Whanganui river the same legal rights as other humans (Kramm 2020). This decision was in deference to the Maori indigenous culture that values deep spiritual ties with the natural environment. It was also in line with the fast-growing global environmental movement that sees legal rights being extended to increasing numbers of forests and bodies of water (CDER, n.d.). Also in 2017, the high court of the Indian state of Uttarakhand granted the Ganges and Yamuna rivers similar rights and legal standing (O'Donnell and Talbot-Jones 2018). While this move met opposition and was eventually overturned by India's Supreme Court (Sen 2019a, 2019b), it was groundbreaking legislation. Even more notable was Bangladesh's declaration in 2019 to recognize as living entities all rivers in the country and grant them legal standing (Ingraham 2020).

Among historians and area studies scholars, Dipesh Chakrabarty is arguably the most eloquent and compelling in pushing for a posthumanist approach. He makes no mention of posthumanism in his landmark and much-debated article "The Climate of History: Four Theses" (Chakrabarty 2009), but his call for species history, planetary and deep or geobiological history reeks strongly of posthumanism. Recently, this article was expanded considerably into a book, *The Climate of History in the Planetary Age* (Chakrabarty 2021). It is empirically richer and more nuanced and philosophically more rigorous but the overall arguments are fundamentally similar to those of the 2009 article. It also makes no mention of posthumanism. It is not clear why. Perhaps the author does not wish to be weighed down by or tied to the contentious debates on contradictions that seem to be built-in in posthumanism.

Chakrabarty (2009, 2018, 2021) endorses the idea of planetary history, which takes the Earth system, life itself, and the whole range of both inanimate and animate subjects as the starting, and perhaps also the end point, of historical analysis and narrative. Its coverage is not exclusive to conventional focus such as countries or civilization and other human creations, but also the other celestial bodies, such as "nature" in other planets, if there is such a thing. Rather than the traditional notion of global or world history as a story of progress of human civilizations, the shift to the idea of a planetary history, takes human history as just one among many strands. Chakrabarty (2009, 2018, 2021) also muses about the notion of species history, which highlights the interconnectedness of the history of humans with those of other species, whereas geobiological history anchors more deeply into the distant past of how life, as the locus of history, evolved on Earth through the aeons. These types of history are truly comprehensive and deep in scope, and they foreground the integral role of pre-humans, humans, plants and animals, as well as the natural environment in historical transformation. Among the key implications is the need to rethink the notion of historical agents, and along with it, historical responsibility to include not just humans but also environmental processes, such as natural disasters. As the book's arguments are multi-layered, wide-ranging and philosophically complicated, and thus require space more than I am allowed here, I can only cherry-pick a couple of key points to illustrate one of the arguments that I wish to put forward in this chapter: resorting to the posthumanist approaches leads to ethical, political and analytical paradoxes which may hinder an effective response to the Anthropocene. It needs to be supplemented or complemented with other approaches.

Chakrabarty seems to believe that by shifting away from the anthropocentric world or global history toward planetary or geobiological history, greater value and concern for the planet and life itself might be generated among people and thus they would be prodded to do much more to help mitigate or reverse the Anthropocene. We are all stakeholders; we are all in this together, is the underlying idea. A philosophical problem here lies in the human-centric conception of the Anthropocene as *the* problem. If viewed from the posthumanist standpoint of the planet or life history itself rather than the human-centric global or world history, the destruction brought about by Anthropocenic conditions may not appear to be a problem at all. Rather than destruction, what would happen may be viewed as mere reconstruction or transformation. Rather than being destroyed, life may just mutate into different forms. In other words, the spectre of apocalyptic destruction that accompanies the Anthropocene would be a problem only if imagined from the current human standpoint; humans would be nudged to act to save the Earth if they consider it as their problem. Chakrabarty's analytic tack requires, rather ironically, conferring anthropomorphic characteristics upon nature, or the planet, to regard the changes in the Earth system as destructive rather than just transformative. By conceiving of planetary or geobiological history, along with its accompanying form of historical consciousness, Chakrabarty in effect paradoxically erases the foundational basis of the problem that he sought to address. His attempt to bake the cake and eat it too, by regarding the Anthropocene as a human problem that can be addressed by non-human-centric, planetary solutions, seems riddled with contradictions that may prove difficult to reconcile within the planetary frame, his preferred platform.

Recalling his earlier book that pushed stridently for "provincializing Europe" (Chakrabarty 2000), it is ironic that Chakrabarty cosmopolitanizes Europe in this new book (Chakrabarty 2021). He draws deeply from European philosophical traditions for support, and ignores the richness of Indian, Chinese and other indigenous thoughts, which are clearly more attuned to environmental causes. He projects European views on philosophy and history into the canvas of the whole of humanity, assuming European modernity seeped thoroughly through the mentality of people across the globe. The Euro–American centric tenor of this book is truly arresting.

Why did Chakrabarty go to great lengths, in around 300 pages of exhilarating philosophical rumination, just to end up cornered in a paradoxical position? I think that he started on what is a problematic premise: that the fundamental roots of the Anthropocene lie in a particular mindset that sees humans as separate from, and can be superior over, nature. It is a mindset that was conditioned for centuries by humanistic hubris. Just as the intellectual roots are deep, so does his push for an equally deep solution—deep history, species history, geobiological history, planetary history. The premise is questionable, so I argue, because the Anthropocene, particularly the Great Acceleration, is likely of a fairly recent origin. Mindful of the debates on when it may have started (Lewis and Maslin 2015; Malhi 2017; Chernilo 2017), I would posit that if it were mainly a mindset problem, it would have occurred several centuries earlier. That it occurred in the recent decades, or in the past few centuries at most, we may infer that that anthropocenic condition was accumulative of countless effects of human behaviours, only a part of which was drawn from certain types of cognitive acts and motivations. As human behaviour is as affective as it is rational in nature, affect or emotion equally, if not more, contributes to and is predictive of human behaviour (Damasio 1994; Haidt 2013;

LeDoux 1998). Causal attributions, therefore, cannot be reduced to a mindset, or a way of thinking. The intellectual fallacy that underpins this approach may have emanated from the rationalist assumption that thinking precedes human behaviour, and such behaviour can be explained as consequential from the conscious choices made by the actor.

The notion of performativity rescues human behaviour from the tyranny of the intellectualist conception that puts the mind separate from, and lording over, the body. By altering the fundamental assumption about human–nature and mind–body relations (from atomistic and adversarial to harmonious and complementary), the direction and the basis for decision-making will, so his approach posits, change as well, and the attendant impact on human behaviour will be more favourable for sustaining and enhancing the habitability of the Earth. The attribution of straightforward causal links between these variables seems very problematic both on ethico-political as well as analytic levels.

From the ethico-political standpoint, this analytic arc undermines efforts to hold morally and politically accountable actors and factors like the government, industrialists, capitalists, religious groups, scientific community, consumerism, colonialism, modernity and structural inequalities on a global scale, as the "unfavourable" mindset is taken as the fundamental cause. If only humans adopted the "right" mindset from the beginning, we would not have reached the Anthropocene. And now, what needs to be done is, following this line of thought, change our mindset, make it, say, more planetary rather than human-centric in orientation. Viewed from this standpoint, it will be easily set aside that over 70% of the world's carbon footprint is contributed by less than 20% of people who are mostly from the developed countries (Center for Global Development n.d).[1] Overshadowed as well is that the "great acceleration" that fast-tracked the Anthropocene is likely to be a post-Second World War phenomenon, which means that greater accountability must be sought from actors and institutions who had the most active involvement in economic development in the past several decades, at most the past two centuries. The deep approach that goes to the heart of Chakrabarty's analysis and his proposed solution cannot but downgrade, if not skirt around, the question of proportionate accountability.

Much of the critical responses to Chakrabarty's 2009 article (and by extension the 2021 book) have to do with this question (Boscov-Ellen 2020; Emmett and Lekan 2016). He himself acknowledges the importance of the question, but his overriding aim to find a solution to the Anthropocene pushed him deep

1 https://www.cgdev.org/media/who-caused-climate-change-historically.

into a philosophical and historical conundrum that can only diminish the role of human choices and sideline the question of human responsibility. For him, any blame-game ought to take a backseat to finding an immediate and effective solution. After all, we have only one Earth, and the immediate task is to save it (Chakrabarty 2009). What good is it to find one at fault if the only home we have is already destroyed, so his logic goes. By decoupling the search for a solution from the question of accountability, he mistakenly assumes that only by flattening responsibility across humanity in all nations, groups and individuals can a decisive action be made to address the planetary challenges. He ignores the possibility that it is precisely the proportionate acknowledgment of blame that may enable a more decisive action. What prompted the US to disengage from the Paris Agreement, for example, may have to do with the lack of proper acknowledgment of its enormous role in bringing about the Anthropocene.

Another problem from the analytic standpoint is that Chakrabarty's approach is ridden with intellectualist bias. This bias emanates from what Bourdieu calls scholastic fallacy (Bourdieu 1990, 2006). This fallacy underpins the tendency among intellectuals to view the world through the supposedly reliable analytic lens, and to regard the analytic standpoint either as the only valid vantage point, or as one that is superior to others. This manifests, among other possibilities, in the exaggerated role attributed to cognition or intellect in bringing out, as well as a solution for, a problem, which Chakrabarty's proposed solution indicates.

Chakrabarty seemed to have assumed that the secularist, human-centric thinking of the scholarly groups and the well-resourced upper classes is the norm shared by most people in the world. For billions of people in the Global South whose everyday life is often characterized by precarity, they remain enthralled by, and at the mercy of, the powerful and sometimes scary forces of nature. They are also psychologically held hostage to fate and faith. Facing typhoons, earthquakes, floods, drought, not to mention corrupt politicians and greedy businessmen on a regular basis, hardly do they have the need for the Anthropocene to know the wrath of nature. Latour's (1993) perceptive claim that "we have never been modern" possibly applies to them, who despite their huge numbers have not had enough tools to destroy the Earth System beyond repair. Also, far from their consciousness is the ability of humans as species to chart and make history. Still operating on the God- or nature-centric cosmology which was taken for granted during the pre-modern times and abandoned by intellectual and economic–political elites in the modern periods, their idea of history has possibly no place for the notion of human agency and accountability. The impending destruction of the planet envisioned in the Anthropocene may be simply too big for them to feel responsible for. History, from their view, appears more as the unfolding of the natural order or of God's plan; humans merely enact God's or nature's "will". Whatever happens is attributable to their design, which they believe that they can only live with, or die of. Seen from this standpoint, one may raise the question of whether the Anthropocene is a problem mainly for those who think otherwise—the intellectuals and other elites who aspire and live through the liberal vision that humans *can*. Paradoxically, this is precisely the idea that forms one of the links in the lengthy and complex chain leading to the Anthropocene.

In short, the Anthropocene is not a fundamentally intellectual problem; it is a problem of multiple, interlocking, and convergent causes. But the change in the mindset envisioned by Chakrabarty entailed the shift to deep or planetary history and presupposes the fundamental role of intellect in causing and addressing the problem. In other words, Chakrabarty is entrapped in the intellectualist fallacy when he overstates the importance of the intellect in diagnosing the problem and in prescribing a solution. This also reflects the apparently exaggerated role that he imputed on the scholarly community in addressing the challenges of the Anthropocene.

Another major challenge to his analysis lies in the fact that many indigenous cultures and religious communities across the globe have for centuries never really abandoned, even continually promoted, posthumanist ideas, such as the unity or interconnectedness of human and non-human ecosystems. "We have always been posthuman", as some scholars have reminded us (Ferrando 2016). In addition, despite the supremacy of European modernity, a lot of people across the globe may have actually "never been modern", to echo the title of Bruno Latour's provocative book (1994). This book underscores the fundamentally non-modernist ideas (e.g., human–nature relations, facts vs values, science vs politics) that most people live by on a daily basis even at the height of modernity. These raise a question as to who Chakrabarty is admonishing to shift towards a posthuman, deep history? Notwithstanding modernity, the fear, awe, and reverential attitude towards nature appear to have remained predominant among most people on Earth, simply because of, as noted above, the persistent precarity of their life. They lack economic and psychological resources to free themselves from the shackles imposed by nature. It seems only those who have enormous economic and intellectual capital, as well as endless desire for progress, have fully embraced and practised the modernist, human-centric, secular view of human–nature relations.

Despite the familiarity with the basic posthumanist ideas on human–nature relations, most people still cut the trees, burn forests, and pollute water and air, which suggests that it requires more than a favourable mindset to prevent harmful behaviour towards nature. The case of the Ganges is instructive. Although long-beheld as sacred by hundreds of millions of Hindus, that did not prevent them from polluting it with an estimated three billion litres of wastewater every day (Sen 2019). Its sacredness even leads people to think that the physical dirt could be neutralized by its spiritual purity (Sen 2019). The need to arrest the river's degradation was the reason for passing a law that granted legal standing to the Ganges and its major tributary, the Yamuna. This point has deep implications for Chakrabarty's proposed shift to a planetary mindset. Despite the reverential attitude to nature among a sizable number of people, which is in line with the planetary or deep history that he aspires for, it seems not enough to forestall environmentally destructive behaviour. One may wonder then, for what use is the shift in mindset that Chakrabarty argues for? Whose mindset needs to shift in the first place? Billions of people appear to remain within, or shifting in and out, of the posthuman mentality. This highlights the need to pay greater attention to disaggregating humans depending on their social–economic classes and to focus on other possibly more fundamental factors, as encapsulated in alternative terminologies that challenge the basic idea of the Anthropocene, such as the Capitalocene and Plantationocene (Haraway 2015; Moore 2016; Moore et al. 2019), among other possibilities.

Another interesting aspect of this case is that resorting to legal measures means that the divine status of the Ganges is being downgraded, now dependent on human legislation for validation or enforcement. If the posthumanist position needs human laws to instantiate and enforce it, does this suggest the uselessness of the posthumanist stance? This brings us back to the core problem with posthumanism: it is being validated through a humanist argumentation and undermines in the process its logical foundation. Carey Wolfe's (2010, XVI) admonition that "(t)he nature of thought itself must change if it is to be posthumanist" sounds very resonant here. How the nature of "thought" can be changed from being uneasy with contradictions to being accommodating of paradoxes is a major challenge to a posthumanist positionality.

2 On Lived Experience and the Embracing of Paradoxes

Neimanis's book *Bodies of Water* (Neimanis 2017) seems to offer a pathway to meet the challenge. Drawing from feminist phenomenology, it offers an alternative, but in many ways, it is also complementary to Chakrabarty's intellectualist approach and aspiration for a macro and deep history. Reconceptualizing the human body as just one among many inter- and intra-acting "bodies of water" across species and "natureculture", the book offers a corrective to the atomistic view of humans by affirming their interdependency and affinity with animals, plants and the natural environment. Just like Chakrabarty, Neimanis seems to believe that the destructive and manipulative impulses of humans towards the natural environment emanates fundamentally from their presumed separation from, and superiority over, nature. By realizing humans as an integral part of the broader natural order, an idea that is very old but was overshadowed by the hegemonic modernist mentality, they may be encouraged to be more respectful and protective of nature.

Bodies of Water reconceptualizes the notions of body and embodiment, which are traditionally tied to the anthropocentric conception of the body as exclusively a human body—singular, autonomous and coherent. The feminist phenomenological approach that Neimanis employs conceptualizes the body as porous and encompassing, and materially (not just conceptually) connected to other bodies through their watery components. This approach enables carving pathways to transcend the separation of a body from other bodies regardless of forms and substance.

The posthuman and feminist phenomenological approach employed in this book allows the idea of the "lived experience" to be extended beyond humans, to include that of the non-human creatures and the physical environment. Their existence, too, constitutes experience, their own experience, thus acknowledging the agency of, say, water, which otherwise is presumed to have none. With water's, and other non-humans', lived experience as legitimate starting points, an alternative imaginaries of inter- and intra-action and connection that permeates the whole life system is rendered intelligible. Neimanis did not discuss explicitly what this exactly means insofar as efforts to deal with the Anthropocene are concerned. My inference is that the recognition of both human and non-human lived experience acknowledges the originary equal value of "life" or existence of the non-humans, and that the existing inequality between life and non-life is due not to nature but to human conception of power and reality. It also legitimizes their "rights" to remain unharmed by humans. In the most generous formulation, humans may exploit the natural environment, but they have no right to do so to the point that it would not be able to regenerate or recover.

In addition, rather than naturalizing and justifying humans' appropriative and destructive impulses toward nature, the idea that the latter is its Other, rather than a part of itself, such impulses need to be viewed instead as self-inflicting destruction. After all, the natural environment that humans injure forms the biggest of all bodies of water that there is. The key point is that we are one water in various forms and locations; thus, by polluting bodies and destroying other aspects of nature, we are harming ourselves. If there is a chapter in this book that links closely with this idea, it is John Charles Ryan's chapter (Ryan 2026). The notion of hydropoiesis foregrounds the role of water in the material creative process involving both humans and non-humans, as well as animate and inanimate entities. By looking into Aboriginal Australian poetries, the rivers are shown to have agencies by letting their voices be heard.

I am not sure if modern humans would be prodded to act more gently toward nature simply by asserting or assuming that we are all interconnected as bodies of water. However, vis-a-vis the intellectualist solution, such as Chakrabarty's, which calls for changing the mindset from the anthropocentric to planetary or geobiological history, people may pay greater heed by being reminded that the poisons and other pollutants that permeate the air, land and oceans get ingested by their own bodies through the food chain and could seriously harm them in the process. In other words, things that could impair their own lived experience may have a greater deterring effect than ones that require a change in mindset to forestall a distant apocalyptic future.

For analytical purposes, the contradiction that besets Chakrabarty's intellectualist approach is rendered inoperative here. Given the phenomenological imagination that views water as the gestational fountainhead of things, the various forms it has taken in different bodies across the same as well as different species, and across animate/inanimate divides, are all embodied manifestations of water. They ought to be embraced as part of a complex phenomenon. The relationship therefore is dialectical and complementary rather than contradictory. Whatever walls separate species from each other and from the physical environment are mere appearances that conceal superficially their unitary essence.

Bodies of Water is striking for embracing the contradictions in feminist, as well as non-feminist, theorizing as a given which we have to live and work around with, rather than deny or pretend to eliminate. The phenomenological approach that takes bodily experience—of anyone or anything—as a legitimate starting point allows contradictions to be seen for what they are, as products of different experiential standpoints and as multiple facets of the naturally complex reality. They are constitutive, rather contradictory elements, of reality. They appear contradictory only owing to the intellectualist standpoint, which breeds the mistaken assumption of the validity of the binary logic that underpins the subject–object, humanist–posthumanist, human–nature divide, which Neimanis's feminist phenomenological approach seeks to make irrelevant. Recast within this frame, Chakrabarty's arguments may be re-formulated by taking the two seemingly contradictory mindsets—the modernist, human-centric history and planetary or deep history—as two among other possible facets of reality. It is not essential for people to shift from one to another, or for the two views to cancel each other out, in an effort to address the challenges of the Anthropocene. A possible implication is that the solution may be explored by embracing diversity in areas other than the mindset, which, to me, is instructive.

Intellectualist bias is another area where Neimanis's posthuman phenomenological approach offers a counterpoint to Chakrabarty's analysis. The expanded notion of experience as a starting point eschews Chakrabarty's assumption of the intellectualist roots of, and solutions to, the problems wrought by the Anthropocene. Rather than offering an analysis of the roots of the Anthropocene, and prescribing possible ways to address it, Neimanis's posthuman phenomenology operates on a different analytic register and is more modest in its aspiration: to describe in detail the configuration of various bodies of water to highlight their connectedness, with a hope that this will help humans promote a healthier relationship with other creatures and the environment. In her words, "Bodies of water … may not be the paddle that will guide us out of this planetary mess. But I am wagering that this figuration might just help us learn to swim" (2017, 26). This aim seems in line with the adaptive more than mitigative approach, which is one of the fault lines in the debates on how to address the Anthropocene.

On the question of accountability, *Bodies of Water* is cognizant of the concerns raised about the potential downside of distributed agency and its implication for human responsibility. The fear, as noted earlier, is that the recognition of the shared agency between humans and non-human agents results in human responsibility being significantly reduced, dissuading people to act. With the presentist nature of experience that grounds Neimanis's phenomenological approach, the deep links between the past, the present and the future that are central to Chakabarty's intellectualist approach, and which compel a dramatic reversal in mindset, are de-emphasized. The acknowledgement of distributed agency does not necessarily mean a reduction in human responsibility. Grounded on the experience of threats of anthropocenic conditions—such as global warming, rising sea levels, more frequent and intense weather extremes—there is no presupposed necessary dilution of human responsibility, as such responsibility is built-in within the demands of self-protection and survival. With everyone or everything assumed to be inter- or intra-linked as bodies of water, whose well-being and survival depend on protective actions in the present, collective responsibility among humans is expected, rather than diminished.

The descriptive character of the phenomenological analysis is well-suited to de-naturalize many fundamental humanist assumptions. By devoting several chapters detailing the inter- and intra-action of watery elements in various bodies across species and environmental actants, Neimanis's book serves to deconstruct the humanist logic, among other things, that for centuries the dominant groups among modern humans have taken for granted as natural. By historicizing or putting in historical context thoughts and behaviour, it becomes clear that the contradictions that seem to be inherent in posthumanistic ideas are due to the limits of the hegemonic humanist rationality, rather than to its own flaws. In other words, posthumanism appears contradictory only because the logic of humanism cannot properly account for the inherently dialectical nature of reality, in which human interests are not the only rightful bases for judgement. Viewed from its own logic, what appears as contradictory are in fact merely different strands that constitute the whole.

3 Conclusion

If the idea is to address the challenges posed by the Anthropocene, the posthumanist stance seems to offer a mixed bag of possibilities. The philosophically fascinating, but contradictory, attempt by Chakrabarty is fraught with analytic and ethical difficulties. With his decision to ignore the huge posthumanist elephant in the room, making no mention whatsoever of posthumanism in his book and article (Chakrabarty 2009, 2021), he missed the opportunity to hold the bull by the horn and prevent it from running wild in his analysis. His proposed shift to a planetary or ecocentric mindset is unlikely to have an appreciable impact, as billions of people may have not actually abandoned it to begin with, and the few but very powerful elites who need this shift are unlikely to do so. It might also gaslight people into thinking that the Anthropocene is fundamentally a mindset problem, that its roots are truly deep in history, and that everyone has caused it. The flattening of contexts and differences is unintended by Chakrabarty but it is arresting nevertheless, justified rather unconvincingly by the idea that we have only one Earth and we are all in this together.

The intellectualist bias in Chakrabarty's analysis might be beneficial for the scholarly community as a whole, affirming their prime importance in the order of things as adjudicators and producers of knowledge. However, the authority endowed upon them by society could enable diverting the public attention away from the possibly more immediate factors like personal lifestyle, capitalism and industrialization. By going very deep—deep history, species history, planetary history—the message that people may pick up is that it is but a logical extension of human proclivities so that we cannot but just live with it, or that the solution is too deeply embedded and thus difficult to achieve, so why bother to do anything at all. Explicitly, he wished to avoid driving people to apathy or inaction, but these are the exact possibilities emanating from his analytic arc.

Critical scholarships often aim to fight for the interest of the general public and the marginalized. Given their own class interests as scholars (Bourdieu 1988; Bourdieu and Wacquant 1989), however, they could be blindsided by the unintended side-effects of their scholarly practices. Sometimes, they can and do harm the groups which they explicitly wish to protect, emancipate or empower. The awareness of the scholastic fallacy or intellectualist bias is necessary for scholars to minimize the potential harm in efforts to do good and to recalibrate scholarship's importance in the scheme of things. Many problems in the world, including environmental issues, are at best only partially or marginally intellectual in nature. Exaggerating the role of the intellect in analyzing and solving problems can have disastrous consequences. Neimanis's feminist posthuman phenomenology reinforces the point by rescuing "knowing" from the customary purely or mainly intellectual anchor point. Factoring in affect (intuition or emotion) and embodied experience, knowing and knowledge go beyond mere representations. Also, the pertinent questions cease to be just about accuracy or veracity, but also about what is meaningful and what will make people safe and happy. One important consequence of the multifaceted "knowing" is the demand for tolerance of diversity in various spheres, be it political or epistemological. It also foregrounds the analysis of power relations that renders explicit the hierarchy and dynamic interaction between different knowledge systems.

The contradictions that plague posthumanism spring from, among other sources, the presumed validity of the logic of humanism and its accompanying intellectualist standpoint. By eschewing this logic, grounding instead in the synthesizing capability of the lived experience, Neimanis's feminist phenomenological approach shows that posthumanism can support the efforts to uphold both human accountability and analytic cogency. The shared agency presupposed by posthumanism does not necessarily mean reduced responsibility for humans, as regardless of standpoint, humans are agents or co-agents, and thus are expected to be accountable. The notion of "reduced" responsibility results from the exceptionalist conception of humans as the sole agent, which characterizes the humanist tradition. It does not apply, thus, within the posthumanist framework where humans are from the beginning assumed to be just one among many. In short, by granting analytic parity between the opposing standpoints, Neimanis's feminist posthuman phenomenology denaturalizes the long-taken-for-granted hegemonic position of humanism and allows us to see alternative possibilities which before were easily dismissed as anomalous or contradictory.

Beyond philosophy, one area where posthumanist ideas and approaches may be useful is in their rhetorical power. I agree with Chris Ingraham (2020, 327), when he argues, referring to India's Supreme Court's decision to overturn the legislation on the Ganges and Yamuna rivers, that "the impracticality of more 'posthuman' laws aligned with Earth jurisprudence is not sufficient to invalidate them", as their importance rests on their rhetorical power to persuade. "(R)hetorics around the rights of nature are imperative even without their successful implementation, because only through the cultivation of such nonanthropocentric thought can we begin to live in more sympathy with the more-than-human with which we are invariably entangled and to which we have an immanent responsibility" (Ingraham 2020, 327). At the end of the day, humans think, act and decide. Decentring them is purely an intellectual sleight of hand, with limited or no impact on people's everyday life.

References

Boscov-Ellen, Dan. 2020. "Whose Universalism? Dipesh Chakrabarty and the Anthropocene." *Capitalism Nature Socialism* 31 (1): 70–83. https://doi.org/10.1080/10455752.2018.1514060.

Bourdieu, Pierre. 1988. *Homo Academicus.* Polity Press in association with Basil Blackwell.

Bourdieu, Pierre. 1990. "The Scholastic Point of View." *Cultural Anthropology* 5 (4): 380–91.

Bourdieu, Pierre. 2006. *Science of Science and Reflexivity.* Polity.

Bourdieu, Pierre, and Loic Wacquant. 1989. "For a Socio-Analysis of Intellectuals: On Homo Academicus: An Interview with Pierre Bourdieu." *Berkeley Journal of Sociology* 34: 1–29.

Braidotti, Rosi. 2019. *Posthuman Knowledge.* Polity Press.

Center for Global Development. n.d. "Developed Countries Are Responsible for 79 Percent of Historical Carbon Emissions." Center for Global Development. n.d. https://www.cgdev.org/media/who-caused-climate-change-historically.

Chakrabarty, Dipesh. 2009. "The Climate of History: Four Theses." *Critical Inquiry* 35 (2): 197–222. https://doi.org/10.1086/596640.

Chakrabarty, Dipesh. 2018. "Anthropocene Time." *History and Theory* 57 (1): 5–32. https://doi.org/10.1111/hith.12044.

Chakrabarty, Dipesh. 2021. *The Climate of History in a Planetary Age.* The University of Chicago Press.

Chernilo, Daniel. 2017. "The Question of the Human in the Anthropocene Debate." *European Journal of Social Theory* 20 (1): 44–60. https://doi.org/10.1177/1368431016651874.

Clark, Nigel, and Yasmin Gunaratnam. 2017. "Earthing the Anthropos? From 'Socializing the Anthropocene' to Geologizing the Social." *European Journal of Social Theory* 20 (1): 146–63. https://doi.org/10.1177/1368431016661337.

Damasio, Antonio R. 1994. *Descartes' Error: Emotion, Reason and the Human Brain.*

Domanska, Ewa. 2018. "Posthumanist History." In *Debating New Approaches to History*, edited by Marek Tomm and Peter Burke, 327–52. Bloomsbury Academic.

Emmett, Robert, and Thomas Lekan. 2016. "Whose Anthropocene? Revisiting Dipesh Chakrabarty's 'Four Theses.'" *RCC Perspectives: Transformations in Environment and Society*, no. 2. https://doi.org/10.5282/rcc/7421.

Ferrando, Francesca. 2016. "Humans Have Always Been Posthuman: A Spiritual Genealogy of Posthumanism." In *Critical Posthumanism and Planetary Futures*, edited by Debashish Banerji and Makarand R. Paranjape. Springer India. https://doi.org/10.1007/978-81-322-3637-5_15.

Gibbard, Philip, Michael Walker, Andrew Bauer, Matthew Edgeworth, Lucy Edwards, Erle Ellis, Stanley Finney, et al. 2022. "The Anthropocene as an Event, Not an Epoch." *Journal of Quaternary Science* 37 (3): 395–99. https://doi.org/10.1002/jqs.3416.

Haidt, Jonathan, ed. 2013. *The Righteous Mind: Why Good People Are Divided by Politics and Religion*. 1. Vintage books ed. Vintage Books.

Haraway, Donna. 2015. "Anthropocene, Capitalocene, Plantationocene, Chthulucene: Making Kin." *Environmental Humanities* 6: 159–65.

Haraway, Donna Jeanne. 1991. *Simians, Cyborgs, and Women: The Reinvention of Nature*. Routledge.

Hayles, Katherine. 1999. *How We Became Posthuman: Virtual Bodies in Cybernetics, Literature, and Informatics*. University of Chicago Press.

Herbrechter, Stefan. 2013. *Posthumanism: A Critical Analysis*. Bloomsbury Academic.

Hollinger, Veronica. 2020. "Historicizing Posthumanism." In *After the Human*, edited by Sherryl Vint, 1st ed. Cambridge University Press. https://doi.org/10.1017/9781108874427.002.

Ingraham, Chris. 2020. "The Human Rights of a River: Codifying the Posthuman." In *Water, Rhetoric, and Social Justice : A Critical Confluence.*, edited by Cassey Schmitt, Theresa Castor, and Christopher Thomas: 313–29. Lexington Books.

Keeling, Diane Marie, and Marguerite Nguyen Lehman. 2018. "Posthumanism." In *Oxford Research Encyclopedia of Communication*, by Diane Marie Keeling and Marguerite Nguyen Lehman. Oxford University Press. https://doi.org/10.1093/acrefore/9780190228613.013.627.

Kramm, Matthias. 2020. "When a River Becomes a Person." *Journal of Human Development and Capabilities* 21 (4): 307–19. https://doi.org/10.1080/19452829.2020.1801610.

Latour, Bruno. 1993. *We Have Never Been Modern*. Translated by Catherine Porter. Harvard University Press.

Latour, Bruno. 1996. "On Actor-Network Theory: A Few Clarifications." *Soziale Welt* 47 (4): 369–81.

LeDoux, Joseph E. 1998. *The Emotional Brain: The Mysterious Underpinnings of Emotional Life*. 1st. Touchstone ed. A Touchstone Book. Simon & Schuster.

Lewis, Simon L., and Mark A. Maslin. 2015. "Defining the Anthropocene." *Nature* 519 (7542): 171–80. https://doi.org/10.1038/nature14258.

Li, Zehou. 2009. "Cconfucian Humanism." In *The Chinese Aesthetic Tradition*, edited by Maija Bell Samei: 39–75. University of Hawai'i Press. https://doi.org/10.21313/hawaii/9780824833077.001.0001.

Malhi, Yadvinder. 2017. "The Concept of the Anthropocene." *Annual Review of Environment and Resources* 42 (1): 77–104. https://doi.org/10.1146/annurev-environ-102016-060854.

Malm, Andreas, and Alf Hornborg. 2014. "The Geology of Mankind? A Critique of the Anthropocene Narrative." *The Anthropocene Review* 1 (1): 62–69. https://doi.org/10.1177/2053019613516291.

Moore, Jason W., ed. 2016. *Anthropocene or Capitalocene? Nature, History, and the Crisis of Capitalism*. Kairos PM Press.

Moore, Sophie, Monique Allewaert, Pablo Gomez, and Gregg Mitman. 2019. "Plantation Legacies." *EdgeEffects* (blog). January 22, 2019. https://edgeeffects.net/plantation-legacies-plantationocene/.

Munoz, Lorna. 2023. "Bolivia's Mother Earth Laws: Is the Ecocentric Legislation Misleading?" *Revista: Harvard Review of Latin America* 22 (2). https://revista.drclas.harvard.edu/bolivias-mother-earth-laws-is-the-ecocentric-legislation-misleading/#:~:text=The%20rights%20of%20nature%2C%20Pachamama,to%20which%20Mother%20Earth%20is.

Neimanis, Astrida. 2017. *Bodies of Water: Posthuman Feminist Phenomenology*. Environmental Cultures Series. Bloomsbury Academic.

O'Donnell, Erin L., and Julia Talbot-Jones. 2018. "Creating Legal Rights for Rivers: Lessons from Australia, New Zealand, and India." *Ecology and Society* 23 (1): art7. https://doi.org/10.5751/ES-09854-230107.

Ryan, John Charles. 2026. "Indigenous Hydropoetics: Voicing Rivers through Aboriginal Australian Poetry." In *Fluid Phenomena*, edited by Iftekhar Iqbal, Hasharina Hassan, and Asiyah Kumpoh. Brill.

Sen, Sudipta. 2019. "Of Holy Rivers and Human Rights: Protecting the Ganges by Law." *Yale University Press Blog* (blog). April 25, 2019. https://yalebooks.yale.edu/2019/04/25/of-holy-rivers-and-human-rights-protecting-the-ganges-by-law/.

Neimanis, Astrida. 2017. *Bodies of Water: Posthuman Feminist Phenomenology*. Environmental Cultures Series. Bloomsbury Academic.

O'Donnell, Erin L., and Julia Talbot-Jones. 2018. "Creating Legal Rights for Rivers: Lessons from Australia, New Zealand, and India." *Ecology and Society* 23 (1): art7. https://doi.org/10.5751/ES-09854-230107.

Wolfe, Cary. 2010. *What Is Posthumanism?* Posthumanities Series, v. 8. University of Minnesota Press.

PART 2

Everyday Life and Livelihoods

CHAPTER 7

Navigating the Blame: an Urban Ethnography of a Riverine Village in Manila, the Philippines

Maria Carinnes P. Alejandria

1 Setting the Stage: Living with Precariousness in a Riverine Settlement

The Pasig river is one of the longest rivers in the Philippines, spanning a total of 25.2 kilometres and crossing major cities in the National Capital Region such as Manila, Pasig, Mandaluyong, San Juan, and Taguig. It is bounded to the north by Manila Bay and in the South by Laguna de Bay. In 2021, Pasig river was identified as the "world's most polluting river for plastic waste" (Meijer et al. 2021, 6). This pronouncement aligned with decades of struggle for governance of the historic river, a struggle that counterposed blame between the precariously tenured residents, city administrators, and industries. More than a mere designation, it starkly underscores the profound and damaging extent of human impact on this urban waterway and its tributaries—a defining feature of the Anthropocene.

This chapter situates the Pasig river as a place of struggle for people residing along its shore. It specifically follows the narratives of residents in the largest precariously tenured settlement in Manila City, where a land mass cornered by the river and bay has created a community that is often marginalized for their lack of legal land tenure. The locale of this study is known as the Baseco Compound, which is a village of approximately 61,000 residents in what is called the Port Area (See Figure 7.1). Unlike other similar settlements in Metro Manila, Baseco Compound sits in a unique geographic location as it is bound by two of the country's largest and most polluted bodies of water: Manila Bay and the Pasig river. This ecological phenomenon creates what is called an estuary.

The compound was not as large as it is today. In the 1940s, the estuary was noticeably larger in area than the land mass. However, as the shipyard developed into one of the country's main thoroughfares in the 1960s, the occupation of the estuary through individual reclamation efforts decreased the size of the estuary and increased the land coverage on which the riverine community of the Baseco Compound now sits. The local government also participated in the reclamation process when it sponsored the building of flood mitigation infrastructure and major roads. These actions have allowed for the initial

 | HTTPS://DOI.ORG/10.1163/9789004758100-009

FIGURE 7.1 Map of Baseco compound 2025

25–30 hectares of land called Engineer's Island to be expanded to its current size, with the addition of Isla Laki, Isla Liit (Big Island and Small Island) and the breakwaters. However, this land area continues to diminish in size as the estuary claims the land that chiefly consisted of concrete debris, sand, broken shells, and most infamously, garbage. This land composition provides an area with a low load-bearing capacity that cannot support massive infrastructure (Steinberg and Lindfield 2011). Even the breakwater structure which was a government project during the term of President Gloria Macapagal-Arroyo—meant to be a barrier against the strong waves caused by weather disturbances—has already been reclaimed by the water. The term that the locals used for this was "Kinain ng Dagat" [swallowed by the sea].

The first recorded residents in the area internally migrated in the early 1960s. The first wave of migrants consisted of the workers of the then National Shipyard and Steel Corporation (NASSCO), which was later renamed Bataan Shipping and Engineering Company (BASECO) after the relatives of late President Ferdinand Marcos Sr., Romualdez, took ownership in 1964 (Murphy 2012). It could be said that Baseco Compound is a migrant barangay (formally called Barangay 649), as most of the residents hail from other parts of the country, creating ethno-linguistic blocks (e.g., Muslim and Visayan migrants) in the community. For example, Block Nine is well-known to locals as a Muslim area. In 1995, the Al-Shatie Mosque (seaside mosque) was finally installed in the area through the concerted efforts of members of the Al-Shatie Livelihood Association, which is the key organization of Muslim residents in the compound (Watanabe 2007). Blocks Six, Seven, and Aplaya are well-known for being predominantly occupied by migrants hailing from the Visayas region. These pockets of ethno-linguistic communities in the compound have created invisible territorial markers that persist today.

As of 2019, the Barangay census reported an estimated 61,000 residents living in the Baseco Compound. Most are with legal documents, as land tenure is not yet available. On 18 January 2002, President Gloria Macapagal-Arroyo signed Proclamation No. 145, series 2002, declaring the Baseco Compound open for disposition to its actual occupants. Through this declaration, a preliminary census was conducted in 2001, such that certain houses were tagged with official numbers and stickers from Barangay to legitimize residents' claims for land ownership. Those within the immediate vicinity of the barangay hall have already received their Certificate of Entitlement to Land Allocation (CELA) together with those that were constructed through the support of non-governmental organizations such as Habitat for Humanity and Gawad Kalinga.

Owing to the precarious residency status of most households in the area, basic city services are nearly unavailable. Blocks adjacent to the barangay hall have access to basic city services, such as potable water, electricity, cemented roads, and garbage collection. Inner blocks have little or no access to these services, especially because most of the houses here were not made part of the 2001 census. In the Aplaya block, I was able to count three direct water lines that were managed by the families that had had them installed. Another waterline was installed in 2020 through a partnership between the University of Santo Tomas and the leaders of the Baseco Women's Organization. This response is a pervading experience among urban poor communities in the country. Manahan et al. (2020, 1) further asserted that "associative water systems were borne out of collective desire to have safe, clean, and affordable water to flow to waterless communities." The rest of the houses in Aplaya are wholly dependent on the shallow wells that the owners of some houses have built, which yield insufficient and contaminated water. Locals purchase gallons of water from direct lines for their consumption, whereas all other water-related household activities use water collected from the wells. During summer, when water from the wells is insufficient to cater to the needs of the residents, households cope by using water from the estuary in order to perform household tasks. However, due to the odour of polluted water, most resort to buying water from direct lines but use it conservatively to the point of not taking showers for several days.

Another challenge faced by the residents in the compound is their inability to access relief support from the local government during periods of flooding, as legally tenured residents are prioritized in the distribution of supplies. Being bound by two of the country's largest bodies of water, the area is at high risk of being affected by environmental hazards, such as typhoons, storm surges, and tidal shifts. Weather variabilities are always expected to have an impact on the daily lives of residents, especially the most vulnerable blocks such as Aplaya, which is the closest to the shore and the inner blocks that have no drainage system. The normalization of flooding in the Baseco Compound has desensitized most of its residents to the risks that it poses.

It is against this backdrop that this chapter discourses on the politics of neglect that the State has operationalized to the disadvantage of the members of this riverine community. Using data from interviews and participant observation of an ethnographic work completed between 2013 and 2020, this chapter documents the forms of negotiations and resistance of the residents to assert their right to live in the city vis-à-vis the myriad of structurally induced challenges ranging from hazard vulnerability to dispossession which are all predicated by a State-sponsored theme of urban renewal and development.

2 Constructing Risk and Vulnerability: Who's Scared of the River?

> *Maam, Langoy tayo!"* [*Let's swim, Maam!*]. *Three children asked me to join as they swam along the shore of the river. Typhoon Maring* [*Tropical Cyclone Doksuri*] *had just passed, and the floodwaters brought the usual: fish that escaped the fish pens, water lilies, and garbage. I asked the children if it was safe to swim while pointing at the overlapping mass of garbage and lilies. They laughed and one teasingly asked, "Takot ka ba, Maam?* [*Are you scared, Maam?*]. Field note, September 2017

The community members who participated in this study often described the community as having fluid membership. This is because of the constant movement of people—coming in and out—within the tentative space. It is common knowledge among residents that their population grows exponentially during elections as *dayo* (visitors) come to stay with relatives to vote. There is a typical insinuation of what is colloquially referred to as flying voters—people who are paid to vote for candidates in a jurisdiction other than where they are registered. Some residents have come to live in the area because of their proximity to business establishments where employment, whether formal or informal, is available. Rose, an elder community member, described this demographic phenomenon as *pagbabaha ng tao* (flooding of people). Like her peers, who have lived in the compound for nearly 35 years, Rose has witnessed how the place has expanded and transitioned into a village that it is now. She recounted, "dati mga barong-barong lang kami dito ngayon may mga sementado na bahay na" ["We used to only have makeshift houses here. Now there are cement houses already"]. I have witnessed what Rose described: In 2014, when I first visited the compound, the houses in *Aplaya* were all made of repurposed materials which were scavenged from the debris brought by floods or the dumpsters along *Divisoria,* an expansive commercial centre. In 2018, right after a road connecting the *Aplaya* to the main road was constructed, cement houses were erected by new residents who alleged having bought the CELA which is colloquially referred to as "rights" of the original residents. This transaction remains illegal as the statutory preconditions of CELA is its non-transferability. It also poses complications in resolving the urban poor housing crisis as the CELA owners are catapulted back to precarious residency status.

The rapid urbanization of Metro Manila has generated challenges in attaining sustainable city life. With a daunting cost of living which is described as the third highest in Southeast Asia despite having a low daily minimum wage for workers, a significant number of Filipinos have decided to take up residence in hazard zones which offer cheap housing options. The banks of the Pasig river have become a prime choice for these groups, not only because of the cheap rental, but also because of the location's general proximity to business districts where they gain employment and other livelihood opportunities. Porio (2011) documented how riverine communities, often due to their residency status, are at high risk of flooding and associated hazards. The severity of vulnerability to hazards is also gendered as women and children experience compounding crises including violence, food insecurity, and displacement (Ramalho 2019). Although the Philippine government recognizes the threat of continued flooding as an effect of climate change on low-lying and structurally at-risk populations, it has remained oriented towards resolving the crisis through infrastructure projects that ignore the key causes of the problem, including excessive groundwater extraction (Rodolfo and Siringan 2006). Issues of flooding in urban spaces are typically ascribed to slums, as they are imagined to be the root causes of the blockage of tributaries through State-sponsored "territorial stigmatization" perpetuated through policies and programmes furthering the consequences of "uneven landscapes of risk" (Alvarez and Cardenas 2019, 242–43). Focusing on the most recent development policy known as the Manila City Ordinance 8119, Precess, Perez, Caguimbal (2015, 87) highlights the uneven development of the city as follows:

> 10 development concepts were formulated for urban renewal. Th ese include: (1) business improvement districts, (2) mixed use development, (3) multi-use development, (4) transportation oriented development, (5) great streets 'Sense of Place,' (6) new urbanism adopting the 'traditional neighborhood design' into the modern urban environment, (7) pedestrianization, (8) histo-cultural conservation, (9) roof gardens and (10) environmental design. These development concepts, however, imply that the main objective of the government is economic growth and histo-cultural conservation and environmental design only come in second.

This exclusionist policy is reminiscent of the 1980s' urban renewal project of the Marcos administration which saw the rise of now iconic infrastructures like the Philippine International Convention Center and the Cultural Center of the Philippines which displaced the then called squatters. Karaos (2016, 84) traced further back and claimed that "administrations [Between Presidents Magsaysay and Marcos] became less interested in the political support which the urban squatters could give them. Instead, they were more concerned with taking advantage of the increased economic payoffs to be derived from the use of urban lands for infrastructure and commercial development."

In 2019, one of the country's largest development corporations, SM Prime Holdings, received approval from the Philippine government to continue the Manila Bay reclamation project with the aim of expanding its lifestyle, retail, and residential units adjacent to its flagship commercial enterprise—The Mall of Asia. This expansion will be in cooperation with China-based firm, China Harbour, and the Dutch company, Royal Boskalis Westminster (Venzon 2019). This project is expected to adversely impact communities along Manila Bay including Baseco Compound regarding their marine-based livelihood, the increased vulnerability to flooding, and land tenure (Valenzuela, Esteban, and Motoharu 2020).

Flooding has become normalized in Baseco Compound to the extent that children are desensitized to the threats, and evacuation protocols are individually implemented depending on the perceived risk (Ancheta, Batan, Balita, Alejandria 2020). In multiple periods of flooding, I realized that my community research partners had a different understanding of flood risk than I did.

Despite years of engagement with the community, I still find myself responding to flood alerts differently than they would. In November 2020, Typhoon Vamco hit the Philippines, directly impacting the riverine communities in the National Capital Region. State agencies provided flood warnings as early as two days before its landfall. Unlike other periods of flooding, that event was different; at that time, the community was about to face an overlap of complex emergencies: the pandemic and the flood. Due to strict quarantine restrictions, the timely delivery of humanitarian support from civil society organizations faced challenges. As the river overflowed into the compound, the residents were left to manage the hazard by digging makeshift drainage for the floodwaters to recede to the river (see Figure 7.3).

> *They kept digging a pathway for the floodwaters to return to the estuary. Rain has been pouring down since the previous night. Although their furniture and appliances have already been moved to higher parts of their houses, rising floodwaters remain a threat that requires community action. They have taken shifts digging the path and scooping buckets of water back into the estuary. When I asked one of the elders if she was okay, she said "sanay na kami! Kinatandaan na namin ito" [We're used to this. We have aged with this experience.]. Field note entry in August 2019.*

This is part of their typical community-initiated response, as the local government continues to ignore the need for sustainable solutions to flooding in this residential area.

FIGURE 7.2 Residents digging makeshift drainage

In recent years, the local government has initiated infrastructure projects targeting the health and educational needs of the residents. To date, two basic education schools and a hospital comprise the infrastructure development projects in the area. A key infrastructure project meant to contribute to flood risk management in the area was the Baseco Evacuation Centre. This four-storey building can accommodate approximately 3,000 evacuees. Noting the number of residents in the area, prioritization is typically implemented to accept evacuees. The first level of screening is based on the documentation status of the individual. Only registered voters in the political jurisdiction can access the evacuation centre. This is followed by proof of a house permit. Residents who have none of these documentations can neither access the evacuation centre nor receive relief packages. It is for these systemic complexities, coupled with personal contexts relating to securing possessions, that the majority of the residents living proximate to the shoreline opt to not evacuate. Instead, these residents rely on their social networks to address issues related to physical safety and food security.

After every flooding, residents must manage a persistent issue that is often attributed to their unwelcomed stay in the city—a river of garbage. Aling Maria, a resident of 30 years, resignedly discussed how it was her routine to clean up the garbage that the flood leaves behind. However, she has noticed that in recent years, the amount of trash that piles up has increased exponentially. She lamented, "tingin ko, hindi naman na amin yan!" [I think that's not our trash anymore!]. This statement highlights the awareness that Aling Maria had regarding the cause of pollution in the river. This was not only because of

FIGURE 7.3 Trash left behind after Typhoon Vamco

them. There are other contributors to the systemic pollution of the Pasig river rooted in the industrial and development policies of the country.

3 River-Based Moral Economy and Shifting Attitudes on Tides

> *The stench was overwhelming. It was surreal to see children rushing to the floating trash, stepping on it as though it was a stable ground. The density of the combined trash and water lilies accumulated by the shore allowed for buoyancy. As the waves folded and the trash layer followed, the children walked unperturbed on top of the washed-up treasures. They looked pleased, seeing that the tides had brought many scraps. Elmer, a 9-year-old boy exclaimed after picking out a large plastic bottle, "may kita kami ngayon!" [We'll have income today!]. Field note: January 2015.*

Flooding is often viewed as a destructive force due to its impact on properties and capacity to disrupt livelihood activities. When tides flood the Compound, accessibility becomes a key challenge. As most areas remain without cemented roads, flooding renders dirt roads impassable for weeks owing to the generation of thick mud. With tricycles as the main mode for transporting people and commodities, flooding exacerbates the pre-existing issues of livelihood and food insecurity by disrupting mobility and access within the compound. One of the most affected informal economic sectors in the compound is the *magbabawang* [garlic peelers] who are typically females performing the role of household providers. Contractors send sacks of garlic cloves to women in the compound who peel them for US$2.00 per 50 kg. Whenever the tides bring in flooding and the roads become impassable, this primary source of income is disrupted and causes food insecurity in their households.

When I asked community members about the best time to evacuate during a flooding event, I received responses related to gendered themes. Presuming that the household is qualified to access the evacuation centre, the following is the community-based evacuation arrangement derived from interviews with household leaders.

Males often have a socially prescribed task of managing the evacuation process within the community, while women are typically expected to manage the needs of the members of the household in the evacuation centre. It was noticeable that women were primarily expected to play the role of caregivers for the household while in the evacuation centre. However, because resources were limited, this role extended to being a primary provider. From the exchange of goods to the promise of services, women negotiated within their networks to

TABLE 7.1 Community-level evacuation arrangement during periods of flooding

Qualitative depth of flood	Response	Member to evacuate
Knee-deep; thigh-deep	Lift all ground items to higher areas.	None
Waist-deep	Prepare important properties and animals for evacuation	Children and pregnant women
Chest-deep	Males will lock their homes and aid in the evacuation of remaining vulnerable individuals	Older Adults

gather food and other supplies for their households. Most older adults refuse to evacuate during periods of flooding because of their perceived need to protect their property from looting. It is for this reason that they are often the last sector to be evacuated in such events. Although the need to evacuate is clearly understood by the residents, some decide to stay in the flooded areas or leave the evacuation centre earlier based on a personal risk–benefit analysis (See Table 7.1). Depending on their risk appetite, some decide to not evacuate due to the potential benefits or what they call *grasya* [blessing] that the floods bring. In this context, the tides that brought the flood are perceived positively, as collected debris directly translates to meals or income for the household. The exchange and consumption of commodities collected from flood debris are socially accepted in a moral economy anchored in the shifting tides of the river. In January 2015, right after Typhoon Amang brought floods to the compound, Tatay Karding, a resident of 25 years, showed me a pail of milkfish while saying, "swerte ang daming nakawala sa Laguna!" ["It's so lucky that so many fish escaped their pens in Laguna!"[1]].

This moral association with tides, as either a bringer of gifts or devastation, has been explored in other riverine communities with an emphasis on the concept of flood cycles (Langill and Abizaid 2020; Bista 2020; Dedekorkut-Howes, Torabi, and Howes 2020). Similarly, residents of the Compound have created a localized understanding of the risks of flooding cycles that determine their responses—big tides, evacuate; small tides, gather blessings. With rising sea levels owing to the impacts of climate change, more coastal communities

1 Laguna de Bay has a thriving fishpen and fishcage industry. Fish often escape their pens when the water level increases due to continuous rain.

will face the hazard of flooding (Neumann et al. 2015). The degree of the risk and vulnerability of these communities is determined by the structural responses designed to reduce the impact of this hazard. In the case of Baseco Compound, sustainable risk reduction programmes remain a gap in State-led and community-based responses. Aling Nene, a resident of 23 years, narrated how programmes for their community seem to be "*ningas cogon*[2]" and that "pampapogi[3] lang madalas mga project dito" [most projects here are only for show]. As flood risk reduction and mitigation measures remain nonviable, residents have learned how to live with the shifting tides of the river.

4 The Pollution Crisis: a Securitized Response on the Margins

> *It felt surreal to see the shoreline so clean. I had not visited the compound since the pandemic began, and I asked Joana what happened as though having a clean shoreline was a bad thing. She said, "nag-rorotation kami ng linis ngayon dyan, maam. May mga pulis at kagawad din na rumoronda para i-check kung naglinis kami." [We are doing rotation cleaning up now. There are police and village officials who go around to check if we are cleaning the area.]. Field note, October 2020.*

An estimated 63,000 tonnes of plastic trash courses annually through the Pasig river into the ocean. This is supported by studies on the presence of microplastics in the river, which indicated "persistent plastic pollution from the river source (Laguna de Bay), its tributaries, as well as the communities and industries situated along the river" (Deocaris et al. 2019, 1). Heavy metals have also been detected in the river (Paronda et al. 2019), prompting policies that discourage the consumption of any food harvested from it. However, the river remained an active site in the foodscape for residents along its shores. News reports of Imelda fish [big head carp] caught by local fisherfolks have drawn varying responses. On the one hand, the idea of having food coming from a highly polluted body of water was considered a public health issue, as pollutants ingested by fish could be hazardous to those who consume them. However, the viability of the river as a healthy food source environment was seen as proof of the success of State-sponsored rehabilitation projects. There are

2 Ningas cogon is a Filipino term used to refer to the zealous attitude of an individual towards a new task which immediately fizzles out thereby leaving the project uncompleted.

3 This notion is also related to a common perception that public officials only start projects to boost their popularity.

indeed multiple stakeholders in the project of rehabilitating the Pasig river: Civil Society Organizations, Community-Based Organizations, residents, and the Philippine Government. On 6 January, 1999, the Philippine government created the Pasig river Rehabilitation Commission through Executive Order No. 54 to govern the renewal and sustainable utilization of the river. One of the key points of action for the commission was the relocation of residents along the bank. This objective was predicated by the predisposition to ascribe blame to what the government labels as "informal settlers" for excessive pollution of the river. This narrative of informal settlers as dirty, unhygienic, corrupters by the city's authorities perpetuated arguments for their eviction and dispossession.

In 2012, the Asian Development Bank published a validation report on the Pasig river Environmental Management and Rehabilitation Sector Development Programme (PREMRSDP). Among its key findings highlighted were the primary sources of the river's pollution which include "population growth, urbanization, and industrial activities" which resulted in the "indiscriminate discharge of untreated municipal and industrial wastewater and solid waste exceeding the river system's absorptive capacity" (ADB 2012, 1). With US$98 million spent on the PREMRSDP, the report concluded that it was unsuccessful in attaining its goals which included "improved environmental management of the Pasig river basin within Metro Manila particularly for wastewater management; and promotion of urban regeneration and renewal in the vicinity of the Pasig river" (ADB 2012, 2). Similar to this programme, succeeding programmes by other organizations to rehabilitate the Pasig river have faced critical challenges in terms of relevance and sustainability. In November 2019, President Rodrigo Duterte abolished the commission through Executive Order No. 93 on the grounds of excessive corruption and failure to achieve its objectives. Aligned with the administration's securitized approach to resolving issues, the pollution crisis in the Pasig river was managed through the intervention by local police units, who provided continued presence as clean-up programmes were implemented by elected village leaders.

In October 2020, I was finally able to return to the Compound after more lenient quarantine restrictions were implemented. However, this compound was unrecognizable. The once trash-filled shores were clean. I was informed later by my community research partners of the securitized environment that they had been living in since the declaration of Enhanced Community Quarantine in March 2020. As a high-density area, a significant number of armed police officers were assigned to the compound under strict orders to arrest anyone violating the quarantine protocols. With a history of experiencing police oppression through the State-sponsored Drug War, the presence of uniformed personnel in the area has generated a panopticon-like situation with residents living cautiously. According to the kagawads [elected village leaders] that I spoke with, the order to have village groups clean up the shoreline on a rotational basis came as an order from the Mayor's office. As early as June 2019, Mayor-elect Francisco Domagoso publicly addressed his goal of "cleaning up the river not only to restore its water quality but also to improve urban life" which he claimed became the failure of the outgoing Mayor Estrada (Hallare 2019, 1). This orientation of improving the living conditions of residents in the Baseco Compound has been adopted by other politicians as they drafted projects that aimed to recalibrate the brand of urban living in the compound. Senator Cynthia Villar launched her Zero Open Defecation for the Baseco Compound Programme, which was popularly dubbed the Toilet Bowl project in 2017. This project aimed to reduce the introduction of human waste into the river. Residents who were part of the 4Ps conditional cash transfer programme of the government were tapped to work on the construction of public toilets in Gasangan, a village block located at the mouth of the estuary. These residents were paid about US$8.00 per day, which was 75% of the minimum wage in the National Capital Region. This project has faced challenges in its implementation. According to the residents, just like any government project in the area, this became yet another example of *a ningas cogon*.

A critical look at the selection of project areas by politicians in the compound highlights the prevalence of disproportionate development plans. The Senator's Toilet Bowl Project in Gasangan was limited to the section of the compound where her mangrove project was situated. Residents from adjacent blocks referred to this area as "Paborito ni Senadora" [Senator's favourite block]. Gasangan is the block that is most proximate to the main road that connects the coastal blocks to the highway. It controls the access to the inner blocks. Every time I passed by the makeshift office of the senator's local coordinators, I would see a pile of undistributed toilet bowls. When I asked a resident why she had not taken a bowl for their own house, Melanie retorted, "saan namin ilalagay yan?" [where will we put that?]. The success of this toilet intervention was compromised from the beginning because of a lack of co-planning with residents. Apart from the limited land area that each household has, which deters them from allocating space for a toilet, the logistical demand for using it is also high. In a community where chronic water insecurity is a reality, wasting water resources to flush toilets is unthinkable. This kind of less inclusive development plan cushioned in the neoliberal agenda of the aesthetic rejuvenation of marginalized pockets within the city exacerbates existing injustices that amplify justifications for discrediting undesired urban dwellers. When these projects fail, the blame automatically shifts to the dwellers instead of the structural gaps that have generated their generational vulnerability. As Nanay Nene lamented, "pati yun kalat ng iba kami ang may kasalanan! Kahit anong linis namin dyan, pagbumaha tambak uli." [Even trash from other areas is made to be our fault! However much we clean up the shoreline, when it floods, the trash piles up again.] The persistent blame against residents along the shores of the Pasig river for polluting the river neglects the structural contexts that generate this issue. From the disposal of industrial waste that remains unchecked, to the inefficient sewerage system in the capital, to the gaps in garbage collection, the external contexts that significantly frame the issue of pollution in the Pasig river remain less emphasized compared to the narrative of blame ascribed to precariously tenured residents. The tentative and often securitized solutions imposed on the community to resolve the pollution problem, that an entire region has been actively contributing to, are reminiscent of the parallel State-perpetuated narrative that associates criminality with this sector without recognizing the accountability of other actors and institutions that play a bigger role in the issue.

5 Reviving the Dead: Afterthoughts of an Ethnographer on the Margins

After seven years of engaged ethnographic work in a community that has been vilified by both the public and the State for the pollution of the historic Pasig river, the precariousness of their lives still haunts me. Although community-based organizations from blocks along the shores of the river actively operates to negotiate with officials and other actors for their rights to live in the city, their land tenure status remain as their utmost source of vulnerability. Perhaps this is the reason why flooding was never seen as a major crisis within the community. Apart from being desensitized to it because of its reoccurrence bordering on normalization, a greater source of insecurity is lodged in their consciousness. By not having land tenure, they are kept in a liminal state of being part of the city, while also being constantly asked to prove their right to stay in it. Whereas residents of Baseco Compound must rely on grassroots resilience to navigate an evolving urban environment due to this limited institutional support, Yong, Azalie, and Hassan's (2026) chapter discussed how climate change has sparked a policy-driven reintegration of traditional water-based living in Brunei. These forms of entanglement between local knowledge and state-capital-development were also documented in the work of Chu and Carnegie (2026) in this book.

With minimal provisions, they are expected to solve the pollution crisis of the city of the region that tides continuously bring to their doorsteps. As Thomas et al. (2020) highlighted in the context of Small Island Developing States, those who often contribute the least to environmental degradation bear the brunt of its consequences. The lived experiences of the residents of Baseco Compound which include facing insufficient and contaminated water, managing a "river of garbage" after flooding, and being unjustly blamed for the river's pollution, exemplify this compounded injustice. Their daily struggles within this significantly altered environment underscore the ethical and political dimensions of the Anthropocene which raise critical questions of environmental justice within urban contexts like Manila.

Should they fail, dispossession looms like an untethered policy that would send them back to their supposed real homes, distant and full of unfamiliarity as they have been born in a city that keeps rejecting them. The community members are expected to thrive independently in a resource-limited and hazard-prone environment as it was their supposed decision to uninvitedly inject themselves into the city's peripheries. This expectation of resilience from a community circumscribed within a politics of neglect relieves the State of the accountability to promote justice through equitable and dignified access to urban life.

Infrastructure projects in the community that resemble attempts by the State to replicate the "good life" that is associated with more affluent sections of the city only serve as band-aid solutions that only "vaccinates citizens and environments so that they can take larger doses of inequality and degradation in the future" (Kaika and Kaika 2017, 1).

In seven years, I witnessed how my older key informants have passed on without attaining their lifelong dream of a land title. They left behind their children and grandchildren the very same stigmatized status by which they had entered the city: an informal settler, a slum, a squatter. Unlike their generation, which had no access to nearby schools or hospitals, their successors had access to these. Perhaps these infrastructures create some semblance of acceptance to the city; perhaps these are just vaccines for intergenerational injustices that their community will keep experiencing until the State recognizes their right to live in the city. As the community struggles to survive in a rapidly developing city, the Pasig river continues to be the proverbial Dead that the community will need to carry on its back as though a ticket that allows them to be in it.

References

Alvarez, Maria Khristine, and Kenneth Cardenas. 2019. "Evicting Slums, 'Building Back Better': Resiliency Revanchism and Disaster Risk Management in Manila." *International Journal of Urban and Regional Research* 43 (2): 227–49. https://doi.org/10.1111/1468-2427.12757.

Ancheta, Arlen, Clarence Batan, Dan Angelo Balita, and Maria Carinnes Alejandria. 2020. "Baha at mga Bata: Exploring the Filipino Children's Perceptions and Experiences of flood as Stories of Vulnerabilities." In *Disaster Archipelago: Locating Vulnerability and Resilience in the Philippines*, edited by Maria Carinnes Alejandria and Will Smith. Lexington Books.

Asian Development Bank. 2012. "Philippines: Pasig River Environmental Management and Rehabilitation Sector Development Program Validation Report Independent Evaluation Department." https://www.adb.org/sites/default/files/evaluation-document/36027/files/pvr-187.pdf.

Bista, Raghu Bir. 2020. "Does Disaster Change Income and Wealth Distribution Toward Extremity of Inequality and Poverty? Analysis of Flood and Landslides in the Vulnerable Locations of Nepal." *Forum for Social Economics*, January: 1–15. https://doi.org/10.1080/07360932.2020.1715810.

Chu, Ta-Wei, and Paul Carnegie. 2026. "Dam Trouble: Transdisciplinarity and the Case of a Riverine Livelihood Study in Northeastern Cambodia." In *Fluid Phenomena*, edited by Iftekhar Iqbal, Hasharina Hassan, and Asiyah Kumpoh. Leiden: Brill.

Dedekorkut-Howes, Aysin, Elnaz Torabi, and Michael James Howes. 2020. "When the Tide Gets High: A Review of Adaptive Responses to Sea Level Rise and Coastal Flooding." *Journal of Environmental Planning and Management* 63 (12): 2102–43. https://doi.org/10.1080/09640568.2019.1708709.

Deocaris, Chester C., Jayson O. Allosada, Lorraine T. Ardiente, Louie Glenn G. Bitang, Christine L. Dulohan, John Kenneth I. Lapuz, Lyra M. Padilla, Vincent Paulo Ramos, Jan Bernel P. Padolina, and Jan Bernel Pradillada Padolina. 2019. "Occurrence of Microplastic Fragments in the Pasig River." *H2Open Journal* 2 (1): 92–100. https://doi.org/10.2166/h2oj.2019.001.

Hallare, Kathrina. 2019. "Isko to Probe Baseco Compound 'Secret Dumpsite.'" *Inquirer*. June 4, 2019. https://newsinfo.inquirer.net/1126689/isko-to-probe-baseco-compound-secret-dumpsite#ixzz7t3PzjPPB.

Kaika, Maria, and Maria Kaika. 2017. "'Don't Call Me Resilient Again!' The New Urban Agenda as Immunology … or … What Happens When Communities Refuse to Be Vaccinated with 'Smart Cities' and Indicators." *Environment and Urbanization* 29 (1): 89–102. https://doi.org/10.1177/0956247816684763.

Karaos, Anna Marie A. "Manila's Squatter Movement: A Struggle for Place and Identity." *Philippine Sociological Review* 41, no. 1/4 (1993): 71–91. http://www.jstor.org/stable/23898158.

Langill, Jennifer C., and Christian Abizaid. 2020. "What Is a Bad Flood? Local Perspectives of Extreme Floods in the Peruvian Amazon." *AMBIO: A Journal of the Human Environment* 49 (8): 1423–36. https://doi.org/10.1007/s13280-019-01278-8.

Manahan, Mary Ann, Enrique Villanueva, and Joseph Edward Alegado. 2020. "(Re) Construir Servicios Públicos Frente a La Gobernanza Neoliberal: Prácticas de Sistemas Asociativos En Torno al Agua En Las Comunidades Urbanas Pobres de

Metro Manila." Relaciones Internacionales-Madrid, no. 45: 205–26. https://doi.org/10.15366/relacionesinternacionales2020.45.009.

Meijer, Lourens J. J., Tim van Emmerik, Ruud van der Ent, Christian Schmidt, and Laurent Lebreton. 2021. "More than 1000 Rivers Account for 80% of Global Riverine Plastic Emissions into the Ocean." *Science Advances* 7 (18). https://doi.org/10.1126/sciadv.aaz5803.

Murphy, D. 2012. *Baseco and Its Proclamation*. Quezon City. http://www.ombudsman.gov.ph/UNDP4/wpcontent/uploads/2012/12/BASECO_PROC1_HTML.pdf.

Neumann, Barbara, Athanasios T. Vafeidis, Juliane Zimmermann, and Robert J. Nicholls. 2015. "Future Coastal Population Growth and Exposure to Sea-Level Rise and Coastal Flooding—a Global Assessment." *PLOS ONE* 10 (3): 1–34. https://doi.org/10.1371/journal.pone.0118571.

Paronda, G. R. A., David, Carlos Primo C., and Dahlia C. Apodaca. 2019. "River Flow Pattern and Heavy Metals Concentrations in Pasig River, Philippines as Affected by Varying Seasons and Astronomical Tides." *IOP Conference Series: Earth and Environmental Science* 344 (1): 012049. https://doi.org/10.1088/1755-1315/344/1/012049.

Porio, Emma E. 2011. "Vulnerability, Adaptation, and Resilience to Floods and Climate Change-Related Risks among Marginal, Riverine Communities in Metro Manila." *Asian Journal of Social Science* 39 (4): 425–45. https://doi.org/10.1163/156853111x597260.

Preecess, Maria, Perez, Julee, and Caguimbal, Rory. 2015. "Urban Renewal of the City of Manila and its Impact to Environment and Physical Design." *Journal of Architecture and Allied Arts in the Philippines* 6: 77–91.

Ramalho, Jordana. 2019. "Empowerment in the Era of Resilience-Building: Gendered Participation in Community-Based (Disaster) Risk Management in the Philippines." *International Development Planning Review* 41 (2): 129–48. https://doi.org/10.3828/idpr.2018.25.

Rodolfo, Kelvin, and Fernando Siringan. 2006. "Global Sea-Level Rise Is Recognised, but Flooding from Anthropogenic Land Subsidence Is Ignored around Northern Manila Bay, Philippines." *Disasters* 30 (1): 118–39. https://doi.org/10.1111/j.1467-9523.2006.00310.x.

Steinberg, Florian, and Lindfield, Michael. 2011. *Inclusive Cities*. Mandaluyong. https://www.adb.org/sites/default/files/publication/29053/inclusive-cities.pdf.

Thomas, Adelle, April Baptiste, Rosanne Martyr-Koller, Patrick Pringle, and Kevon Rhiney. 2020. "Climate Change and Small Island Developing States." *Annual Review of Environment and Resources* 45: 1–27. https://doi.org/10.1146/annurev-environ-012320-083355.

Valenzuela, Ven Paolo Bruno, Miguel Esteban, and Motoharu Onuki. 2020. "Perception of Disasters and Land Reclamation in an Informal Settlement on Reclaimed Land: Case of the BASECO Compound, Manila, the Philippines." *International Journal of Disaster Risk Science* 11 (5): 640–54. https://doi.org/10.1007/s13753-020-00300-y.

Venzon, Cliff. 2019. China's island builder has US embassy in Manila surrounded. *Nikkei Asia.* https://asia.nikkei.com/Business/Construction/China-s-island-builder-has-US-embassy-in-Manila-surrounded.

Watanabe, Akiko. 2007. "The Formation of Migrant Muslim Communities in Metro Manila." *Kasarinlan: Philippine Journal of Third World Studies* 22(2): 68–96.

Yong, Gabriel, Izni Azrien bin Noor Azalie, and Noor Hasharina Hassan. 2026. "Riverine Lifelines of the Brunei Sultanate." In *Fluid Phenomena*, edited by Iftekhar Iqbal, Hasharina Hassan, and Asiyah Kumpoh. Brill.

CHAPTER 8

Making Kin along the Mekong River: Exploring Multispecies Entanglements in the Anthropocene

Maya Dania

The control of the river system by human-related forces under global hydropower development is now a major feature of the Anthropocene era. Freshwater environments are one of the most vulnerable points in the Earth's ecosystem. Although it is scalable, the transformation of the river system utilizing hydropower technology has drawn artificial boundaries that exert a powerful influence on nearby human communities. The construction of dams across rivers offers a case of environmental transformation that holds the latency to modify the alluvial and stratigraphic record (Ortega et al. 2014). Water is rapidly becoming scarce as humans increasingly depend on energy production to serve agricultural irrigation, hydroelectric power generation, water supply for industrial and domestic use, and flood control (Rodhe and Killingtveit 2020). Hydropower dam construction seems to be a modern solution to produce abundant and reliable energy, including building reservoirs for water storage and other urban uses. Moreover, large dams symbolize modernization, national prestige, and human dominance over nature.

Human actions are changing the world, and the world may be changing humans mutually in a tenuous assemblage of delicate bonds. Donna Haraway (2015) has provided a starting point of the "species turn" to relate non-human and human collaboration through the co-construction of the niche and the entanglement of "intra-acting", coevolutionary histories and co-becoming in her concept of "making kin". Haraway's relationality concept has been particularly significant, providing conceptual tools for structuring multispecies research, notably regarding the interrelated notions of becoming companion species and nature cultures. In contrast to the modern Western ontology of separation and independence, relationality is an ontology in which entities (or beings) come into existence and are transformed through relations. It posits interconnection, co-becoming, and mutual complementarity as key existential principles.

This chapter aims to look at how the lifeways and livelihoods of human co-inhabitants in the Chiang Rai wetland livelihood are changing due to the feral

 | HTTPS://DOI.ORG/10.1163/9789004758100-010

effects of anthropogenic intervention on the Mekong river ecologies. This study focuses on Chiang Khong district in Chiang Rai province in the northernmost part of Thailand that shares the Mekong river as its natural border to the Lao town of Huay Xao. The local communities have a long history of engaging in riverine activities, including fishing. This study will mainly assess the central concept of the Anthropocene in river landscape derived from participant observations, interviews, archival research, and discourse analysis in the qualitative method. Therefore, this research proposes a multispecies Anthropocene framework using the trope of "making kin" to analyze how the frontiers of China's BRI projects in the Lower Mekong Basin (LMB), manifested in the large-scale hydroelectric dams, are new forms of anthropogenic landscapes that accelerate the long-term destruction of the biosphere, where diverse beings are dependent on complex entanglements with social, political, economic, and cultural relations.

This chapter explores the potential application of tentacular thinking in "making kin" that recognizes the interdependencies of humans in a web of relationships encapsulating a narrative of *natureculture* that move beyond a human-centric view. In this study, Chiang Khong is a multispecies field that acknowledges the interconnectedness and inseparability of humans and other life forms and seeks to extend ethnography beyond the human realm. By using Donna Haraway's concept, this study of social phenomena is attentive to the agency of the more-than-human world at the intersections between ecological relations, political economy, and cultural representations.

1 "Making Kin": Haraway's Multispecies Studies in the Anthropocene

As Haraway stated (2015), human actions always involve other abiotic processes and biotic organisms, and they do not terraform the world independently. Although there is still debate on whether the Anthropocene marks a new geological period, what is generally accepted is that the Anthropocene is a particular juncture in time where planetary impacts caused by human-induced changes have exceeded unprecedented levels. In many ways, the Anthropocene is a strikingly resonant iteration of the problem articulated forcefully in Donna Haraway's Cyborg Manifesto (2013), which sees the human, non-human, culture, and nature as inextricably linked and warns that attempts to dominate human and non-human nature can be both disastrously successful and productively perverse. The Anthropocene, according to Haraway, considers both "too big and too small" concepts to explain the perilous state of our planet; she believes that "making kin" with the more-than-human world is urgent for humanity in a relational universe. Kin-making means reclaiming human relationships with nature and unseen forces that shape life to co-create the future of all things in the Anthropocene. Further, Haraway offers an alternative name for the Anthropocene, the Chthulucene, an age of "trouble" that requires making kin with, rather than self-making, to stay with the trouble of living together with something other than entities bound by ancestry. The Anthropocene period is conceptualized as the Chthulucene to describe interconnected human and more-than-human nature in tentacular practices to build liveable futures.

Haraway (2003) describes nature and culture as already bound up in complex relationships called *'natureculture'* to emphasize that nature and culture are always entanglements in human lives, social practices, and roles and responsibilities. Haraway (2015) has also provided a starting point of the "species turn" to relate non-human and human collaboration through the co-construction of the niche and the entanglement of "intra-acting", coevolutionary histories and co-becoming in her introduced concept of "making kin". Haraway's posthumanism theory has been particularly significant, providing conceptual tools for structuring multispecies research, notably regarding the interrelated notions of becoming companion species and *naturecultures*. In contrast to the modern Western ontology of separation and independence, relationality is an ontology in which entities (or beings) come into existence and are transformed through relations. It posits interconnection, co-becoming, and mutual complementarity as key existential principles. Local theory in relational ontologies emphasizes material, spiritual, ecological, and cultural interconnection: all are inseparable in *natureculture*.

Through the multispecies lens, this chapter explores how the assemblages of organic species are actors "making kin" (Haraway 2015) involving collaborations of interspecies relationships (Tsing 2015) to thrive together in the Anthropocene. In Haraways's "making kin" concept, the term 'multispecies' includes the arrangements of relative interdependency and mutual relationships between humans and non-humans made possible through the inter-species collaboration of gain-making. Multispecies studies understand our world through immersing ourselves in forms of the liveliness of humans and non-humans, the non-living, that entangle in diverse relationships of knowing and living together. Interconnectedness relationships emerge from coevolutionary histories and co-becoming. The narrative involves multiple entanglements in the social and cultural landscape as human and more-than-human nature become vibrant actors of Chiang Khong place-making.

2 Situating the Mekong River in Regional Development

The Mekong is the longest river in Southeast Asia, rising in the glaciated mountains of Tibet and flowing nearly 3,000 km through tributaries in the ASEAN countries (Myanmar, Lao PDR, Thailand, Cambodia, and Vietnam) before entering the South China Sea in the delta. From a multispecies perspective, the Mekong is the world's second most biodiverse river (after the Amazon river). It has a local carrying capacity for migratory fish in each upstream habitat. The LMB is also home to the world's largest inland fishery. Ziv et al. (2012) discovered 877 fish species in the Mekong river Basin. The LMB is one of the most biodiverse areas globally, abundant with diverse native and migratory species. The basin is rich with a diversity of freshwater river fauna, with nearly 500 species of fish that provide a large variety of fisheries as the second largest and the most significant in the world after the Amazon river (Valbo-Jørgensen et al. 2009). Downstream, much development is taking place along the Mekong's tributaries. Major proposals are constructing tens of huge dams and reservoirs in Laos, Vietnam, Cambodia's tributaries, and 11 mainline dams in Laos and Cambodia (Urban et al. 2018).

In recent years, Chinese dam builders and investors have been active in neighbouring countries along the LMB areas. The projects are located at Pak Beng, Luang Prabang, Xayaburi, Pak Lay, Sanakham, Pak Chom, Ban Koum, Phu Ngoy (Lat Sua), Don Sahong, Stung Treng, and Sambor. The Xayaburi Dam was completed in May 2017 to pursue renewable and low-carbon energy in the region. Damming rivers to create hydropower can help alleviate the world's energy crisis and lower the danger of global climate change, yet, damming can have massive detrimental consequences on the environment and ecosystems (Fan et al. 2022). Dam activities have considerably influenced the natural flow regimes of most of the world's great rivers. The LMB is mostly a wetlands ecosystem, with three major elements influencing productivity: hydrology, floodplain quality, and fish migration patterns (Baran and Myschowoda 2009).

The Mekong river Basin's water and ecosystem production and integrity are inextricably related to the river's seasonal flow pattern, characterized by wet season flow. Flood flows deposit silt on the river's banks, which offers an extremely fertile base for horticulture and agriculture with the arrival of the dry season (Olson and Morton 2018). This natural cycle of flooding and receding of the Mekong's waters is critical to the river's ecosystem. The dam-induced changes in streamflow have led to the extinction of a flagship species from the river due to its potential to cut off the long-distance migration route of the Mekong Giant Catfish and affected many aspects of the riverine ecosystem in the Mekong, including the loss of flood pulses, floodplain habitat, wetlands, and sediment transport (Piman and Shrestha 2017).

According to Baran and Myschowoda (2009), the inland capture fisheries in the Mekong Basin are very important. Their enormous size is only equalled by their economic importance, critical involvement in rural lives, and important role in regional food security. The basin's fisheries also influence the population's dietary demands. The majority of the catch is gathered by impoverished part-time and subsistence fishermen who utilize fishing as part of a diversified livelihood strategy (Kura et al. 2017). This diversification contributes to food security, earns money with minimum capital investment, provides cash in times of need (for example, to buy rice seeds at the end of the dry season), and "insures" against crop failure (McKenney and Tola 2002). The existence of fish is critical to food security and livelihoods in the region. Fish also contribute calcium and important amino acids to the diets of the inhabitants in the region (Jensen 2001). In his book "The river of Life", Santasombat (2011) explicitly aims to develop a better understanding of transnational megaprojects, particularly the construction of hydropower dams on the Mekong, and their impact on changing ecosystems and the lives and livelihoods of marginal communities in the Lower Mekong Basin through an ethnoecological perspective that engages resource use, class differentiation, lifestyles, and cultures, to identify serious health, economic, and environmental consequences.

Riverine wetlands in Chiang Khong are involved with Thai history, national identity, religion, and the way of life. The discovery and analysis of historical settlements of ethnic groups and archaeological sites indicate that people have lived in the wetlands of northern Chiang Rai since the eighteenth century (Phuwanatwichit and Srirat 2014). In the highlands of Northern Thailand, wetlands provide a vital and diverse range of ecosystem services that support local livelihoods. Wetland ecosystems significantly benefit humans, plants, and wildlife. They temporarily store runoff and serve as fresh markets, food sources, and other resources for local people. For many generations, wetlands have provided important sociocultural and spiritual benefits to local people (Hempattarasuwan et al. 2021).

3 Multispecies Anthropocene in the Narrative of the Mekong Giant Catfish in Chiang Khong

In Northern Thailand, the impacts of the dams are mainly experienced by the local people in the Chiang Khong district (Phurisamban 2019). Local people who live in this Mekong village (Chiang Khong) are known as the descendants of the Tai-Lue ethnic group that shared similar cultures to people in Luang Prabang (Lao PDR) and Sipsong Panna (Xishuangbanna) along Yunnan's border. Fishing is a livelihood for the local villagers in the district. It involves certain non-living objects embodied in spirits to support fishing practices and bring good luck. The fishing activities reflect their economic income sources and the entwined ecological and cultural beliefs and knowledge. The decline of fish in Chiang Khong is an obvious indicator of a transformed river that also alters the local people's socioeconomics.

Research by Jakkrit Sangkhamanee (2010) demonstrates the involvement of local cultures and subjectivities in the ethnology of hydraulic knowledge, which gives crucial insights into the linkages between culture and the sustainability of water resources across the region. Dams will benefit the region's economies greatly, but they will severely influence other industries that rely on river flows, such as fisheries. Several academics have recognized the limits of an ontology centred on the dichotomy between people and nature, which is widespread in many Western industrialized nations, in tackling the concerns of the Anthropocene. Some scholars' alternative ontologies are present in non-Western countries. For example, Raffles (1999) introduces the interconnectivity of nature and culture in reading "locality" with a case study of an Amazonian place that he calls Igarapé Guariba in the northern Brazilian state of Amapa. In a more specific way, Inoue and Tickner (2016) emphasize the many modes of knowledge of many worlds and natures in one planet used by numerous indigenous peoples in Brazil. Western epistemology frequently dismisses stories, poetry, myths, dreams, written texts, and songs, but these manifestations help the Yanomami of Brazil to grasp and cohabit with the non-human world.

Chiang Khong district, located on the upper reaches of the LMB, highlights the area's physical proximity to the Mekong river. It is the home of the aquatic ecosystem where the Mekong Giant Catfish migrates to spawn in the river covered by wetlands and forest in the ancestral Boon Ruang community lands. Chiang is a northern dialect word that translates literally as 'town', while Khong refers to the Mekong river. Thus, the deterioration of cultural assets in Chiang Khong district demonstrates a clear case for possible danger to community identity. The Mekong Giant Catfish is a source of pride for the community, especially for Ban Hat Krai residents who live along the village's riverside. The Mekong Giant Catfish (Pangasianodon gigas, in Thai: ปลาบึก—pla buek) is native to Chiang Khong, and the species is now considered critically endangered due to accelerating habitat loss. The species is widely recognized as an indicator of the ecological integrity of the Mekong ecosystem due to its long-distance migrations to find suitable spawning grounds. It is three metres in length and over 300 kg in weight, and has good longevity (Lorenzen and Sukumasavin 2007).

Ban Haad Krai is a village located in the Chiang Khong district. This village is the main place for the Mekong Giant Catfish to spawn because the village is the site of the banks that attract Catfish to lay eggs and guard their nests for the young to mature. With the abundant numbers of Mekong Giant Catfish, Chiang Khong district chose the Catfish as the town logo with the motto "the city known for catfish". The village recorded the largest freshwater Giant Catfish with an average weight exceeding 200 kilogrammes in the 1990s. As fishing has become the major livelihood source, rivers and water are sacred entities for the local people and serve as life-givers. The local people build their knowledge and memory from the river culture. In the process of hunting the Catfish, the local people perform an ordination ceremony for the godfathers of the Catfish on 18 April every year.

Originally, there was a ritual before catching the Giant Catfish together, which was called the "Wai Phong Ceremony". Only professional catfish hunters participate in this ceremony that has been passed down from their ancestors. There is a small shrine built for organizing a sacrifice of Phee Ploung (ghost owner of Giant Catfish) or Phaya Nak to fortune-tell the number of Catfish that will be possibly caught (the number of Catfish allowed by the angels to be caught each year). The ceremony also includes offerings to the ghost of Mae Ya Nang (แม่ย่านาง). The ghost of Mae Ya Nang is a female deity who watches over the natural power of the river to protect the fishing boats. The entity is similar to the Mae Pho Sop ghost, who conquers the soil that protects the growing fertility of rice for the Thai people. With the offerings to Mae Ya Nang, fishermen receive protection for their boats every time they go sailing to catch the fish.

The type of wood allowed to make the boat is from the Takian Tree and must also be taken from the deep forest with some rituals to ask for the Forest God's permission before cutting the trees. With the Takian boat, four to five fishermen can share a space to catch the fish. On each journey, two men row the boat, and two to three people spread the nets to catch the fish. In the past, fishing was a male responsibility to obtain income and fulfil the family's food security. People also believed that "Pla Khong Beng", or eating the Giant Catfish, would make people smart and have charisma and merit. Before the development of the dams, the local people had already considered the Giant Catfish a commodity. When the Mekong Giant Catfish became a commodity, the fish meat price was higher than the original selling price of fish in the market. In late 1977, the price increased to 100 baht per kilogramme, and in late 1987, the price increased to 250 baht per kilogramme and continued to rise until it was about 500 baht per kilogramme in the late 1990s.

With the increased popularity of Giant Catfish hunting in the late twentieth century, fishing activity has been transformed into a male hunting pastime as the activity of Giant Catfish hunting increased. The village used to host the tourism promotion of Giant Catfish hunting to attract tourists to visit the village from April to May every year. This attraction also grew the tourism trends in Chiang Khong as the business of hotels and restaurants expanded, with the intensity of fish hunting increasing from the 1980s to the late 1990s. Many tourists came to see the catfish hunting, causing more than 200 men in the village, including children, teenagers and adults, to become catfish hunters and caught 69 catfish, which was the highest record in fishing.

However, in the early 2000s, the Giant Catfish was rarely seen in the Mekong river, and the World Wildlife Organization (WWF) of Thailand attributed the threat to the Catfish to the impacts of overhunting. The worse environmental condition was also created by the riverbank explosion of the waterway development project for commercial channel navigation due to the commercial agreement on the Lancang-Mekong river on 20 April, 2000, at Tachilek Province (the part bordering Chiang Rai Province). The explosion of the river banks has destroyed the natural spawning area and breeding ground for the Catfish to mate and lay eggs. With the dam construction on the Mekong river, downstream aquatic habitats experience water turbidity and sedimentation changes. The river flows slowly, starving fish of the nutritious sediments eliminated by the regulated water outflow due to mainland China's present upstream hydropower dams. The presence of the Mekong Giant Catfish is an important indicator of the health of the river and its ecosystem. Hence, the endangered status of the fish shows the fragile ecosystem of the Mekong river that affects not only the livelihoods of people living in the basin but also the regional environment for a sustainable future.

4 Kin Making with Entangled Lives along the Mekong River in Chiang Khong

Aside from making a policy to prohibit catfish hunting, regional diplomacy efforts are also being made to mitigate the negative impacts of the dam construction in the upper part of the Mekong river. In the early 2000s, the local community worked with the local government and NGOs to organize a conservation project to deliver the Giant Catfish back into its habitat in Ban Haat Krai. In 2020, The Mekong river Commission (MRC) held the 8th Regional Stakeholder Forum to gather recommendations on the drafted Mekong Basin Development Strategy (BDS) (2021–2030) with stakeholders from the four Member States: Cambodia, Laos, Thailand, and Vietnam (Suhardiman and Gehab 2022), as well as the State of the Basin presented at this forum indicating that the Mekong river has experienced seasonal changes to its flow. According to the MRC (Mekong river Commission) report, by 2040, the sediment in the Mekong river will decrease at an alarming rate (Soukhapon et al. 2021). The MRC has revised the evidence to allay the concerns of lower Mekong countries by creating Mekong monitoring to monitor the damage due to sediment loss.

The local legend of the Mekong Giant Catfish is now revitalized to cultivate environmental awareness beyond a human-centric view. Even though catfish hunting has been cancelled for 20 years now due to the extinction of the Giant Catfish, causing Chiang Khong fishermen to turn to other fish for their livelihood, the sacrificial ritual is still held every year and has become an annual event (festival) of the people of Chiang Khong. Aside from holding ceremonies for the river Ghost, the local people in Chiang Khong also consider the Mekong river as the home of Naga (the half-serpent being in Thai mythology). The belief in Naga has been with Thai people for a long time, which is seen in many tourist attractions that mention Naga or the temples where the serpent statues appear. Naga is considered a sacred entity with supernatural power that is involved in the history of Buddhism for Thai people due to its role of protecting Buddha from rain and sun heat while praying. The role of Naga related to Buddhism also represents the acculturation of the local belief in animism with Buddhism as a new religion. Especially in Chiang Khong, the district's history is close to that of Naga because the legend of the birth of the Mekong river was derived from two Nagas named Suthonakarat (Sri-Sut-Tho Naga) and Suwannakharat (Suwannaga) who resided in the Chiang Khong area and fought each other until God Indra (Lord Indra) had to come down to stop the battle by proposing that the two Nagas should compete for each other to build a river. If either party could build a river that reaches the sea first, God Indra would reward the winning Naga, the Giant Catfish, to live in the river and remain the companion of the Naga.

The two Nagas accepted the proposal from God Indra and started the competition to build the longest river to reach the sea. Sutthonakharat chose to build a river to the eastern side of the town, but due to his impatience, he built a long curved river until it was called the "Khong river" (Mekong river). On the other side, Suwannakharat was calmer and more detailed and created a straighter river than the Mekong river, known as the Nan river. Even though both rivers similarly reach the sea, Sutthonakharat completed the river's creation before Suwannakharat. Then, God Indra gave the Mekong Catfish as a prize to live only in the Mekong river in the Chiang Khong area. For this reason, the local people in Chiang Khong that live near the Mekong river strongly connect with the myth of the Naga.

Until today, the local people in Chiang Khong believe that the Naga lives in a large cave under the Mekong river water. As in the legend, they built Doi Mae Ya Mon Temple, located at the mouth of the Mekong river, next to the border with Laos. This river spot is a three-way area with a whirlpool which is active all year round, and is considered one of the most dangerous spots on the Mekong river. The elders forbid their children to swim in this area because, in addition to being afraid of the vortex's danger, it also includes the legend of the Naga that lives in that area. According to legend, it is said that at this temple, there is a Naga hole beneath the Doi Mae Ya Mon Temple. Originally, at the mouth of the Mekong river near the temple, the Naga family created a secret route to travel back and forth between the Naga world and the human world. The Naga father has a fierce personality, and he likes to use the route to steal a look at boatmen in the Mekong river, often capsizing the boats and rafts of the local fishermen.

However, one day, this ferocious Naga father accidentally became stuck in a net fishing trap and could not escape. Worrying about the father, the Naga daughter swam to the human world to ask for help, and she met Ya Mon, an old woman who went out from her home to pick mushrooms by the Mekong river. Ya Mon felt pity for the Naga daughter and helped the Naga father to be released from the net trap by using a small knife. Ya Mon then asked the Naga father to practise the Buddhist precepts and stop his actions from making trouble in the human dimension. The Naga father agreed, and he closed the hole to the path and the cavern connecting the human world to the city of Naga on the Mekong river. Before sending Ya Mon back to the riverbanks nearby the village, the Naga father gave her a basket of vegetables that turned into gold leaves. With the gold, Ya Mon asked the villagers to build layers of relics to close the Naga route, and the location of the Naga route is located in the basement of Doi Mae Ya Mon Temple in Chiang Khong. With this Naga belief, the local people in Chiang Khong built a strong connection to their land and resist the development projects supported by Chinese investors.

The Mekong area in Chiang Khong is also a geographically targeted economic corridor territory for development by China's Belt and Road Initiative (BRI) megaprojects, especially in relation to water resources and agricultural production. As a major emerging country, China is a crucial player in the transformation of the global energy structure. Chiang Rai Province is becoming an increasingly important economic city, with the potential for border trade and tourism and as a logistic hub because Chiang Rai links Thailand with three neighbouring countries: the Republic of the Union of Myanmar, the Lao People's Democratic Republic, and the People's Republic of China (southern China).

In 2015, the National Committee on Special Economic Zone Development approved Chiang Rai province as one of five areas in Thailand designated as "Special Economic Zones Phase 2", aiming to develop the border areas, increase their competitive ability, and solve security problems. The Chiang Rai Special Economic Zone (CRSEZ) is a government-initiated development supported under the larger framework of Special Economic Development Zones (SEZs) which covers three districts of Chiang Rai Province in Chiang Khong, Chiang Saen, and Mae Sai, which are seen as conducive to trade and investments due to their strategic location on the North–South Economic Corridor, connecting Northern Thailand to Southern China via two routes: Chiang Khong-Laos and Chiang Saen-Myanmar (Kaewkanta et al. 2022).

Chiang Khong's neighbouring district, Chiang Saen, has now transformed into a SEZ with its unique geographical location known as the Golden Triangle to connect Thailand with Myanmar and Laos through a natural borderline drawn by the Mekong river. In the opposite part of Chiang Saen, Bokeo district of Laos, King Romans Casino, a private casino supported by Chinese investors, is fully constructed with the possibility of extending other development and land acquisition. Bokeo also shares a border with Chiang Khong district on another side, with the meeting point located nearby Doi Mae Ya Mon Temple. Although Chiang Saen and Mae Sai have now become the SEZ area in Chiang Rai, the Chiang Khong community rejected the plan and built-up community resistance to the development by implementing local traditions and the cultural myth of the river.

Mae Nam Khong Local Knowledge Institute, a local school in Chiang Khong, is led by a teacher named Kru Tee (Mr. Niwat Roikaew), who always opens his home to young Thais, academicians, and other communities who want to learn about the history and natureculture of Chiang Khong. It is a two-storey wooden building located on the banks of the Mekong river. Rak Chiang Khong is the name of the environmental movement group that Kru Tee also leads. In Thailand, the efforts by Kru Tee and the local community in Chiang Khong to keep their homeland free from land acquisition and damaged riverine ecosystems due to dams being constructed along the Mekong river have inspired the younger generations to protect the river and nature for a sustainable future. They share three main important things about the Mekong river, which are: the history, the ecology, and the culture. On 26 May, 2022, Kru Tee received the Goldman Environmental Award as the representative of Asia for his work related to protecting the Mekong river.

Chiang Khong people are becoming kin with the river through diverse ways of practices to aspire bonds and intimacies with more-than-human nature. Through kin-making, the re-establishment of both material and spiritual cosmos, such as Naga mythology and cultural folklore, has never been more critical in the Anthropocene to sustaining ecological connections in the dynamic relationships between humans and more-than-human entities. The local people in Chiang Khong still hold their belief in Buddhism–animism which can be traced back to their shared belief with Lao culture. Beyond the human-centric view, the river is a sacred entity as the home of sacred animals, like Nagas and their offspring, the Mekong Giant Catfish. The revival of the narrative of the Mekong river through history, ecology, and culture has resulted in the Chiang Khong locality sharing in complex ways of embeddedness in natureculture. The exceptional and daily practices, the construction of multispecies entanglements and the making of the Mekong Giant Catfish as a locational identity tied up with the environmental context through the stories that people repeatedly tell reinforce a personal connection to the riverine landscape. As the locals in Chiang Khong travel with their stories, they narrate their place with its associated histories to proliferate the meanings of multispecies entanglements.

5 Conclusion

The way of life of the people of the Mekong river has long been associated with the river. So from the Chiang Khong people's perspective, the Mekong is not just a river but refers to 'life'. The Mekong river comes from a diverse ecosystem. Some areas are surrounded by sandbars formed by the deposition of river sediments. Many fish come into this area. Including in the middle of the river, there are many fish species; it is a place to lay eggs for some birds as well, and there are various insects that are food for fish. The Mekong river, therefore, is the blood vessel that nourishes people because their life depends on the Mekong river, whether for consumption, transport, fishing, or cultivation. It is also a source of civilization and customs of the people in the community.

Perhaps the change of consciousness—the change within ourselves—to spirituality, culture, and how we think will have the most significant impact. After all, as Donna Haraway (2003) reminds us, nature and culture are always already bound up in complex relationships called 'natureculture'. Haraway's term 'natureculture' emphasizes that ideas are embodied in social formations and are both material and ideological; nature and culture are always already natureculture. Local theory in relational ontologies emphasizes material, spiritual, ecological, and cultural interconnection: all are inseparable in natureculture. Not surprisingly, this rationality has caused a cascade of socio-ecological changes, and in turn, habitat fragmentation and biodiversity loss in areas where human control has simplified and disrupted nature. Through the multispecies lens, this chapter has tried to narratively deliver how the assemblages of organic species are actors "making kin" (Haraway 2015) involving collaborations of interspecies relationships (Tsing 2015) to thrive together in the Anthropocene.

References

Baran, E., and Myschowoda, C. 2009. Dams and fisheries in the Mekong Basin. *Aquatic Ecosystem Health & Management*, *12*(3): 227–34.

Fan, P., Cho, M. S., Lin, Z., Ouyang, Z., Qi, J., Chen, J., and Moran, E. F. 2022. Recently constructed hydropower dams were associated with reduced economic production, population, and greenness in nearby areas. *Proceedings of the National Academy of Sciences*, *119*(8), e2108038119.

Haraway, D. J. 2015. Anthropocene, capitalocene, plantationocene, chthulucene: Making kin. *Environmental Humanities*, *6*(1): 159–65.

Haraway, D. J. 2013. A cyborg manifesto: Science, technology, and socialist feminism in the late twentieth century. In *The transgender studies reader*: 103–18. Routledge.

Haraway, D. J. 2003. *The companion species manifesto: Dogs, people, and significant otherness*: Vol. 1, 3–17. Prickly Paradigm Press.

Hempattarasuwan, N., Untong, A., Christakos, G., and Wu, J. 2021. Wetland changes and their impacts on livelihoods in Chiang Saen Valley, Chiang Rai Province, Thailand. *Regional Environmental Change*, *21*: 1–15.

Inoue, C., and Tickner, A. B. 2016. Many worlds, many theories? *Revista Brasileira de Política Internacional*: *59*.

Jensen, J. G. 2001. Managing fish, flood plains and food security in the Lower Mekong Basin. *Water Science and Technology*, *43*(9): 157–64.

Kaewkanta, C., Suma, Y., Kiattisaksiri, P., and Eaktasang, N. 2022. Municipal Solid Waste Management in Special Economic Zone (SEZ) in North Thailand. *Multicultural Education*: *8*(5).

Kura, Y., Joffre, O., Laplante, B., and Sengvilaykham, B. 2017. Coping with resettlement: A livelihood adaptation analysis in the Mekong River basin. *Land Use Policy*, *60*: 139–49.

Lorenzen, K., and Sukumasavin, N. 2007. Mekong giant catfish. *Catch and Culture*, *13*(1): 22.

McKenney, B., and Tola, P. 2002. *Natural resources and rural livelihoods in Cambodia: A baseline assessment* (No. 23). Cambodia Development Resource Institute.

Olson, K. R., and Morton, L. W. 2018. Polders, dikes, canals, rice, and aquaculture in the Mekong Delta. *Journal of Soil and Water Conservation, 73*(4): 83A–89A.

Ortega, J. A., Razola, L., and Garzón, G. 2014. Recent human impacts and change in dynamics and morphology of ephemeral rivers. *Natural Hazards and Earth System Sciences, 14*(3): 713–30.

Phurisamban, R. 2019. *Against the Current?: Using Local Ecological Knowledge to Negotiate Impacts from Large-scale Mekong River Development*. PhD Diss., Geographisches Institut der Universität Zürich.

Phuwanatwichit, T., and Srirat, S. 2014. Sustainable tourism planning and management preparing for the ASEAN Economic Community (AEC): Wiang Nong Lom wetland, Chiang Rai province. Chiang Mai Univ J Soc Sci Human 15 (2): 132–64.

Piman, T., and Shrestha, M. 2017. Case study on sediment in the Mekong River Basin: Current state and future trends. *Stockholm Environment Institute: Stockholm, Sweden.*

Raffles, H. 1999. Local theory: Nature and the making of an Amazonian place. *Cultural Anthropology, 14*(3): 323–60.

Rodhe, A., and Killingtveit, Â. 2020. Catchment hydrology. In *Geochemical Processes, Weathering and Groundwater Recharge in Catchments*: 77–107. CRC Press.

Sangkhamanee, J. 2010. The hydraulics of power and knowledge: Water management in Northeastern Thailand and the Mekong region.

Santasombat, Y. 2011. *The river of life: Changing ecosystems of the Mekong Region.* Mekong Press.

Suhardiman, D., and Geheb, K. 2022. Participation and politics in transboundary hydropower development: The case of the Pak Beng dam in Laos. *Environmental Policy and Governance, 32*(4): 320–30.

Soukhaphon, A., Baird, I. G., and Hogan, Z. S. 2021. The impacts of hydropower dams in the Mekong River Basin: A review. *Water, 13*(3): 265.

Tsing, A. L. 2015. The Mushroom at the End of the World. In *The Mushroom at the End of the World.* Princeton University Press.

Urban, F., Siciliano, G., and Nordensvard, J. 2018. China's dam-builders: Their role in transboundary river management in South-East Asia. *International Journal of Water Resources Development, 34*(5): 747–70.

Valbo-Jørgensen, J., Coates, D., and Hortle, K. 2009. Fish diversity in the Mekong River basin. In *The Mekong*: 161–96. Academic Press.

Ziv, G., Baran, E., Nam, S., Rodríguez-Iturbe, I., and Levin, S. A. 2012. Trading-off fish biodiversity, food security, and hydropower in the Mekong River Basin. *Proceedings of the National Academy of Sciences, 109*(15): 5609–14.

CHAPTER 9

The Many Meanings of Water: an Exploration of the Life-World of Two Cities on the Murray River

Heather Downey

The IPCC (2022) posits that global warming caused by human impacts, notably intensified resource extraction and agriculture, has disrupted the social and physical global systems that sustain life. This scenario has resulted in constructions of the Anthropocene as an era of crisis (Robin 2016, 27; Christoff 2016; Davies and Lawrence 2019; Groenfeldt 2019). While not disputing the extreme threat facing the Earth, Head (2007, 837) suggests that focusing on 'human impacts' reinforces the dualisms of 'natural and human environments' and positions humans outside nature rather than part of "the Earth and its processes".

In Australia, particularly in the southeastern states, global warming, characterized by climate variability and extremes, has caused a series of environmental disasters. El Niño seasons of severe drought and water scarcity followed by the black summer bushfires of 2019/2020, then multiple La Niña seasons of heavy rains and flooding (L. Cox 2022) have placed the natural environment, particularly rivers and water, at the fore of the nation's psyche. Consequently, awareness of the inextricable links between people and the natural world and concern for the Earth's capacity to support current and future generations have increased (Jackson 2022). Those already experiencing disadvantages, such as those residing in rural Australia, where health and welfare outcomes and life expectancy are far poorer than in metropolitan centres (AIHW 2018), are most keenly affected by the effects of the Anthropocene (Dominelli 2014). In this geographic context, the injustice of marginalisation has been exacerbated by water scarcity, river overflows and marked changes to place and community, including loss of ecosystems, livelihoods, homes, and lives, and displacement following environmental disasters (IPCC 2022). However, Robin (2016, 16) argues that the Anthropocene must still "be lived in hope", and this opens the way for us to traverse the Nature–Society binary and understand rivers in creative imaginations.

This chapter aims to contribute to the development of a "planetary imaginary" or "our systemic entanglements" (Clarke 2015, 151), by drawing on green social work that incorporates Indigenous knowledge to question the

 | HTTPS://DOI.ORG/10.1163/9789004758100-011

dominance humans have assumed over the natural world (Dominelli 2014) to understand the multidimensional meanings of the Murray river for two rural Australian river communities, Mildura and Albury/Wodonga (Downey et al. 2022a, 2022b). Considering multiple worldviews in this way develops Habermas' (1987) theoretical concept of the life-world or the behaviours and perspectives that can be understood. There are five parts to this chapter: (1) Murray river imaginaries and the research context, (2) theoretical frame, method and analytic frame, (3) descriptions of the study sites, (4) a comparison of findings from the two regional cities, Mildura and Albury/Wodonga, and (5) a discussion of the findings and conclusion.

1 Murray River Imaginaries

In Australian Indigenous worldviews, rivers are central elements of Country, a term that comprises people, animals, plants, air, the Earth and water. Country is a sentient entity infused with spirit, a place of belonging that "expects its people to maintain its integrity … And Country has an obligation in return—to nourish and sustain its people" (Rose 1992, 109). Consequently, rivers are holders of profound knowledge, enablers of cultural practices, and central within the ways of being (Ryan 2023), with river health and the health of Indigenous Peoples intertwined. Country and Dreaming, or stories of how the universe came to be, are inextricably linked (Ryan 2023). As there are myriad Australian Indigenous Peoples, there are diverse stories of creation, including the creation of rivers. Traditional custodians of the Albury/Wodonga region, the Wiradjuri, Waywurru, and Dhudhuroa Peoples, for example, posit that at the beginning of time, the big fish, Waa Wee, swam through the landscape, making channels in the Earth. Rain followed, and the channels filled to create the Murray river (Burraja Cultural Centre n.d).

Australian Indigenous values were the dominant culture for millennia prior to colonisation (Marshall 2017). However, settler colonialism imposed new legal, social, and political systems that bore no resemblance to Indigenous lore and customs (Marshall 2017). The legal construction of the Murray river as the boundary demarcating the states of New South Wales and Victoria, a territorial definition imposed under s 1 of the Australian Constitutions Act 1850, conflicted with Indigenous People's beliefs that each of the approximately 500 Australian Indigenous groups had a defined territory from which they had been created (Carney 2013). This study is located in the Australian context which has unique characteristics. However, there are also some similarities with other colonized nations that saw rivers as a means to open up the interior (Arnold

2026; Van Schendel 2026). In Australia, colonists sought to eliminate Indigenous societies through forcible removal from traditional lands and waters that were appropriated for industry, primarily agricultural production (Berry and Jackson 2018; Marshall 2017).

Irrigation systems, vast feats of engineering introduced in response to the extreme drought conditions experienced in Australia's southern Murray Darling Basin (MDB) during 1877–1881 became a central tenet in settler-colonial identity and were fundamental to constructions of a "white water imaginary" (Wolfe 2006, 395) that excluded Indigenous Peoples. Informed by Western values of exploiting and owning natural resources, modernist irrigation systems irreversibly altered Country (Marshall 2017) and reconfigured natural flows (Davies and Lawrence 2019). Thus, the powerful role of colonization and modernity transformed rivers, with some suggesting that they are now a form of hybrid socio-cultural-economic ecosystem, produced naturally, socially, culturally, economically, and technologically (Jackson 2022; Thoms, Rose, and Dyer 2020; Davies and Lawrence 2019).

2 The Research Context

In the contemporary context, settler-colonial Australia continues to be influenced by a disabling story that has historically constructed the nation as "a sunburnt country, a land of … droughts and flooding rains" (MacKellar 2013, 224). However, the experience of catastrophic and incremental environmental events has resulted in significant water policy implementation, particularly in Australia's food bowl, the Murray Darling Basin. Technical responses, notably, "recovery of 2,750 gigalitres of water from irrigation farming (*Basin Plan 2012* (Cth))" are central tenets of the Murray Darling Basin Plan (MDBP), the mechanism underpinning the implementation of the Water Act (2007) which aims to restore the ecosystem while protecting all beneficiary water users (Downey et al. 2022a, 557). At the same time, market-based instruments, such as complex water trading arrangements for irrigators that aim to encourage efficient water use, have served to construct water as private property (Christoff 2016). Critically, the strength of the agrarian culture, specifically the privilege afforded the irrigation agricultural sector's voice in water policy debates, has marginalized the views of the Indigenous and non-Indigenous residents of rural river towns and cities and the environment (Downey et al. 2022a).

However, environmental catastrophes affect all rural residents, and there is increasing recognition of the diversity of water values. First, Indigenous People's tenacity has resulted in their customary connections to water remaining central to individual and group identity and health (Downey et al. 2022a). Environmental extremes continue to cause collective grief as Indigenous Peoples' reciprocal relationships with rivers are further threatened (Downey and Clune 2020). Second, experiences of water scarcity and flooding impact green and blue spaces such as rivers, wetlands, parks, and gardens. These spaces are important sources of pleasure, and enablers of social connection, physical activity, and individual and community health (D.T. Cox et al. 2017) for those who live in rural towns and cities.

The latest Australian climate projections suggest that Australia faces a future of frequent and severe climate extremes with decreases in average rainfall and MDB river flows expected to continue (CSIRO 2022). This chapter explores emerging river imaginaries, specifically, their dynamic, multidimensional meanings for the residents of two unique rural Australian river cities.

3 Theoretical and Analytic Frame

Green social work incorporates "a socially just, culturally respectful and sustainable world with Indigenous wisdom paving the way" (Ellis, Napan, and O'Donoghue 2018, 535). Critically, green social work draws on Indigenous knowledge to challenge the hierarchical relationship that has developed between humans and the non-human world (Curaming 2026) by understanding the connectedness between everyone and everything and our responsibility to care for the Earth today and into the future (Dominelli 2014). Green social work understands that disadvantaged groups, such as those residing in rural Australia, are further marginalized by the impacts of anthropogenic climate change (Noble 2016). Consequently, this worldview calls for the "adoption of lifestyles that can be applied universally and are possible to sustain without injustice to fellow humans and other species" (Noble 2016, 18).

In this chapter, concepts relating to green social work serve as a frame for interpreting the findings from a mixed methods study of the multidimensional meanings of water in the Australian rural river communities of Mildura and Albury/Wodonga. Specifically, the chapter explores diverse aspects of hydrosocial relationships and the ways that these contribute to emerging river imaginaries.

4 Method

This research compares results from analogous, cross-sectional, convergent mixed-method online questionnaires (Creswell and Creswell, 2017) from two unique rural communities in Northern Victoria, Australia. One, Mildura, was emerging from an intense, prolonged drought when the study was conducted from July to August 2020. The other, Albury/Wodonga, was experiencing a La Niña event, with associated flooding during questionnaire administration from September to October 2021 (Department of Environment n.d.). Both sites were subject to COVID-19 isolation restrictions that permitted residents to leave their homes for one hour per day for exercise. This research project was approved by the La Trobe University Human Research Ethics Committee, HEC20081, Mildura, and HEC21260, Albury/Wodonga. The questionnaires were informed by the Social and Economic Long Term Monitoring Program (SELTMP) for the Great Barrier Reef (Marshall et al. 2019). Questionnaires comprised 45 items and three open-ended questions (see Table 9.1 for open-ended questions), presented in five sections: (1) meanings of the Murray river, (2) meanings of blue and green spaces, (3) water availability, shortages, and restrictions, (4) people's ideas about water, and (5) sociodemographic data. The estimated completion time for the questionnaire was 25 minutes. Community residents in both sites were invited to participate via local print and online local media and local government social media channels, with no incentives. The inclusion criteria required respondents to reside in either Mildura or Albury/Wodonga and be aged at least 18.

I am a middle-class female academic with colonial settler heritage. I reside in a rural, riparian community located on the Murray river. This chapter provides a novel analysis as little is known about the topic. First, data from each study site were analyzed separately. Deductive descriptive analysis of quantitative data was performed using the Statistical Package for the Social Sciences software (SPSS). Qualitative data were managed using NVivo software. Qualitative data analysis is an inductive process, specifically, a thematic analysis of answers to open-ended questions and respondents' observations using green social work as a lens to examine emerging water imaginaries. The use of respondents' verbatim comments supported confirmability (Liamputtong 2013). The convergent mixed-method design permitted the integration of quantitative and qualitative data from each study site separately to form one data set for Mildura, and one for Albury/Wodonga (Creswell and Creswell 2017). This analytic

TABLE 9.1 Open-ended questions

What are the first words that come to mind when you think of the Murray River?
If water was not restricted, what would you use it for?
Is there anything more that you would like to tell us about the ways that you use or interact with water and/or the Murray River?

approach has been used previously to explore domestic violence service delivery during COVID-19 (Cortis et al. 2021). Using each data set as a whole enabled a comprehensive understanding of the thick, rich, contextual findings in holistic narrative form for each study site. Importantly, using the two data sets in this way also allowed comparisons of the findings between the two regional cities.

5 Study Sites

Study sites were selected for their location, industry, and climatic characteristics. Both rural cities are located on the Murray river, in the Murray Darling Basin, that produces food and fibre worth $22 billion annually and is imagined as Australia's food bowl (MDBA 2022). While Mildura is in the remote northwest of Victoria, Albury/Wodonga is located in the northeast of the state. Mildura is the site of Australia's first irrigation colony and is rich in non-Indigenous and Indigenous history. In the contemporary context, Mildura's industries include horticulture, healthcare, welfare, and education (ABS, n.d.). Originally known as Bungambrawatha, Albury/Wodonga was settled by cattle graziers in the 1830s, and the Murray river was used for irrigation purposes since the construction of the Hume Dam in 1919. Albury/Wodonga's significant industries now include health, human services, and defence (ABS 2021). Commonly imagined as a desert oasis, Mildura has the driest and hottest climate in Victoria, with mean annual precipitation of 285.4 mls and summer temperatures regularly in the low 40 degrees C (Bureau of Meteorology 2021). Albury/Wodonga's climate is slightly more temperate, with summer temperatures regularly exceeding 30 degrees C and annual rainfall of 643.1 mls (Bureau of Meteorology n.d.). The Murray river is the water source for both cities, and State-mandated permanent water-saving rules apply to residents. These laws permit the use of buckets, hand-held hoses, or watering cans at any time, for any purpose, yet restrict using watering systems to water residential or commercial gardens and lawns outside the hours of 6 p.m.—10 a.m. (Victorian Government 2019).

6 Findings

The findings are presented under the themes of (1) participants' characteristics, (2) connections to the Murray river, (3) recreational use and enjoyment of the Murray river, (4) connections to blue and green spaces (5) Murray river health, and (6) water scarcity. Attention is paid to the similarities and differences within and between Albury/Wodonga and Mildura community residents' perspectives on the meanings of the river.

7 Participants' Characteristics

Mildura's population is approximately 32,738, and Albury/Wodonga's population is 97,793 (ABS 2021, 2020). In Mildura, 33 people responded to the online survey, with 23 completing the demographic questions. One hundred and sixty-three Albury/Wodonga residents submitted the survey, with 133 providing demographic information.

Self-reported information on gender, age, length of residency in Mildura and Albury/Wodonga, cultural heritage, education level, and annual household income was collected (see Table 9.2).

Of those providing demographic information in Mildura, the proportion of males (52.18%) and females (47.82%) differed slightly from that of the official Mildura population: male (48.10%) and female (51.90%). In Albury/Wodonga, the proportion of males (32.34) and females (67.66) differed considerably from that of the Albury/Wodonga population: male (48.5) and female (51.5). In both sites, the respondents' ages (Mildura, range 19–70 (mean age of 50 years) and Albury/Wodonga, range 19–82 (mean age of 46.63 years) exceeded the median age of residents, i.e., 39 years in Mildura and 38 years in Albury/Wodonga (ABS 2021, 2020). The proportion of respondents in Mildura born in Australia was 86.95%, and for Albury/Wodonga, the figure was 85%, which are both higher than the official figures for the Mildura and Albury/Wodonga populations, i.e., 77.00% and 82.5%, respectively. Of these, 4.35% of Mildura and 3.75% of Albury/Wodonga respondents identified as Indigenous, similar to the Indigenous populations of both sites, i.e., 4.60% of Mildura and 3.70% of Albury/Wodonga residents. In Mildura, the proportion of respondents with either a Bachelor or Post Graduate qualification was 60.86%, far higher than that of the Mildura population with a figure of 6.40% (ABS 2020). Similarly, in Albury/Wodonga, those with a Bachelor or Post Graduate qualification, 53.37%, far exceeded the rate of the Albury/Wodonga population of 17.7%. In Mildura, 79% of respondents reported an annual pre-tax income of $50,000 to over $100,000, exceeding Mildura's median annual income of $53,196 (ABS 2020), while in Albury/Wodonga, 66.67% of residents reported an annual pre-tax income of $50,000 to over $100,000, also exceeding the median of $75,400.

TABLE 9.2 Comparative household income in Mildura and Albury/Wodonga

Mildura			Albury/Wodonga		
Characteristics	**N = 23**	**%**	**Characteristics**	**N = 133**	**%**
Gender			*Gender*		
Female	11	47.82	Female	90	67.66
Male	12	52.18	Male	44	32.34
Cultural Heritage			*Cultural Heritage*		
Settler Colonial Australian	20	86.95	Settler Colonial Australian	114	85.00
Indigenous Australian	1	4.35	Indigenous Australian	5	3.75
Other	2	8.70	Other	15	11.25
Age	*Range*	*Mean*	*Age*	*Range*	*Mean*
	19–70	50.33		19–82	46.63
Years of residency in Mildura	*Range* 1–65	*Mean* 16	*Years of residency in Albury/Wodonga*	*Range* 1–82	*Mean* 25.91
Education level	*N = 23*	*%*	*Education level*	*N = 133*	*%*
Less than Year 12 or equivalent	2	8.70	Less than Year 12 or equivalent	8	6.01
Year 12 or equivalent	3	13.04	Year 12 or equivalent	21	15.78
Diploma	4	17.40	Diploma	33	24.84
Bachelor's degree	7	30.43	Bachelor's degree	40	30.07
Postgraduate degree	7	30.43	Postgraduate degree	31	23.30
Household annual income (before tax)			*Household annual income (before tax)*		
Up to $50,000	3	13.04	Up to $50,000	32	24.06
$50,000 – $100,000	8	34.78	$51,000 – $100,000	41	31.58
Over $100,000	10	43.48	Over $100,000	48	36.09
Preferred not to answer	2	8.70	Preferred not to answer	11	8.27

8 Connections to the Murray River

When listing the first words that came to mind when thinking about the Murray river, residents in both Mildura and Albury/Wodonga prioritized

FIGURE 9.1 Canoeing on the Murray river

"life" and "lifeblood". Here, the term 'life' encompassed other than human life to include trees, gardens, wetlands, wildlife, and native fish. Both cohorts frequently mentioned climate change impacts, notably "flood" and "drought". However, Mildura residents also chose the terms "crisis" and "stress". Importantly, the river was also imagined as a threat and source of pleasure in both sites. A few residents chose descriptors such as "scary" and "dangerous", yet it was more common for residents to observe the river as "beautiful", "peaceful", "magnificent", and "mighty". Further, respondents from Mildura and Albury/Wodonga drew attention to the ways that the river provides space for aquatic pursuits, including boating, kayaking, canoeing, swimming, and fishing, and other recreational activities such as camping, walks, and social connection, particularly time with grandchildren, family, and friends.

9 Recreational Use and Enjoyment of the Murray River

Reflecting the connection between people's wellbeing and the river, 81.9% of Mildura respondents and 91% of Albury/Wodonga respondents had visited the Murray river for recreational purposes in the past year. While there was not a lot of variation, summer was clearly the most popular season, with 27% of Albury Wodonga respondents and 21% of Mildura respondents indicating that they visited the river on more than 30 days during December, January, and February. Residents in both sites most commonly visited the river for non-aquatic exercise such as jogging, dog walking, walking, sightseeing, picnicking, and barbecuing. They went alone or in the company of friends, family, or their dog.

10 Connections to Blue and Green Spaces

Mildura and Albury/Wodonga residents noted the importance of having access to green and blue spaces and a garden and provided reasons for their answers. Both sites value the access to green spaces, with 96% of Mildura residents and 97% of Albury Wodonga residents observing that green spaces were extremely or very important to them. Curiously, access to blue spaces was extremely or very important to 92% of Mildura residents and only 62% of Albury/Wodonga residents. Gardens were extremely or very important to 72% of Mildura residents and 79% of Albury/Wodonga residents.

10.1 *Green Spaces*

Both cohorts suggested the ways that green spaces enabled opportunities for physical activity, particularly for children, and rejuvenation, as illustrated by comments such as "green spaces are calming, and they encourage deep thought" and "a city that has no green space has no heart. Places like that recharge citizens' minds and souls." Comments including, "it's important to have green spaces as it encourages interaction with others" through "picnics and barbecues," "parties," and "weddings" highlighted that green spaces provide opportunities for informal and formal social interaction. The very few respondents at each site that noted the importance of access to green spaces for maintaining wellbeing and social interaction during COVID-19 isolation periods observed that "Covid made me appreciate green spaces even more and how important they are to our wellbeing physically, emotionally and spiritually" and "I loved being able to walk in green spaces during lockdown. I saw people and their dogs."

Critically, Mildura residents' comments about green spaces, such as "out here the land clearing has been phenomenal, tragic, heartbreaking and continuing. The dust storms that result from all the open country can only be reduced by regenerating the country with parks and nature reserves," strongly suggest a sense of environmental degradation and the city's arid, dusty surrounds, a characterization of a place that was absent in Albury/Wodonga responses. However, overall, respondents' comments did not explicitly recognize the Murray river as the water source and enabler of green spaces.

10.2 *Blue Spaces*

The importance of access to blue spaces, such as rivers and swimming pools, was markedly different in both sites, with 92% of Mildura respondents and just 62% of Albury/Wodonga respondents observing that these spaces were extremely or very important. Interestingly, a striking contrast between the groups was a strong preference in Mildura for unaltered nature, namely rivers, while overall, Albury/Wodonga residents preferred to access the built environment, specifically swimming pools. The few Mildura residents that preferred swimming pools constructed the river as a threat—a consistent theme noted in Albury/Wodonga residents' responses. Pools were imagined as accessible, supervised, safe spaces for families and those with health issues to exercise. For example, one respondent described that "pools are better to take kids to because it's easier to watch them; they are less likely to be bitten by wildlife and the water is arguably cleaner", and another noted, "local swimming pools are ideal for people like me who carry long-term injuries. I can practise deepwater jogging and the like without jumping into a cold river with unseen obstacles and hazards underfoot". Both cohorts constructed a similar picture of the seasonal benefits that swimming pools provide via observations such as

FIGURE 9.2 Natural waterscape: the Murray river at Mildura

"swimming pools are important for providing year-round swimming opportunities if they are heated" and "my whole family swim in the local pool about once a week and during the summer, swim in the river whenever we can."

Across the two cohorts, the respondents constructed their cultural connections to the river, yet did so in very different ways. For example, one Albury/Wodonga respondent with Indigenous heritage was the only person to view the river as cultural heritage, as evidenced by the statement, "it is important to me because I have spiritual ties to the river, going back generations in my family. It holds significance to my identity and who I am, making me feel connected to my ancestors". In Mildura, the river was seen as a respite from a harsh climate, for example: "It is the river which gives us the connection with being an oasis. All water is a representation of being soothed from the heat in our dramatic summers", and as pleasures that may be transferred intergenerationally, "water sports and activities are things that I and my children grew up with, and now I want my grandchildren to enjoy the river".

10.3 *Gardens*

The respondents' observations about the meaning of their garden were similar between Albury and Mildura. The respondents grew a wide range of plants and trees in their gardens. Albury/Wodonga residents most frequently grew native plants (106 people), followed by flowers (95 people), vegetables (89 people), flowering bushes such as camellias (86 people), native trees (87 people), and fruit trees (82 people). Mildura respondents preferred edible crops over plants that provided only aesthetic and sensory value, and grew vegetables (n = 19), fruit trees (n = 19), native plants (n = 17), flowers (n = 15), flowering bushes such as camellias (n = 15), and native trees including grevillea (n = 13). Comments from one Mildura resident encapsulated the satisfaction of growing fresh produce: "the sheer joy of growing my own food and being able to feed myself is a pleasure". However, the benefits of gardens extended to individual wellbeing, relationships with nature, and opportunities for creativity, play, privacy, shade, exercise, and social interactions.

First, the relationship between gardens and emotional wellbeing was a common theme, with both groups noting that the garden is "somewhere to escape from the world for a little while", "a place to sit, read, and sleep" and "a coping mechanism for me when my anxiety flares up and things get tough". Second, both cohorts viewed gardens to be enablers of connections with the natural world through the creation of "their own micro-climates" that have cooling effects and create habitats for native animals, birds, and insects. Finally, observations including "having gatherings with friends in my garden", and "a nice entertaining area" and "a gathering place for the family", constructed gardens

FIGURE 9.3 The Sheer Joy of growing my own food.

as enablers of social relationships. Importantly, the health and extent of gardens are enabled by water from the Murray river and nearly three-quarters of the respondents in both sites made this link.

11 Murray River Health

Eighty-two percent of Albury/Wodonga respondents and 73% of Mildura respondents expressed concerns about the Murray river's health. People in Albury Wodonga were most concerned about the future of the Murray/Darling river system and climate change impacts on the Murray river, with people wanting to do more to improve the water quality of the river. Mildura residents' concerns related to the impacts of blue-green algae on the Murray river, overuse by large irrigators upstream, and the health of freshwater fish in the region.

It was common for people in both sites to note that Australia is a dry continent, with droughts occurring more frequently. Residents' comments pointed to a need for corporate and government responses to this scenario, as illustrated by: "we are rapidly using up the river through overpopulation and the greed of the big agribusinesses" and "using our water for cotton and almond production is absolutely ludicrous in Australia as the amount of water required is enormous." In an arid context, the management of environmental flows, a term referring to the quality, quantity, and timing of flows necessary to support fish, wetlands, and regeneration of native plants such as reeds (Hemming et al. 2019), was considered to be extremely or very important by 97% of Mildura respondents and 91% of Albury Wodonga respondents. Aquatic ecosystems were regarded as important elements of community identity, in comments including, "Mildura wouldn't be Mildura without the fish and wetlands.

They're a part of who we are" and "I love everything about Albury and care what happens to it; without environmental water, many wetlands in our local area would have died/have died". Critically, observations such as "without a healthy natural environment, our lives are impoverished. We have neglected to support the water needs of native flora and fauna due to our own demands. It is time to reverse that before we cause more extinctions" constructed a strong sense that some see themselves as but one element of ecosystems, one with an obligation to redress the impact of anthropogenic climate change. However, a few responses suggested that concerns may be driven by self-interest, as illustrated by: "If fish are dying what does this tell us about the quality of the water which we are also drinking, bathing and washing in and using to water the food that we eat?" and "our family loves to fish."

12 Water Scarcity

Seventy-three percent of Mildura respondents and 74% of Albury/Wodonga residents strongly agreed that there is a connection between the water that they use and the Murray river. Unsurprisingly, 93% of Albury/Wodonga residents and 88% of Mildura residents observed that using water in sustainable ways was extremely or very important to them. Observations including "climate change is scary, and it's important for me to be able to do my part to help" strongly suggested a sense of personal responsibility for sustainable water usage. Both cohorts described domestic water-saving practices that included saving greywater for garden usage, only watering gardens in the evening, capturing roof run-off, and installing "water-saving shower heads, dish and laundry washing machines with water-saving settings and timers on our garden hoses". Although some had installed rainwater tanks, some others described the challenges that impeded this strategy: "I would like to get a rainwater tank but live in a unit and only have a courtyard so am concerned about the space it would take up". It was common for people to frame water-saving behaviours in generative terms, for example: "one drop less I use today is another drop my kids can use in the future". However, two residents, both from Albury Wodonga, appeared to adopt water-saving strategies due to "cost" and to "avoid paying high bills". Critically, just over a third of Mildura respondents, 36%, and 24% of Albury/Wodonga residents agreed very strongly that they would consider moving away from their communities to areas in close proximity to reliable water if water became even more scarce than it is now.

Despite the considerable concern for water scarcity, only 30% of Mildura respondents and 21% of Albury/Wodonga respondents felt that there would be no change in their usage if water was not restricted. In this imagined scenario,

Mildura residents suggested that they would use additional water for gardens, pools, and domestic cleaning, including washing the car and "cleaning the dust off everything". Albury/Wodonga residents also preferred water for garden use, specifically for filling fish and turtle ponds, growing more food, installing garden water features, and watering lawns twice daily. However, in contrast to Mildura residents, water as a pleasure was a consistent theme for Albury/ Wodonga residents, who mentioned longer showers, water for their dogs to play in, and the freedom to "put a sprinkler on for my kids to play under the whole day on a summer's day."

The findings here raise an interesting tension. On one hand, the respondents' comments about water use adaptation illustrate an awareness of the precarity of water in the context of anthropogenic climate change and how their choices impact the environment that they live in. Yet, the majority of the respondents suggested a nostalgic longing to return to a modernist era where people had a sense of entitlement to use water from the natural environment as they pleased.

One-third of the Mildura respondents (33%) and 30% of the Albury/ Wodonga respondents took the trouble to provide additional comments on the ways they use or interact with the Murray river. Overwhelmingly, Mildura residents focused on river management in observations including, "the people that live along the river need it to be safe, sustainable and a resource for all, not just for businesses that need it to make money", "it is not a climate issue but a management issue; they keep selling it in large amounts to companies who can afford it", and the "health of the river should be legislated and not sold". Here, the reference to river health suggests the ways that humans may anthropomorphize, or transcribe human characteristics, onto waterbodies (Downey Threlkeld and Warburton 2013). In Albury Wodonga, management was only one of several themes, including tourism and the river's recreational and spiritual meanings. Some comments about management from Albury Wodonga reflected fierce nationalism and that 11 per cent of Australian water entitlements are now owned by international entities, primarily from Canada, China, and the United States (Productivity Commission 2021): "multinational companies that are profit-run should have no access to our water. The same goes for foreign countries; they should have no ownership of our water". However, most of the respondents voiced region-centric views, for example, "I support the people of Albury through their local council having control over the usage and distribution of water in our city". Others suggested that local government could do more to protect the river as illustrated by: "I love our Murray river and hate what speed boats are doing to the banks. They should be banned". A few residents pointed out the economic benefits that river-based tourism provides. However, others voiced environmental concerns, expressed in sentiments such as "I am worried about the effects of more visitors on platypuses". Overwhelmingly, Albury-Wodonga respondents focused on recreational use of the river and cited benefits, including opportunities for sailing, waterskiing, kayaking, bike riding, walking, and photography. However, some called for "more bicycle and walking tracks", "more river frontage access", and "more parks with easy access for swimming". Yet, some saw the river as a spiritual and cultural place, observing that, "when I am submerged in the water, I feel whole. When I am observing the river, I feel at peace" and, importantly, "my family is Indigenous, and losing that connection to Country by not being able to access the waterways would be devastating". Such comments reflect a key principle of Indigenous worldviews, i.e., that of selfhood, where self-growth is enabled by attachment to and interdependence between all entities, human and non-human (Townsend and McMahon 2021). Critically, some comments from those with settler-colonial heritage reflected increased awareness of and openness to Indigenous worldviews: "the realization of the importance of Country and the care of Country have become a focus in my later life", and "we could all learn more from our local Wiradjuri elders."

This chapter has used a green social work lens to compare data sets from two rural riparian communities, namely Mildura and Albury/Wodonga. This approach permitted the identification of points of congruence and difference within and across both sites concerning respondents' river imaginaries.

Respondents from both sites emphasized the river's life-generating characteristics. This important finding has congruence with Strang's (2015, 41) argument that "living water: the water that flows through the world enabling life" is a notion that is central to every culture. Critically, some respondents' deep concern for other than human life and desire to improve the river's health is consistent with a non-hierarchical worldview of ecological systems, and suggests a human capacity for environmental good that may counter destruction.

Mildura residents' view of the river as an entity under threat may reflect the town's positioning on the edge of the desert and the community's experience of severe drought and environmental degradation (Bureau of Meteorology 2021). Mildura is located downstream from the Darling river, the site of unprecedented fish deaths in 2018/2019. Here, residents' concerns focused on blue/green algae, fish health, and upstream irrigators' water misappropriation. Evidence suggests that these three issues may be interlinked and underpinned by climate change and changes to irrigation water allocations (Jackson and Head 2020). The Australian Academy of Science (2019) pointed out that in 2018/2019, large blue/green algae blooms, driven by low flows, resulted in insufficient dissolved oxygen in the water for fish to survive. Overall, the concern for present and future water conservation was palpable, with the majority of the respondents in both sites having implemented water-saving strategies and devices. However, only one-third of the respondents would maintain the water conservation behaviours if water was more available, with Mildura residents prioritizing the use of additional water for cleanliness. In contrast, Albury/Wodonga residents prioritized water for pleasure.

Critically, a considerable number of residents in both sites suggested that they would relocate if water scarcity intensified which may affect social connections.

Seeing the river as a source of fear may be deep-seated and stem from early socialization. Such constructions are consistent with observations that children and young adults commonly experience a fear of deep water and drowning (Sorin 2003; Newsome, Zimmerli, and O'Kelly 2001). Findings that showed residents' frequent and year-round engagement with the river for a range of aquatic and non-aquatic activities that permit both social interaction and quiet solitude supported the notion of the river being for pleasure. The centrality of the river to individual and community identity was palpable. However, while intergenerational relationships were an explicit theme within the responses, an Indigenous person was the only respondent to explicitly describe the river as a songline enabling deep ancestral connections.

In both locations, public green spaces were seen as providing opportunities for exercise and rejuvenation, with importance for social health. The latter observation is consistent with the literature that has previously explored the ways that green spaces may enable a sense of connection and community identity (Baur, Gómez, and Tynon 2013). Importantly, Mildura residents, who experience a far drier climate, also consider green spaces to be a respite from arid conditions. The findings concerning the meaning of gardens were consistent with the literature highlighting that these private green spaces are highly valued for providing multiple benefits, including connecting with the environment and others (van Holstein 2020) and food production (Downey, et al. 2022a). Despite findings that strongly suggested the recreational, spiritual, and environmental importance of the Murray river to both Mildura and Albury/Wodonga respondents, access to blue spaces was overall less important than access to green spaces and, curiously far less important to Albury/Wodonga residents than to those residing in Mildura. The strong preference for swimming pools in Albury/Wodonga reflected the considerable concerns about river safety. Overall, the lack of comments relating to access to green and blue spaces during COVID-19 isolation periods may suggest that restrictions caused little change to people's interactions with natural spaces and were consequently unnoteworthy.

Many respondents' additional observations concerning river management at macro and meso levels appear to reflect an "an understanding of humans as subjects" and the river "as their ontologically distinct object" (Bäckstrand and Lövbrand 2015, 11). Critically, these observations contrasted markedly with the few who constructed a deep spiritual connection with the river and those who expressed a desire to learn from Indigenous ontologies.

13 Conclusion

In one of the first studies to explore and compare rural riparian Indigenous and non-Indigenous community residents' river imaginaries, this study aimed to disrupt the dominant discourse that sees rivers as resources to be exploited for production. The differences between Mildura and Albury/Wodonga residents' concerns may be attributable to the markedly different climate in each site, as well as geographic location. In sum, the findings suggest that while many residents of these two Murray river cities may be ensnared by modernist ideas of a human–nature binary, others appear to recognize people's deep entanglement in the natural system. A limitation of this study is the absence of diverse river imaginaries beyond those of Indigenous and non-Indigenous river city residents. Further research is needed to explicitly explore the river practices and imaginaries of ethnic minority migrants in an Australian riparian city context.

References

ABS. 2020. *2016. Quickstats Mildura (RC)*. Australian Bureau of Statistics. https://www.abs.gov.au/census/find-census-data/quickstats/2016/SSC21671.

ABS. 2021. "Albury – Wodonga 2021 Census All persons QuickStats." Australian Bureau of Statistics. Accessed January 13, 2023. https://www.abs.gov.au/census/find-census-data/quickstats/2021/1001.

AIHW. 2018. *Australia's health 2018*. Australian Institute of Health and Welfare.

Arnold, D. 2026. "Rivers as Representational Tropes in the Tropicalization and Modernization of India." In *Fluid Phenomena: The River and Anthropocene Life-World in the Asia-Pacific*, edited by Iftekhar Iqbal, Hasharina Hassan. And Asiyah Kumpoh. Brill.

Australian Academy of Science. 2019. *Investigation of the causes of mass fish kills in the Menindee Region NSW over the summer of 2018–2019*. Australian Academy of Science (Canberra). https://www.science.org.au/supporting-science/science-policy-and-sector-analysis/reports-and-publications/fish-kills-report.

Bäckstrand, Karin, and Eva Lövbrand, eds. 2015. *Research handbook on climate governance.* Edward Elgar Publishing.

Baur, Joshua, Edwin Gómez, and Joanne F. Tynon. 2013. "Urban Nature Parks and Neighborhood Social Health in Portland, Oregon." *Journal of Park & Recreation Administration* 31 (4).

Berry, Kate A., and Sue Jackson. 2018. "The making of white water citizens in Australia and the Western United States: Racialization as a transnational project of irrigation governance." *Annals of the American Association of Geographers* 108 (5): 1354–69. https://doi.org/10.1080/24694452.2017.1420463.

Bureau of Meteorology. 2021. "Climate statistics for Australian locations." Accessed December 15, 2022. http://www.bom.gov.au/climate/averages/tables/cw_076031.shtml.

Bureau of Meteorology. n.d. "Climate statistics for Australian locations." Bureau of Meteorology. Accessed January 15, 2023. http://www.bom.gov.au/climate/averages/tables/cw_072160.shtml.

Burraja Cultural Centre. n.d. "Fish & The River People." Burraja Cultural Centre.

Carney, Gerard. 2013. The Story behind the Land Borders of the Australian States – A Legal and Historical Overview. High Court of Australia.

Christoff, Peter. 2016. "Renegotiating nature in the Anthropocene: Australia's environment movement in a time of crisis." *Environmental Politics* 25 (6): 1034–57. https://doi.org/10.1080/09644016.2016.1200253.

Clarke, Bruce, ed. 2015. *Earth, life, and system: evolution and ecology on a Gaian planet.* Fordham University Press.

Cortis, Natasha, Ciara Smyth, Kylie Valentine, Jan Breckenridge, and Patricia Cullen. 2021. "Adapting service delivery during COVID-19: experiences of domestic violence practitioners." *The British Journal of Social Work* 51 (5): 1779–98. https://doi.org/10.1093/bjsw/bcab105.

Cox, Daniel TC., Danielle F. Shanahan, Hannah L. Hudson, Richard A. Fuller, Karen Anderson, Steven Hancock, and Kevin J Gaston. 2017. "Doses of nearby nature simultaneously associated with multiple health benefits." *International journal of environmental research and public health* 14 (2): 172. https://doi.org/10.3390/ijerph14020172.

Cox, Lisa. 2022. "With another La Niña under way, experts warn Australia's east coast remains at high risk." *The Guardian Australia*, September 25, 2022. https://www.theguardian.com/australia-news/2022/sep/25/la-nina-weather-experts-warn-australia-east-coast-high-risk-rain-flood-flooding.

Creswell, John W., and J. David Creswell. 2017. *Research design: Qualitative, quantitative, and mixed methods approaches.* Sage publications.

CSIRO. 2022. *State of the Climate.* Commonwealth Scientific and Industrial Research Organisation. Commonwealth of Australia.

Curaming, R. A. 2026. "Ethico-political and Analytic Ambivalence of a Posthumanist Approach to Environmental History." In Fluid Phenomena: *The River and Anthropocene Life-World in the Asia-Pacific,* edited by Iftekhar Iqbal, Hasharina Hassan, and Asiyah Kumpoh. Brill

Davies, Peter, and Susan Lawrence. 2019. "Engineered landscapes of the southern Murray–Darling Basin: Anthropocene archaeology in Australia." *The Anthropocene Review* 6 (3): 179–206. https://doi.org/10.1177/2053019619872826.

Department of Environment. n.d. "Victorian Water Accounts A Statement of Victoria's Water Resources 2019–2020." Victorian Government. Accessed November 8, 2022.

Dominelli, Lena. 2014. "Promoting environmental justice through green social work practice: A key challenge for practitioners and educators." *International Social Work* 57 (4): 338–45.

Downey, Heather, Guinever Threlkeld, and Jeni Warburton. 2013. "How are Ageing Farmers Constructed in the draft Murray Daling basin plan and immediate media reports." *Rural Society* 23 (1): 87–100.

Downey, Heather, and Tim Clune. 2020. "How does the discourse surrounding the Murray Darling Basin manage the concept of entitlement to water?" *Critical Social Policy* 40 (1): 108–29.

Downey, Heather, Evelien Spelten, Katie Holmes, Sean MacDermott, and Prue Atkins. 2022a. "A green social work study of environmental and social justice in an Australian river community." *Social Work Research* 47 (3): 207–19. https://doi.org/10.1093/swr/svad013.

Downey, Heather, Evelien Spelten, Katie Holmes, Sean MacDermott, and Julia Mynott. 2022b. *The Multidimensional Meanings of Water for the Albury Wodonga Community.* La Trobe University.

Downey, Heather, Evelien Spelten, Katie Holmes, and Julia Van Vuuren. 2022. "A Rapid Review of Recreational, Cultural, and Environmental Meanings of Water for Australian River Communities." *Society & Natural Resources*: 1–19.

Ellis, Lynsey. M., Ksenija Napan, and Kieran O'Donoghue. 2018. "Greening social work education in Aotearoa/New Zealand." In *The Routledge handbook of green social work*: 535–46. Routledge

Habermas, Jurgen. (1987). *The theory of communicative action: Lifeworld and system* (Vol. II). Beacon Press.

Head, Lesley. 2007. "Cultural ecology: the problematic human and the terms of engagement." *Progress in Human Geography* 31 (6): 837–46. https://doi.org/10.1177/0309132507080625.

Hemming, Steve, Daryle Rigney, Simone Bignall, Shaun Berg, and Grant Rigney. 2019. "Indigenous nation building for environmental futures: Murrundi flows through Ngarrindjeri country." *Australasian Journal of Environmental Management* 26 (3): 216–35. https://doi.org/10.1080/14486563.2019.1651227.

IPCC. 2022. *Climate Change 2022: Impacts, Adaptation and Vulnerability.* Intergovernmental Panel on Climate Change. Cambridge University Press.

Jackson, Sue. 2022. "Caring for Waterscapes in the Anthropocene: Heritage-making at Budj Bim, Victoria, Australia." *Environment and History*: 1–21. https://doi.org/10.3197/096734022X16384451127393.

Jackson, Sue, and Lesley Head. 2020. "Australia's mass fish kills as a crisis of modern water: Understanding hydrosocial change in the Murray-Darling Basin." *Geoforum* 109: 44–56.

Liamputtong, Pranee. 2013. *Qualitative research methods.* 4th ed. Oxford University Press.

MacKellar, Dorothea. 2013. "My Country." In *Australian Poetry Since 1788*, edited by Lehmann. G. and R. Greay, 224. University of New South Wales Press.

Marshall, Virginia. 2017. *Overturning Aqua nullius: Securing Aboriginal Water Rights.* Aboriginal Studies Press.

MDBA. 2022. "Values of the Basin." Murray Darling Basin Authority. Accessed January 15, 2023. https://www.mdba.gov.au/rivers-worth-it/values-basin.

Newsome, John A, William Zimmerli, and Christine O'Kelly. 2001. "Teaching Young Adult Non-swimmers How to Swim." *Strategies* 14 (5): 28–29.

Noble, Carolyn. 2016. "Green social work-the next frontier for action." *Social Alternatives* 35 (4): 14–19.

Productivity Commission. 2021. *Register of Foreign-owned Water Entitlements: Productivity Commission inquiry report.* Productivity Commission.

Robin, Libby. 2016. The view from off-centre. In *Nature, Temporality and Environmental Management: Scandinavian and Australian perspectives on peoples and landscapes,* edited by Lesley Head, Katarina Saltzman, Gunhild Setten, and Maria Stenseke. Routledge. https://doi.org/10.4324/9781315597591-6.

Rose, Deborah Bird. 1992. *Dingo makes us human: Life and land in an Aboriginal Australian culture.* Cambridge University Press.

Ryan, J. C. 2026. "Indigenous Hydropoetics: Voicing Rivers through Aboriginal Australian Poetry." In *Fluid Phenomena: The River and Anthropocene Life-World in the Asia-Pacific*, edited by Iftekhar Iqbal, Hasharina Hassan, and Asiyah Kumpoh. Brill

Sorin, Reesa. 2003. "Reconceptualising fear in young children." *Australian Research in Early Childhood Education* 10: 121–34.

Strang, Veronica. 2004. *The Meaning of Water.* Berg.

Thoms, Martin, Teresa Rose, and Fiona Dyer. 2020. "Riverine landscapes, water resource development and management: A view from downunder." *River Research and Applications* 36 (4): 505–11. https://doi.org/10.1002/rra.3622.

Townsend, Annie, and Mishel McMahon. 2021. "Covid-19 and BLM: Humanitarian contexts necessitating principles from First Nations World views in an intercultural social work curriculum." *The British Journal of Social Work* 51(5): 1820–38.

van Holstein, Ellen. 2020. "The relational emergence of community and technology through the politics of water use." *Social & Cultural Geography* 21 (3): 314–35. https://doi.org/g/10.1080/14649365.2018.1489977.

Van Schendel, W. 2026. "A Monsoon River and its Humans: The Karnaphuli in India and Bangladesh." In Fluid Phenomena: The River and Anthropocene Life-World in the Asia-Pacific, edited by Iftekhar Iqbal, Hasharina Hassan, and Asiyah Kumpoh. Brill.

Victorian Government. 2019. "Permanent water saving rules." Victorian Government. Accessed January 15, 2023. https://www.water.vic.gov.au/liveable/using-water-wisely/advice-and-rules/permanent-water-saving-rules.

Wolfe, Patrick. 2006. "Settler Colonialism and the Elimination of the Native." *Journal of genocide research* 8 (4): 387–409. https://doi.org/10.1080/14623520601056240w.

CHAPTER 10

River Frontier: Nexus of Land, Water, and People in the Bangladesh Delta

Mohammad Tareq Hasan

The relationship between people and rivers has always been an inquiry of interest; in 1906, Arthur Hayes published The Source of the Blue Nile, describing the people of Abyssinia. In that trajectory, by the middle of the twentieth century, water and rivers became more prominent in anthropological investigation, delineating the relationship of water to agriculture and political organization. In 1949, Julian Steward published an article that predated Wittfogel's later, more significant work on "hydraulic societies," in which he discussed the importance of irrigation in the growth of "ancient civilizations" in China, Mesopotamia, Peru, and Mesoamerica (Wittfogel 1956, 1957). Irrigation system ethnography has gained prominence since the 1950s, with studies starting to see water as a social construct that holds mythical, symbolic, political, and economic significance (e.g., Geertz 1972). Over time, this perspective has led to rethinking rivers and freshwater as a "scarce resource" and a "commodity" (Bakker 2003; Strang 2004; Mehta 2005; Wagner 2012), across a variety of historical contexts (e.g., Davis 1996, 2013; Donahue and Johnston 1997; Mosse 2003; Strang 2004, 2009; Baviskar 2007; Lahiri-Dutt and Samanta 2013; Wagner 2013; Barnes 2014; Hastrup and Hastrup 2016). Climate change and the focus on the Anthropocene add yet more intensity and urgency to the anthropological study of water (Crate and Nuttall 2009; Hastrup 2009; Hastrup and Skrydstrup 2013; Orlove 2009) and converge with studies that focus on human rights (Sultana and Loftus 2012; Hossen 2014), health (Whiteford and Whiteford 2005), power and inequality (Demos 2017), and governmental and technical assemblages (Ogden et al. 2013), i.e., how the world is made, understood, explained and remade (see Gibson and Venkateswar 2015).

Building on such a trajectory, the research (a river ethnography) is centred on the river and the port rather than a specific community or issue. It focuses on how diverse individuals and communities live in the vicinity. One of the main goals is to answer the question, how does the river, through its flow, pervade the imagination of these people? I ask this question because, on the one hand, a river and its port facilitate mobility; on the other hand, the production of the spaces where this mobility occurs, the river port, is not characteristically

 | HTTPS://DOI.ORG/10.1163/9789004758100-012

bounded within an area. Further, though the construction, reconstruction, and use of a port are now motivated mainly by economic interests, we can trace how the environmental and economic frontier translates into the social frontier. This approach aligns with the anthropological concerns that emerged in response to the Anthropocene. Though the Anthropocene is sometimes characterized as a new geologic epoch with the planetary scale of anthropogenic influences on the composition and function of the Earth's ecosystem, this has generated analytical and methodological perspectives dismantling the rifts that separate humans from other life forms or rejecting the boundary between life and non-life (see Mathews 2020; cf. Gibson and Venkateswar 2015, 5; see also Ingold 2000; Harris 1998; Hage 2012, 305).

In this chapter, I depict a river port (of Chandpur) in Bangladesh and the perspectives unique to the region. The study, along with its concerns with the Anthropocene, at the analytical level, also aligns with critical discourses concerning highways found in Marc Auge's Non-Places (1995), as well as earlier works by Henri Lefebvre like The Production of Space (1991) and David Harvey's The Condition of Post-modernity (1989). As Marc Auge (1995) argued, while facilitating high mobility, highways are typical non-places—alienating spaces lacking social significance. According to Lefebvre (1991) and Harvey (1989), highways [port or the river in this case] are spaces controlled by the economic powers of capitalism and governance. A river and/or a river port may appear to be such non-places or alienating spaces—due to their characteristics as not holding anything but rather as spaces of mobility. However, suppose that one reconfigures the analysis at the microscale of ethnographic research. In that case, the river port emerges not merely as a space of alienation and domination but also as a space whose uses contribute to dealienation.

As I will describe below, the port has been a binding site for understanding new articulations of life and governance. Thus, river ports can be considered a frontier—in terms of mobility from and through linking different parts of the country into new territorial formations (see Cons and Eilenberg 2019, 2; see also Tsing 1993; Cresswell 2019, 201). As such, river ports can be called mobile places representing emplaced mobilities (cf. Salazar 2023, 1). Overall, Chandpur river port as a frontier allows us to understand people's relationship to the river and land, people's relationship with the state, and initiatives for sustainable development, which are radically reconfigured. The river port in Chandpur has positioned the locale as a frontier in a dual sense. Firstly, it enables the mobility of people. Secondly, the idea of the frontier lets us see that a process not bounded within a specific area can mould a region's resourcefulness, possibilities, and dangers. As for Chandpur, discourses of developing/modernizing the people, socio-economic development, saving the environment, better use of resources, etc., have espoused policies that have produced paradoxical results (cf. Tsing 2005, 32).

The river port analysis will allude to the analysis of power and inequality from a relational perspective (cf. Massey 1994). Therefore, the paper critically examines the relationship between mobility and place (cf. Salazar 2023). I explore the river port and the historical elements of its governance in varying configurations without prioritizing any specific event. Hence, taking a conjunctural approach—"peeling back layers of meaning and practice, and tracking relations across different spans of space and time" (cf. Li 2014, 16)—this chapter probes how multiple forces produce temporal social dynamics at one point of time in history. The river port as a frontier could help us conceptualize places as provisional and emergent, having networks of local to global linkages (cf. Escobar 2008) similar to what Tsing et al. (2020) conceptualize as feral effects emerging through relations among people and other living and non-living entities in a landscape.

Following Strang (2014, 135), I urge "a theoretical shift towards less anthropocentric visions of human-environmental interactions" and "a more egalitarian bioethics of relationality." The argument is about conceptualizing rivers ontologically inclusively from a political, economic, and legal standpoint. Thus, it would not reflect a concrete state of being but a process of becoming (Braidotti 2013, 66). Hence, with a lesser anthropocentric perspective, the paper reflects how the people with differential positionalities, the river, the port, and everything else are interconnected in the collective becoming and forming the total social fact, i.e., being integral, even essential, to the domains or institutions of society (cf. Orlove and Caton 2010; Ingold 2010; Krause and Strang 2016). By doing this, the paper moves beyond the concept of humans as a confined entity. It explores their entanglement with other life (material and non-material) forms (cf. Kohn 2007, 4; see also Tsing 2012, 142).

1 The River Port

Chandpur in Bengali means the place of the moon. It is one of the oldest river ports in the region, and there are several tales about its name. It is attributed to Chand Roy, the Zamindar (Landlord) of Bikrampur. However, some argue that it is named after Chand Fakhir or Shah Ahmed Chand, an administrator from Delhi in the fifteenth century (Khan, Hossain, and Sultan 2014, 23). The Chandpur river port is situated at the confluence of the Padma, Meghna, and Dakatia rivers in south-central Bangladesh. During the early twentieth century, the river port was a vibrant trading centre because it was used to bring in raw materials for the European and Indian jute mills. Nevertheless, with the partition of the Indian sub-continent in 1947, most of these businesses were closed down. Additionally, erosion has submerged some parts of the town into riverbeds. Nonetheless, this port remained an essential centre for jute and grain trades. In addition, many other businesses, including shipbuilding, salt refinery, and fishing, continue to thrive because of this port.

Today, the river port has features of its own. Disembarking from a river vessel, one is greeted by a host of street vendors/sellers who try to persuade people to buy commodities like bread, fruits, and other food items. Many work the whole night and sleep during the day, and others cater to the people travelling throughout the day. To get out of the terminal, one must pass these people (Figure 10.1). Those who travel via the port always look out for the Hilsa of Chandpur. Hilsa is

FIGURE 10.1 An entry-exit to the river port.

the national fish of Bangladesh and one of the most sought-after fish in Bangladesh and Kolkata, India. Many Hilsa sellers embark on small boats and sell their fish to travellers on the ships when they are docked to drop off/board passengers (Figure 10.2). As people are interested in Hilsa fish from Chandpur, sellers always say that they have just caught the fish, but locals also commented, "Fish from Chandpur are never sold here; they go to Dhaka. The fish sold at the terminal are basically from Barisal; people do not understand, so they buy these at higher prices." Later, I talked with a fellow traveller about why he bought Hilsa here. He replied, "These Fish are good; I will have to buy fish anyway; why not from Chandpur? I am returning to Dhaka, and my family would be happy." As the comment indicates, the Chandpur river port, the river, the people travelling through, the fishermen, and the Hilsa fish are all connected. The connection has generated a unique perception of the fish and the place.

At the exit of the terminal, one meets another group of people who run vehicles to take people to their destinations. As many people from neighbouring districts pass through the port, several separate sections of public transportation are designated in terms of their destination. Just after the terminal for public transport, we find shops and restaurants. Then, one finds squatters hosting thousands of people of lower income strata working as transportation labourers, daily labourers, or workers of the fishery jetty (Figure 10.3). These people contribute to the daily functioning of the port and fish processing.

FIGURE 10.2 Vendors selling Hilsa fish to the passengers on a launch.

FIGURE 10.3 Squatters beside the river port

Though the squatters lack basic supplies, the river makes it convenient to use the river water for drinking, cooking, and washing.

The snippet of the port indicates how the port and the people are entangled. Thousands of people's lives are formed and influenced by the river and the port. On the one hand, it facilitates the mobility of the people of southern Bangladesh toward the eastern part of the country. The river also forces mobility on the people via riverbank erosion. On the other hand, the mobility along the river and its resources allow people to make a living. The paper's focus is not mobility but the social ensemble where mobility is an element. The living arrangements that one will see in this chapter are only possible near the port and the river. Consequently, the area has enabled a particular social arrangement among the people. While the relations found in the vicinity of the port are unique, those, in turn, give the port and the river characteristic features. The entanglements of people and the river could be seen as the creation of a thick place—that enhances the sense of meaning, belonging, and connectedness—in contrast to the thin places—that lack the specificity of thick places (Van Schendel 2015, 99; see also Casey 2001; Duff 2010).

2 Three Stories

Rahima has been living on the riverbank for eight years now. Before that, she lived in Dhaka, migrating from a nearby char. Now, she makes a living working as a labourer with the support of her sons. She said that she went to Dhaka after her home was eroded and later came to live in the squatter huts near the port once life in Dhaka became difficult. In Dhaka, she used to work as a housemaid, but her husband thought that they could make a better life at the port as his cousins were already here and working in the fishing jetty. Her son, who used to work as a transportation mechanic, is now working as a driver of an electric rickshaw. Though life is also difficult, she said, "There are many more opportunities here." Besides, the cost of living near the river is comparatively lower than in Dhaka; they do not have to pay rent (Figures 10.4 and 10.5).

FIGURE 10.4 An informal market on the street

FIGURE 10.5 Housing on the riverbank

This situation and perspective blur the distinction between the development paradigms' centre and periphery. The capital city, Dhaka, is usually believed to attract people from poorer regions, but the dominant urban centre has limitations. Devoid of social relations, the people migrating from different regions face hard times when encountering crises. Because of the socio-ontological perspectives, the ways that we think of ourselves and the socio-economic resources that help us sustain our lives in some ways contradict the dominant narratives of development and progress. For example, when the first case of COVID-19 was detected in Dhaka, and a nationwide lockdown was declared, we found that thousands of people were hurrying back to rural areas to attend to their families' needs and better manage the crisis with the help of the broader social supports of the village (Hasan 2020); also, the perception about the possible outcome of the disease and their purported vulnerability contradicted the modernist medical views reflecting alternative ontological dimensions of life, death, and the afterlife (Hasan 2022a). The blurring of existing geographic, social, and economic distinctions and the emergence of new fault lines indicated the paper's concerns about finding entangling relationships among people, places, and values. The following two stories further elaborate the claims.

Badshah was born near the Chandpur port. He used to run a small business but always aimed to migrate to the Middle East. At the age of 18, he travelled to Qatar. He lived in Qatar for seven years and returned a year ago. He is planning to go again but, in the meantime, has started a small business (stall for selling snacks) on the riverbank, which has gradually become a tourist spot. In establishing the snack stall, he has loaned from his cousins. He mentioned that when he needs money, he asks his cousin for money whom he had helped before. Even though his cousin is not pleased, he continues to help his cause. He also mentioned that when he first went to Qatar, he spent money loaned from his relatives for the first few months. Badshah's case also reflects his tendency to return to the place and people he considers his own. Like Rahima, he also survives the difficult times by acquiring support from the people he is related to by kinship. This may seem true for everyone, but in these cases, one crucial element is the place, i.e., the river and river port. Their existence near the river has shaped everything that they do and how they perceive their life.

Karim runs a small grocery shop on the riverbank. He makes a living out of it, serving both the people who live in the vicinity and those who come to see the confluence of the three rivers and visit the char (popularly known as the mini Cox's Bazar). He said he started to live near the port around 15 years ago after the river eroded their homestead lands. They were displaced thrice, once before Bangladesh's independence and twice afterward. His paternal cousins arrived here first, and later when they came to live near the port, his cousins helped them to rent a home and eventually financed his business. However, it took him five years to return their loan.

The commonality in all their life was the river. On the one hand, it made their life tough and forced them out of their home, but on the other hand, they were making a living around the river and the port. It shows how the river could be destructive and creative simultaneously.

3 The New Kinship

The river and port have allowed the informal sector workers' economic survival. There is a history of employment distribution through kin networks in the port area. To avoid using the vocabulary of evolving workplace circumstances, this study instead discusses the significant—yet sometimes underappreciated—relationship between kinship and contemporary societies. This emphasis on kinship would add to the scholarly literature that has argued for the continued importance of kinship in labour usage and organization in a capitalist economy (Bear 2013, 2015; Lazar 2018; McKinnon and Cannell 2013; Mollona 2005; Narotzky 2015; Shever 2012, 2013; Yanagisako 2002, 2018). Furthermore, I argue that kinship—expressed through relationships, sentiments, and idioms—is a generative force for the capitalist economy (cf. Shever 2012; Hasan 2022b). Anwarul—a fisherman, told me,

> The river washed away our home when I was a child. Many of my cousins no longer fish, but I still go on fishing trips. However, it is not easy to continue fishing now. We can hardly profit from fishing. I do not want my children to enter this profession. […] What we can earn from fishing is insufficient to sustain a family. Sometimes, we have to work as day labourers to complement our income. I have no alternative; we must survive somehow.

For fishing trips, Anwarul depends on his kin network. He said that his cousins and nephews accompanied him during the fishing trips. Some others help with cash to arrange boats, fuel, and nets. He further said that some of his cousins, who were not involved in fishing anymore, worked as small shop owners, auto-rickshaw drivers, or day labourers and helped him by lending him small amounts to arrange fishing trips.

Primarily, out of formal financial support, they use micro-credit and personal loans to survive economic hardship or revive economic opportunities.

Ahmed et al. (2021) identified that fishers in the Meghna estuary are dependent on financial support for their fishing trips—60 per cent of the Hilsa fishers borrowed money from Non-governmental Organizations (NGOs), 20 per cent from money lenders, 16 per cent from relatives and neighbours, and only 4 per cent from formal banks. As the data show, Ahmed et al. (2021) found that the financial support from kinship networks was comparatively lower. However, it should be noted that besides giving cash, they help each other by mediating/guaranteeing loans from NGOs or more lenders. So, kinship was crucial for accumulating cash for fishing trips to the river and financing small businesses. Supporting each other in everyday life was much more common. This is also evident in Badshah's comment, "I have run many businesses in my life. I can still survive because of my family. My cousins and relatives always help each other with money and labour."

When I visited the fishery jetty and market (Figure 10.6), I found a similar process where kinship was a crucial factor in work distribution, where apprentices are hired based on kinship. Shafiq worked in the fishery jetty and helped process the fish. He said he came from Munshigonj to work along with his maternal uncle. Shafiq's uncle Majid, who had worked in the jetty for 20 years, told us that he started to work with his father when he was young. He further said that everyone here knows everyone and those who work together are some sort of relative. He continued,

FIGURE 10.6 The fish market

> We, the river people, always helped each other. Still, we help others on fishing trips and give money to people in need. Previously, due to the possibility of river erosion, we could not survive without others' help. Since the riverbanks erode one side at a time, we live by reciprocity. We are always ready to help since maybe tomorrow I might need help.

The experience of living by the riverbanks generated a rigorous form of reciprocity as well as competition. This also indicates that the river port as a place was created by the people, with their experience, use, and perception (cf. Van Schendel 2015, 99).

Another area feature is marital relations with other regions connected via the river. It was repeatedly said by many that usually, they had affinal connections with people living upstream or downstream of the river. As for Rahima's case, she was married off in Munshigonj (two hours away from Chandpur). It can be argued that mobility and kinship networks gain a larger spatial range as enabled by the river. Besides enabling inter-region marriage, the river always helps/encourages and forces displacement/migration on different scales (as the cases of Badshah and Karim show). In these displacements and adjustments, kinship plays a vital role. Hence, the river or the port cannot be analyzed devoid of the social aspects that have developed in the multiplicities of the living and non-living elements (cf. Van Schendel 2015; Braidotti 2013; Tsing et al. 2020).

The forms of associations and alliances among the people indicate that although the river port may appear to be a place that bears the closest, most apparent, and most tangible effects of globalization's abstract and distant processes, people who live nearby reconstruct it through their daily practices and discourses as a common theme of their narratives; they use it, conceptualize it, and perceive it in their everyday terms, challenging the doctrinal principles of the global political economy of ports, resource management, and governance. However, as I show in the next section, there is still a contradiction: the very act of making small, everyday deals helps to reproduce the inequalities, creating a complicated, concurrent process that can be represented ethnographically in terms of socio-spatial particularities (see Marcus 1995).

The river port is generative of a particular life and life-ways as in the case of other mobility-facilitating infrastructures. This feature of the river is also manifested in Maya Dania's paper (this volume), where she traces the process of making kin along the Mekong river—the web of relationships that the river forges (see also Haraway 2015). Hence, one could find ontologies that represent alterity opposing independence and separation with diverse forms of relationalities. Mobility is integral to the place—forming the river port; as I will discuss in the following section, daily mobilities constitute the place (cf. De Certeau 1984; Pandya 1990).

4 Uncertainties, Politics, and Possibilities at the River Frontier

The river port is famous for the Hilsa fish. Many people living in the vicinity depend on fishing in the river. Many people work as labourers, while some organize the labour and equipment to fish (Figure 10.7). However, fishing has become an unpredictable venture. Minhaz, a man of around 50 years, said,

> When I was young, we caught an abundance of fish. However, we could not sell at a higher price. On the contrary, now the price of fish is high, but we cannot catch enough fish. [...] Even if we catch a good number of fish, we have to incur interest on our loans, and the cost of everything increases too much. So, our profit is decreasing every day.

From the discussions with the fishermen, it was clear that, sometimes, because of the syndicate of people who buy the fish in lots and preserve and supply them across the country, they offer prices that allow fishermen to live only marginally (see Alam and Yousuf 2024).

FIGURE 10.7 Fishermen preparing nets

However, besides the low price and other logistical shortages, one of the challenges that they face is the prohibition on fishing during the fish's breeding season. Locally termed "Obhijan", the police raids during the prohibition period became a problem for them to survive. In a year, fishing is prohibited for almost four months. The prohibition on fishing arguably ensures the sustainable use of the river resources. Statistically, Hilsa production has increased after increased management efforts, such as sanctuary management and ban periods, along with incentives for the fishers (Ahmed et al. 2021, 7–10).

Hilsa fish production increased from 2.89 lac tons in 2008–2009 to 5.71 lac tons in 2022–2023—approximately a 92 per cent increase (The Business Standard 2023). However, especially during the Hilsa's breeding season, the fishers who invest money and organize fishing do nothing or rarely engage in other work. The labourers tend to engage in daily labour in the area or migrate to cities for a short period. Ahmed et al. (2021) identified that 28 per cent of fishers were compelled to reduce their meal frequency to two meals per day and consume less expensive food items to withstand the income shortage during the ban period. With the rising cost of living and shorter time for fishing, people involved in the occupation have to diversify their income source. On the contrary, the port area and the river have become a place of opportunity. Since the confluence of three rivers has become a tourist attraction, the local government has also taken the initiative to build a public park for people where many shops have been allotted (Figure 10.8).

FIGURE 10.8 A tourist spot near the confluence of the Meghna-Dakatia-Padma

Hence, the area is not only a place of business or transit but also a destination for people looking for fun. The situation of the fishermen in the port area indicates how the more extensive process of governance and sustainable resource use policies create uncertainties and show us the multiplicities of the port, whereas new possibilities are emerging. However, the permission to set up a shop and do business is outside the jurisdiction of government policies, determined by the local kinship-based alliances. For instance, Badshah said that the allotment he received to establish the snack stall was only possible because his relatives mediated with the local government administration. Additionally, many street sellers told me that they started working with someone with affinal or consanguineal relations and gradually became independent sellers. Nonetheless, the inequalities and politics of resources are an everyday aspect of the lives of the people along the river in terms of the price of the fish in the fishery jetty that eventually big business owners control.

Since the river supports mobility and provides fish resources, over the years, the port area has experienced growths in population and economic activities. As part of the government initiative to develop the port and the region, the government, over the years, has taken on several projects, including maintaining the water depth and reducing siltation, establishing developed port facilities, setting up a river-centric tourist spot (CPA News 2021), and building an international river port (The Business Post 2023). However, since the economic resources are accumulated in the hands of the elites, the people still need more basic amenities (as described in the earlier section). To counter such lack, the government, over the years, has implemented infrastructural projects and the regulation of fishing rights that allow wealth accumulation by the elites (e.g., New Age 2023; Marof 2023; Alam and Yousuf 2024). As the fishermen of Chandpur revealed, the government is not interested in supporting them but in implementing policies that only harm them. Selim reported, "The government does not allow us to fish for a few months, but they do not do anything in terms of keeping the cost of living down or implementing policies so that we may obtain interest-free loans or get better prices for fish from the mohajon—the people who lend money for fishing trips, and buy the fish (in advance in some cases). He further commented,

> Everyone who lives by the riverbank has relocated their homes a few times because of river erosion. We have lost our homes, land, and everything. However, we have always got on with life. We do not fear uncertainty. We arrange fishing trips every year, hoping for a better fish catch or price. We hardly survive but never lose hope. We are saving cash, obtaining loans, and investing in fishing trips. Some also try to send one sibling abroad to

> the Middle East. Although now we do not worry about river erosion, we face uncertainties every day.

This comment reflects that the search for a better future is like the river—the people living near the river are constantly venturing into new uncertainties and possibilities. As Tsai (2019) showed, hope and possibilities emerge within uncertainties, and it becomes evident when we become attentive to the practices and relations rather than grand narratives of development (cf. Escobar 1995).

Another looming crisis is the encroachment over the Dakatia river, which runs through the city. Economic and political power holders construct commercial multi-storey buildings on both sides of the river, narrowing its flow. Yunus, a businessman, said, "Much of the river is already encroached upon, but they ignore such illegal structures along the riverbanks. Government officials are more into restricting the fishermen from fishing." However, as already discussed above, the fishermen who depend more on the income from the fishing trips possess a value system that encourages them to seek and provide support to others in their kin networks. While these supports are primarily directed to the survival of the marginal people, in turn, the survival also ensures that the profit-seeking practices of the power holders can continue.

One way of looking at the issues of government initiatives could be termed as what Maria C. P. Alejandria called "band-aid" solutions (this volume). Still, alternatively, considering the idea of the frontier, one could see how these diverse forces enable the co-becoming of the region and its people—for instance, dependence on kinship and the continuous search for hope. Consequently, we could see how the river port is a place that is inherently moving (cf. Redmond 2001). As such, one could also argue that all places are formed with networks that stretch beyond each such place (cf. Sheller and Urry 2006). Similarly, activities are not separated from the place comprised of the river and the port (cf. Sheller and Urry 2006; Anderson 2012). What characterizes the frontiers are relational movements; the places and the people are always on the move. Historically speaking, the river port can be analyzed as a moment.

5 Concluding Remarks

The life and living conditions around the Chandpur river port reveal that the port is a site of resource and depletion, various policies, and strategies of the people coping with vulnerabilities and looking out for opportunities. The paper reflects how people continually respond to and practically engage with their specific surroundings (cf. Harris 1998, 66). Hence, as humans, our being is always in relation to our social or physical environment (cf. Ingold 2000, 2010).

The paper is structured around the idea that rivers, ports, or other places have social lives and require in-depth ethnographic inquiry and description. The river is social because it permeates and breathes life into human existence, not just at the molecular level but also as a total social fact—being central to economic conduct, the livelihood of the people, mobility, social sentiments, governance, and politics (cf. Orlove and Caton 2010; see also Krause and Strang 2016). It is evident that Chandpur's rivers and river ports influence every aspect of social activity and are crucial for creating class and kinship. Along the process, people build and reconstruct their ties with the river. As a result, the river port transcends its status as a fixed geographically peripheral place or a venue for culturally structured negotiations to become a zone of constant becoming. Analyzing the nexus of land, water, and people as an entangling frontier would allow a better understanding of the Anthropocene since, and as we have seen throughout the paper, configurations of living and non-living elements are always on the move.

Acknowledgement

I thank the people of Chandpur for allowing me to explore their worldviews. Special thanks to Dr. Iftekhar Iqbal for the current shape of the paper. My wife, Bani, travelled to Chandpur once and supported me in every other way as I completed the research. All errors remain mine.

References

Ahmed, Maruf, Sabrina Jannat Mitu, Petra Schneider, Masud Alam, Mohammad Mojibul Hoque Mozumder, and Md. Mostafa Shamsuzzaman. 2021. "Socio-Economic Conditions of Small-Scale Hilsa Fishers in the Meghna River Estuary of Chandpur, Bangladesh." *Sustainability* 13 (22): 1–18. https://doi.org/10.3390/su132212470.

Alam, Mohammad Shaheen, and Ahmed Yousuf. 2024. "Fishermen's Community Livelihood and Socio-economic Constraints in Coastal Areas: An Exploratory Analysis." *Environmental Challenges* 14 (1): 1–12. https://doi.org/10.1016/j.envc.2023.100810.

Anderson, Jon. 2012. "Relational Places: The Surfed Wave as Assemblage and Convergence. *Environment and Planning D: Society and Space* 30 (4): 570–87. https://doi.org/10.1068/d17910.

Auge, Marc. 1995. *Non-Places*. Verso.

Braidotti, Rosie. 2013. *The Posthuman*. Polity.

Bakker, Karen. 2003. *An Uncooperative Commodity: Privatizing Water in England and Wales*. Oxford University Press.

Barnes, Jessica. 2014. *Cultivating the Nile: The Everyday Politics of Water in Egypt*. Duke University Press. https://doi.org/10.1215/9780822376217.

Baviskar, Amita, ed. 2007. *Waterscapes: The Cultural Politics of a Natural Resource*. Permanent Black.

Bear, Laura. 2013. "'This Body is Our Body': Vishwakarma Puja, the Social Debts of Kinship, and the Theologies of Materiality in a Neoliberal Shipyard." In *Vital Relations: Modernity and the Persistent Life of Kinship*, edited by Susan McKinnon and Fenella Cannell, 155–78. School for Advanced Research Press.

Bear, Laura. 2015. *Navigating Austerity: Currents of Debt along a South Asian River*. Stanford University Press.

Casey, Edward S. 2001. "Beyond Geography and Philosophy: What Does It Mean to Be in the Place-World?" *Annals of the Association of American Geographers* 91 (4): 683–93.

Cons, Jason, and Michael Eilenberg. 2019. Introduction. In *Frontier Assemblages: The Emergent Politics of Resource Frontiers in Asia*, edited by Jason Cons and Michael Eilenberg. John Wiley & Sons Ltd.

CPA News Desk. 2021. "Chandpur is the First River-centric Tourist Destination in the Country." *CPA News: Maritime Publication of Bangladesh*. December 25, 2021. https://cpanewsbd.com/2021/12/25/chandpur-is-the-first-river-centric-tourist-destination-in-the-country/#:~:text=Chandpur is the first river-centric tourist destination in the country,-By&text=Chandpur is the meeting place, to see its immense beauty.

Crate, Susan A., and Mark Nuttall, eds. 2009. *Anthropology and Climate Change: From Encounters to Actions*. Left Coast Press.

Cresswell, Tim. 2019. *Maxwell Street: Writing and Thinking Place*. University of Chicago Press.

Davis, Wade. 1996. *One River: Explorations and Discoveries in the Amazon Rainforest*. Simon & Schuster.

Davis, Wade. 2013. *River Notes: A Natural and Human History of Colorado*. Island Press.

De Certeau, Michel. 1984. *The Practice of Everyday Life*. University of California Press.

Demos, T. J. 2017. *Against the Anthropocene: Visual Culture and Environment Today*. Sternberg Press

Donahue, John, and Barbara Rose Johnston, eds. 1997. *Water, Culture and Power: Local Struggles in a Global Context*. Island Press.

Duff, Cameron. 2010. "On the Role of Affect and Practice in the Production of Place." *Environment and Planning D: Society and Space* 28 (5): 881–95.

Escobar, Arturo. 1995. *Encountering Development: The Making and Unmaking of the Third World*. Princeton University Press.

Escobar, Arturo. 2008. *Territories of Difference: Place, Mobilities, Life, Redes*. Duke University Press.

Geertz, Clifford. 1972. "The Wet and the Dry: Traditional Irrigation in Bali and Morocco." *Human Ecology* 1 (1): 23–39. https://doi.org/10.1007/BF01791279.

Gibson, Hannah, and Sita Venkateswar. 2015. "Anthropological Engagement with the Anthropocene: A Critical Review." *Environment and Society: Advances in Research* 6: 5–27. https://doi:10.3167/ares.2015.060102.

Hage, Ghassan. 2012. "Critical Anthropological Thought and the Radical Political Imaginary Today." *Critique of Anthropology* 32 (3): 285–308. https://doi.org/10.1177/0308275X124491.

Haraway, Donna J. 2015. "Anthropocene, Capitalocene, Plantationocene, Chthlucene: Making Kin." *Environmental Humanities* 6 (1): 159–65. https://doi.org/10.1215/22011919-3615934.

Harris, Mark. 1998. "The Rhythm of Life on The Amazon Floodplain: Seasonality and Sociality in a Riverine Village." *Journal of the Royal Anthropological Institute* 4 (1): 65–82. https://doi.org/10.2307/3034428.

Harvey, David. 1989. *The Condition of Postmodernity*. Blackwell.

Hasan, Mohammad Tareq. 2020. "Homecoming from the City: Covid-19 and Dhaka." *Medical Anthropology at UCL*, April 15, 2020. Accessed 17 April, 2020. https://medanthucl.com/2020/04/15/homecoming-from-the-city-covid-19-dhaka/.

Hasan, Mohammad Tareq. 2022. "Ontologies of a Pandemic: Polarisations Exposed by COVID-19 in Bangladesh." *South Asia: Journal of South Asian Studies* 45 (3): 543–59. https://doi.org/10.1080/00856401.2022.2000143.

Hasan, Mohammad Tareq. 2022b. *Everyday Life of Ready-made Garment Kormi in Bangladesh: An Ethnography of Neoliberalism*. Palgrave Macmillan.

Hastrup, Kirsten B., ed. 2009. *The Question of Resilience: Social Responses to Climate Change*. Danish Royal Society of Sciences and Letters.

Hastrup, Kirsten B., and Frida Hastrup, eds. 2016. *Waterworlds: Anthropology in a Fluid World*. Berghahn Books.

Hastrup, Kirsten B., and Martin Skrydstrup. 2013. *The Social Life of Climate Change Models: Anticipating Nature*. Routledge.

Hossen, A.M. 2014. "Water Policy and Governance for the Empowerment of River Basin Communities in Rural Bangladesh." PhD diss., University of British Columbia.

Hayes, Arthur J., 1906. *The Source of the Blue Nile: A Record of a Journey Through the Soudan to Lake Tsana in Western Abyssinia, and of the Return to Egypt, by the Valley of the Atbara, with a Note on the Religion, Customs of Abyssinia*. Smith, Elder & Co.

"Hilsa Production Increased by 92% in 15 Years: Fisheries Minister." *The Business Standard*. October 11, 2023. https://www.tbsnews.net/bangladesh/hilsa-production-increased-92-15-years-fisheries-minister-716298#.

"Int'l Standard River Port to be Set Up in Chandpur: Khalid." *The Business Post*, May 29, 2022. https://businesspostbd.com/economy/intl-standard-river-port-to-be-set-up-in-chandpur-khalid.

Ingold, Tim. 2010. "The Textility of Making." *Cambridge Journal of Economics* 34 (1): 91–102. https://doi.org/10.1093/cje/bep042.

Ingold, Tim. 2000. *The Perception of the Environment: Essays on Livelihood, Dwelling and Skill*. Routledge.

Khan, Shamsuzzaman, Md. Altaf Hossain, and Amir Rahman Sultan. 2014. *Bangladesh Lokojo Somskriti Gronthramala: Chandpur* (Present State of Folklore in Chandpur District). Bangla Academy.

Krause, Franz, and Veronica Strang. 2016. "Thinking Relationships through Water." *Society & Natural Resources* 29 (6): 633–38. https://doi.org/10.1080/08941920.2016.1151714.

Kohn, Eduardo. 2007. "How Dogs Dream: Amazonian Natures and the Politics of Trans-species Engagement." *American Ethnologist* 34 (1): 3–24. https://doi.org/10.1525/ae.2007.34.1.3.

Lahiri-Dutt, Kuntala, and Gopa Samanta. 2013. *Dancing with the River: People and Life on the Chars of South Asia*. Yale University Press. https://doi.org/10.12987/yale/9780300188301.001.0001.

Lazar, S. 2018. "A 'Kinship Anthropology of Politics'?: Interest, the Collective Self, and Kinship in Argentine Unions." *Journal of the Royal Anthropological Institute* (N.S.) 24 (2): 256–74. https://doi.org/10.1111/1467-9655.12809.

Lefebvre, Henri. 1991[1974]. *The Production of Space*. Blackwell.

Li, Tania M. 2014. *Land's End: Capitalist Relations on an Indigenous Frontier*. Duke University Press.

Mathews, Andrew S. 2020. "Anthropology and the Anthropocene: Criticism, Experiments, and Collaborations." *Annual Review of Anthropology* 49: 67–82. https://doi.org/10.1146/annurev-anthro-102218-011317.

Marcus, George M. 1995. "Ethnography in/of the World System: The Emergence of Multi-sited Ethnography." *Annual Review of Anthropology* 24: 95–117. https://www.jstor.org/stable/2155931.

Marof, Mehedi Hasan. 2023. "Fishing Ban in Bangladesh is a Bilateral Issue." *The Daily Star*, May 29, 2023. https://www.thedailystar.net/opinion/views/news/fishing-ban-bangladesh-bilateral-issue-3331831.

McKinnon, Susan, and Fenella, eds. 2013. *Vital Relations: Modernity and the Persistent Life of Kinship*. School for Advanced Research Press.

Massey, Doreen B. 1994. *Space, Place, and Gender*. University of Minnesota Press.

Mehta, Lyla. 2005. *The Politics and Poetics of Water: Naturalizing Scarcity in Western India*. Orient Longman.

Mollona, Mao. 2005. "Gifts of Labor: Steel Production and Technological Imagination in an Area of Urban Deprivation, Sheffield, UK." *Critique of Anthropology* 25 (2): 177–98. https://doi.org/10.1177/0308275X05052022.

Mosse, David. 2003. *The Rule of Water: Statecraft, Ecology, and Collective Action in South India*. Oxford University Press.

Narotzky, Susana. 2015. "The Payoff of Love and the Traffic of Favors: Reciprocity, Social Capital and the Blurring of Value Realms in Flexible Capitalism." *In Flexible Capitalism: Exchange and Ambiguity at Work*, edited by Jens Kjaerulff. Berghahn Books.

"65-day Ban on Netting Fish in Sea Starts Today." *New Age*, May 20, 2023. https://www.newagebd.net/article/202123/65-day-ban-on-netting-fish-in-sea-starts-today.

Ogden, Laura, Nik Heynen, Ulruch Oslender, Paige West, Karim-Aly Kassam, and Paul Robbins. 2013. "Global Assemblages, Resilience, and Earth Stewardship in the Anthropocene." *Frontiers in Ecology and the Environment* 7: 341–47. http://dx.doi.org/10.1890/120327.

Orlove, Ben. 2009. "The Past, the Present and Some Possible Futures of Adaptation." In *Adaptation to Climate Change: Thresholds, Values, Governance*, edited by W. Neil Adger, Irene Lorenzoni and Karen O'Brien. Cambridge University Press. https://doi.org/10.1017/CBO9780511596667.

Orlove, Ben, and Steven C. Caton. 2010. "Water Sustainability: Anthropological Approaches and Prospects." *Annual Review of Anthropology* 39: 401–15. https://doi.org/10.1146/annurev.anthro.012809.105045.

Pandya, Vishvajit. 1990. "Movement and Space: Andamanese Cartography." *American Ethnologist* 17 (4): 775–97. https://doi.org/10.1525/ae.1990.17.4.02a00100.

Redmond, Anthony. 2001. "Places that Move." In *Emplaced Myth: Space, Narrative, and Knowledge in Aboriginal Australia and Papua New Guinea*, edited by Alan Rumsey and James F. Weiner, 120–38. University of Hawaii Press.

Salazar, Noel B. 2023. "Mobile Places and Emplaced Mobilities: Problematizing the Place-mobility Nexus." *Mobilities* 18 (4): 582–92. https://doi.org/10.1080/17450101.2023.2226358.

Shever, Elana. 2012. *Resources for Reform: Oil and Neoliberalism in Argentina*. Stanford University Press.

Shever, Elana. 2013. "'I am a Petroleum Product': Making Kinship Work on the Patagonian Frontier." In *Vital Relations: Modernity and the Persistent Life of Kinship*, edited by Susan McKinnon and Fenella Cannell. School for Advanced Research Press.

Sheller, Mimi, and John Urry. 2006. "The New Mobilities Paradigm." *Environment and Planning A: Economy and Space* 38 (2): 207–26. doi.org/10.1068/a3726.

Steward, Julian H. 1949. "Cultural Causality and Law: A Trial Formulation of the Development of Early Civilizations." *American Anthropologist* 51, no. 1: 1–27. doi.org/10.1525/aa.1949.51.1.02a00020.

Strang, Veronica. 2004. *The Meaning of Water*. Berg.

Strang, Veronica. 2009. *Gardening the World: Agency, Identity and the Ownership of Water*. Berghahn Books.

Strang, Veronica. 2014. "Fluid Consistencies: Material Relationality in Human Engagements with Water." *Archaeological Dialogues* 21, no. 2: 133–50. https://doi:10.1017/S1380203814000130.

Sultana, Farhana, and Alex Loftus, eds. 2012. *The Right to Water: Politics, Governance and Social Struggles*. Earthscan.

Tsai, Yen-Ling. 2019. "Farming Odd Kin in Patchy Anthropocenes." *Current Anthropology* 60, no. S20: S342–53. https://doi.org/10.1086/703414.

Tsing, Anna L. 1993. *In the Realm of the Diamond Queen: Marginality in an Out-of-the-Way Place*. Princeton University Press.

Tsing, Anna L. 2005. *Friction: An Ethnography of Global Connection*. Princeton University Press.

Tsing, Anna L. 2012. "Unruly Edges: Mushrooms as Companion Species." *Environmental Humanities* 1, no.1: 141–54. https://doi.org/10.1215/22011919-3610012.

Tsing, Anna L., Jennifer Deger, Alder Keleman Saxena, and Feifei Zhou. 2020. *Feral Atlas: The More-than-Human Anthropocene*. Stanford University Press.

Van Schendel, Willem. 2015. "Spatial Movements: Chittagong in Four Scenes." In *Asia Inside Out: Connected Places*, edited by Eric Tagliacozzo, Helen F. Siu, and Peter C. Perdue, 98–117. Harvard University Press. https://doi.org/10.4159/9780674286320-003.

Wagner, John R. 2012. "Water and the Commons Imaginary." *Current Anthropology* 53, no. 5 (October): 617–41. https://doi.org/10.1086/667622.

Wagner, John R., ed. 2013. *The Social Life of Water*. Berghahn Books.

Whiteford, Linda, and Scott Whiteford. 2005. *Globalization, Water and Health: Resource Management in Times of Scarcity*. School of American Research Press.

Wittfogel, Karl A. 1956. *The Hydraulic Civilizations*. University of Chicago Press.

Wittfogel, Karl A. 1957. *Oriental Despotism: A Comparative Study of Total Power*. Yale University Press.

Yanagisako, Sylvia. 2002. *Producing Culture and Capital: Family Firms in Italy*. Princeton University Press.

Yanagisako, Sylvia. 2018. "Reconfiguring Labor Value and the Capital/Labor Relation in Italian Global Fashion." *Journal of the Royal Anthropological Institute*, (N.S.) 24, no. S1: 47–60. https://doi.org/10.1111/1467-9655.12798.

CHAPTER 11

Freshwater River Fisheries in Borneo: the Case of Ikan Semah in Ulu Temburong, Brunei Darussalam

Magne Knudsen and Misa Juliana Minggu

Borneo is an island of rivers. Rivers connect people and enable exchange (Langub and Ishikawa 2016). Historical and ethnographic studies demonstrate the central importance of rivers and the trading networks which they facilitate in the formation of social groups and kingdoms on the island (Boomgard 1998; Eghenter 2001; King 1993). River fisheries, however, have received little attention. Accounts of inland groups' 'traditional' ways of making a living focus on swidden cultivation, land-based hunting and gathering, cash-cropping, and upriver–downriver trade (Appell 1976; Cramb 1989, 2007; Freeman 1955; Rousseau 1989; Sellato 1984; Sutlive 1992; Wadley and Mertz 2005). Although scholars point out that fish have been a vital source of animal protein for some interior Borneo communities (King 1985, 45), fishing is typically constructed as a minor sideline activity, something that people do in their spare time. The livelihood, food security, and socio-cultural importance of freshwater river fisheries deserves greater attention (Funge-Smith and Bennett 2019).

To begin filling this gap in the literature, the chapter looks at freshwater river fisheries in Temburong District, Brunei Darussalam. People who speak and identify as Iban are the main resource users in the freshwater sections of the river system (King and Knudsen 2021). Based on ethnographic fieldwork, the chapter zooms in on the knowledge and skills that fishers draw on to catch Ikan Semah, a large-bodied Masheer fish, in Ulu Temburong. It shows that Iban in Temburong have developed advanced fishing skills and possess in-depth knowledge of fish species, fishing grounds, and freshwater ecology. They know much about how, where, and when Ikan Semah feed, migrate, and spawn. Moreover, we argue that the riverine territory of Iban in Temburong is significantly defined by the habitat of Ikan Semah. Put another way, where Semah swim, Iban follow.

In the last section, we explore why Temburong Iban value the fish so highly. Freshwater river fish and other aquatic species are central to their food culture, and Ikan Semah stands out as a particularly delicious food to them. We pay attention to what people say about the qualities of the fish, including its taste,

 | HTTPS://DOI.ORG/10.1163/9789004758100-013

texture, and aroma, and briefly document how the fish are preserved, shared, and cooked.

1 A River-Based Cultural Ecology Framework

Our argument for a river-based cultural ecology framework that includes fish and fisheries draws upon previous work on cultural ecology and environmental anthropology (Ellen and Fukuyi 1996; Ingold 2000) and more recent ethnographic work on rivers (Wagner and Jacka 2018). Broadly understood, cultural ecology is the study of human–environment interaction. The focus is on how people use, shape, perceive, classify, and know their environment (King and Wilder 2003, 231). Rather than the mechanistic transmission of knowledge and skills, people learn from and adapt to their environments in an active, innovative manner, by putting their knowledge to use (Ingold 2000). Enskilment takes place in a field of relations, involving master–apprentice and community relations (Pálsson 1994), and interactions with other communities within a wider socio-economic and political context (Hviding and Bayliss-Smith 2000). To further explore the depth and complexity of human–environment interaction, some anthropologists are now paying renewed attention to the kinds of relations that humans maintain with 'other-than-human' beings, including animals (Lien and Pálsson 2021; White and Candea 2023). By following people who follow Ikan Semah up a river on Borneo, we add new insights on how human lives and "the lives of rivers" are closely interwoven (Wagner, Jacka, Hviding, Mawyer, and Muru-Lanning 2018, 5).

Anthropologists and ecologists have long studied the resource use patterns, material culture, and knowledge systems of small-scale fishers in tropical saltwater environments of the Asia–Pacific (Fabinyi and Barclay 2022; Firth 1966; Hart 1956; Hviding 2006; Johannes 1981). The literature shows that fishers depend on many different kinds of knowledge and skills to ensure success at sea. Small-scale fishers often know how to make boats and fishing gear suitable for different kinds of locations and fisheries. By being out fishing, they develop the bodily skills to efficiently operate them. Fishing gear and boats are not external to their bodies, but operate as a "bodily extension in quite a literal sense" (Pálsson 1994, 910). Small-scale fishers tend to have considerable place-specific knowledge of fishing grounds and bottom conditions (Randall 1977). Through naming practices and storytelling, they systematize and share this knowledge with others. The temporal dimensions of small-scale fishing are important. Fishers know much about how the movement of the moon and the sun affects the tides, winds, currents, and behaviour of the target species

(Knudsen 2022). Skilful fishers possess detailed knowledge of fish biology and their behaviour, including how, where, and when different kinds of fish feed, group together, and spawn (Johannes and Hviding 2000). We suspect that many river-oriented Borneo groups have equally profound knowledge of fish and how to catch them. As there is hardly any literature on the topic, it remains to be studied how widespread and important freshwater river fisheries are for inland communities across the island, and the kinds of knowledge that groups have of fish, fishing grounds, and river ecology.

Simon Funge-Smith and Abigail Bennett (2019, 1177) argue that freshwater capture fisheries are highly significant for inland communities worldwide, including in terms of catch levels, income, livelihoods, food security, and cultural services. According to them, inland fisheries contributed an estimated 12.7 per cent of total global capture fisheries in 2017. Yet, inland fisheries remain overlooked in both research and policy discussions. This lack of attention may have to do with the nature of much inland capture fisheries, being small-scale, dispersed, seasonal, and part-time (Funge-Smith and Bennett 2019, 1185).[1] As men, women, and children are involved in diverse fishing practices at irregular intervals and multiple locations, researchers may easily fail to grasp the extent to which people are involved in fishing activities and their importance for local communities.

Another reason why inland capture fisheries have received little attention among anthropologists working in Southeast Asia is the presence of a rather entrenched narrative about the nature of traditional economies, one where upland or 'hill' people are automatically defined as farmers and hunters and gatherers, and coastal dwellers and maritime-oriented peoples are quickly assumed to be fishers (Eder 2003). In northwestern Borneo, British colonial officials' interest in the commodity value of standing tree crops, and their concern for the "problems of shifting cultivation", resulted in support for research on agriculture (Leach 1950). The impressive work of Derek Freeman on "Iban Agriculture" (1955) set the tone and encouraged subsequent generations of Borneo scholars to examine similar topics (Appell 1976).

A similar tendency marks the literature on the traditional economies of Brunei. While scholars point out that fish have been a key source of food and income for Brunei's inhabitants (Chua et al. 1987), and fishing in estuarine environments is mentioned by a few studies (Abas 2021; Maxwell 1980, 108–12), the existing literature is largely focused on fishing in Brunei Bay and offshore environments (Anderson, Headly, and Tubb 1952; Lim 1986; Cinco, Teh, Zylich, and Pauly 2015). When mapped into ethnic categories, the result is one where

1 These characteristics are commonly observed in small-scale fisheries in marine spaces as well.

Brunei-Malays are constructed as the fishing-oriented ethnic group, while Kedayan, Bisaya, Dusun, and others are perceived as agriculturalists focused on the production of rice, vegetables, and fruits.

More recently, in response to Borneo rivers becoming subject to large-scale dam constructions, and in response to the negative effects of industrial logging, mining, plantation agriculture, and climate change, impact studies on water flow, river ecology, and indigenous communities have flourished (Baumgartner and Wibowo 2018; Douglas et al. 1993; Shirley and Word 2018; Sovacool and Bulan 2011). Biologists, too, have scaled up their mapping of the diverse species living in Borneo's rivers (Choy 1996; Kahar, Ahmad, and Arai 2023). In these accounts, a fuller picture of human interaction with and dependence on rivers is missing. Clearly, humans interact with, use, classify, and know rivers in a variety of ways (Anderson et al. 2019; Wagner and Jacka 2018; Iqbal, Hassan, and Kumpoh 2026), which include fisheries.

While human-induced environmental crises have dramatically escalated since the mid-20th century – perhaps warranting the designation of the current epoch as the Anthropocene – many proposed solutions to these crises remain grounded in a rigid nature-culture dichotomy (Fletcher et al. 2021). Challenging the conservationist narrative that portrays biodiverse tropical landscapes as pristine and untouched by human activity, Fletcher et al. (2021) highlight a growing body of evidence demonstrating how local communities have, over millennia, actively shaped the biodiversity and ecological complexity of tropical rainforests. They have accomplished this, as the current volume argues, by being respectful of the needs and desires of non-human and other-than-human agents, thus demonstrating the "epistemic duality" embedded in the idea of the Anthropocene (Iqbal, Hassan, and Kumpoh 2026, 2). Nevertheless, conservation practices that seek to preserve "untouched" environments continue to proliferate and marginalize the knowledge systems and resource practices of Indigenous groups (Fletcher et al. 2021, 2). The creation of Ulu Temburong National Park exemplifies this tension. Within the park, most human activities—such as hunting and tree cutting—are classified as illegal. Yet, as our case study shows, traditional resource users have successfully negotiated limited access for freshwater river fisheries. However, the timing of fishing trips is becoming increasingly complex, with climate change emerging as a contributing factor to these shifting dynamics.

2 Background to the Study and Methods

When the first author visited Temburong as a tourist in 2015, on a daytrip, he noticed that people were fishing with traps, lines, nets, and other fishing gear.

During an overnight stay in an Iban longhouse in Temburong, he noticed that the residents stored a lot of fishing gear in sheds near the river and around their longhouse units. His host family served freshly caught river fish for lunch. They had more fish stored in their freezer. These initial observations led to further inquiry and, eventually, to the formulation of a research project called "The life of the Temburong river: An Ethnographic Perspective" (2018–2019).

The second author, Misa Juliana Minggu, joined the project as a research assistant. She is from Temburong, is fluent in Iban, Malay, and English, and holds a Bachelor of Arts degree in Sociology and Anthropology. In the project, we documented past and present human uses of the Temburong river and some of its tributaries. We joined local experts on fishing trips in the saltwater and freshwater sections of the river, and we interviewed twenty respondents, both men and women, who were or had been involved in fishing and gleaning. Our questions circled around what people catch and collect from the river, what they do with the catch, what technologies they use, and how they name, classify, and know the riverine landscape.

In casual conversations and interviews with Iban men, we began hearing stories of fishing trips to a hard-to-reach area of Ulu Temburong called Wong Wan ('waterfall wait'). During follow-up fieldwork and visits to Temburong from 2020 to 2022, we heard more stories of fishing trips to Wong Wan. The stories imbued the place with near-mythical qualities, a place where large-sized Ikan Semah are plentiful and local spirits are powerful, posing a threat to the safety of those going there.

In September 2022, the first author joined two Iban men, a father and his son, on a week-long fishing trip to Wong Wan.[2] The father, who we refer to as Aki (grandfather) for anonymity, is locally well-known for his fishing and boat-navigating skills in upriver Temburong. He first went to Wong Wan in 1970, when he was 15 years old. His son, Ombay (pseudonym), has been on one or two fishing trips to Wong Wan over the last twenty years. In addition to some of the data from our previous work, data from the Wong Wan fishing trip will be presented in some detail, to lend support to the chapter's argument.

3 Iban Settlement Pattern and Resource Use in Temburong

The Iban are the most populous ethnic group in Sarawak (about 840,000 in 2020). In Brunei, they are a minority of about 20,000 (King and Knudsen

2 We obtained permits and funding for the trip to Wong Wan through the project "Small-scale fishing in Brunei Darussalam" (UBD/RSCH/1.2/FICBF(b)/2021/028).

2021, 11, footnotes 5 and 6).[3] The ancestors of the people who later became known as Iban are believed to have originated from western Borneo (Freeman 1992), in what is now West Kalimantan, Indonesia. Through migration, shaped by a distinctive 'journeying' (*bejalai*) complex (Kedit 1993), Iban expanded across northwestern Borneo, through what is today the Malaysian state of Sarawak, and then into Brunei around the turn of the twentieth century (King and Knudsen 2021; Sercombe 1999). In Brunei, Iban are not recognized as a *puak jati* (indigenous population) by the state. Still, many have become citizens of the country over the last fifty years.

Iban settlements and livelihoods in Temburong have historically been strongly associated with the freshwater sections of the river system. Kedayan Malay are the main group of resource users in the estuarine sections of the river. Although fewer in number, Murut, the Brunei-Malay label for the original pre-Malay populations of Temburong (Maxwell 1980), are found in both ecological zones (King and Knudsen 2021).[4]

In the Temburong river, saltwater reaches about four kilometres inland from Bangar town, the district capital. From Bangar, a road follows the Temburong river up to Batang Duri, located 18.6 kilometres south of the town. From Batang Duri to the juncture of Temburong and Belalong rivers (Nanga[5] Belalong in Figures 11.1), there are no settlements today. The entry point to Ulu Temburong National Park, established in 1991, is at Nanga Belalong. The park covers nearly 50,000 hectares and is densely forested.

Iban traditionally used rivers and streams to access land for agriculture, and to hunt, gather, and fish (Freeman 1992; Sutlive 1992). Figure 11. 1 shows the approximate location of some Iban resource uses in Temburong from just below Nanga Belalong to *ulu ai*,[6] the headwaters in the southern-most part of the district. Swidden cultivation was practised until the early 1980s. The maps

3 In 2021, the total population of Brunei was estimated to be 440,715 (Department of Economic Planning and Statistics 2022).

4 Murut in Temburong often refer to themselves as *lun bawang* (people of the country) (Maxwell 1980, 23). Kedayan and Iban moved into Temburong in larger numbers in the early twentieth century. Murut seem to have intermarried quite extensively with both Kedayan and Iban over the years.

In 2010, the ethnic composition of Temburong District, according to an official source, was: 5,852 Brunei-Malays (Kedayan Malay and Brunei Malay classified together), 2,092 Iban, 811 Murut, 206 Chinese, 21 Dusun, and 211 'Others' (Temburong District Office [Information Department] 2011). In 2018, Temburong had an estimated population of 10,900 people (Department of Statistics 2019).

5 *Kuala* is the Malay name for a place where rivers meet (as in Kuala Lumpur). On the map, we use the place names of our Iban informants, including *sungai* for river.

6 *Kepala ai* is another term for 'head waters.'

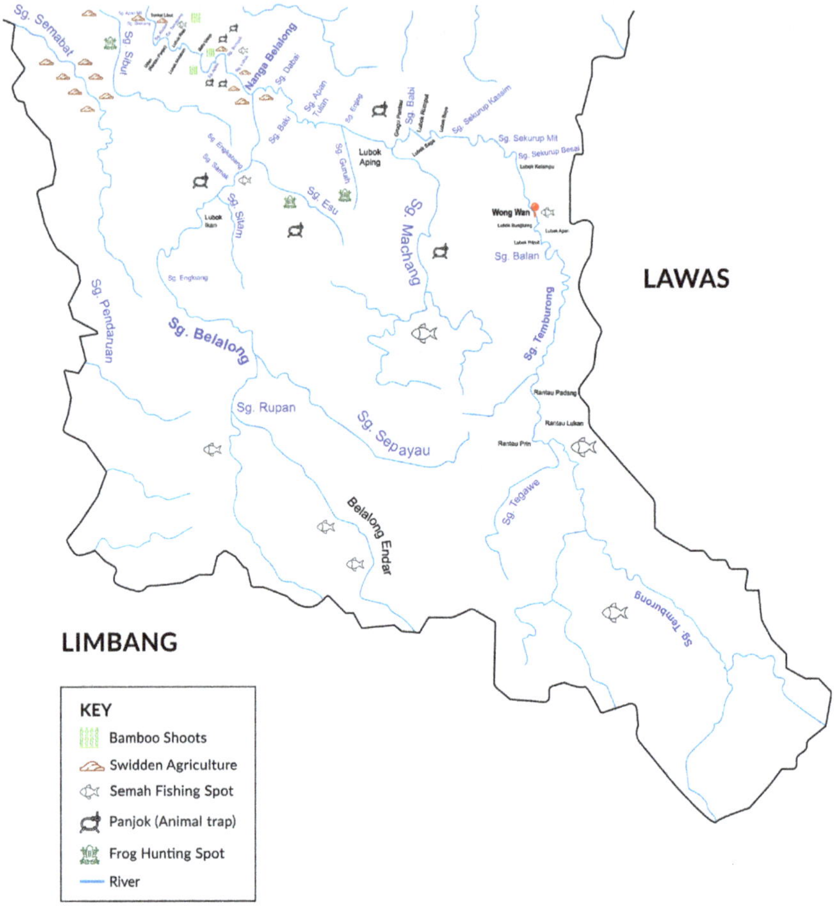

FIGURE 11.1 Resource use map. Approximate location of Iban resource use until the 1980s (Temburong District).
SOURCE: AUTHORS. THE PLACE NAMES WERE PROVIDED TO US BY LOCAL IBAN SPEAKERS

show that swidden fields were located close to rivers. Today, fruit trees grow on many old swidden plots. Popular hunting spots were also located near rivers and streams, where wild boar and deer cross or come to drink. The approximate location of some Ikan Semah fishing spots are indicated on the resource use map (Figure 11.1), mainly in the upriver areas of Sungai Belalong and Sungai Temburong. Farming, hunting, and gathering are today only practised outside the Ulu Temburong National Park. The park, under the control of the Forestry Department and Ministry of Primary Resources and Tourism, strictly regulates resource use (banning land-based hunting, tree felling, and other activities). At the same time, small-scale river fishing is tolerated inside the park.

Locals use many different hook-and-line (*nginti*), trap (*bubu*), and net (*pukat*) fishing techniques to catch a variety of fish and other aquatic species, including freshwater shrimps, snails, turtles, and crabs. The casting net (*nyala*) is frequently used. Frogs are caught at night in smaller streams with a torch light and spear. When the water is clear, diving (*nyelam*) is another technique, which is to find fish among rocks in deeper pools. In the past, until the 1970s, *tubai* (derris root) was used to catch fish. This kind of fishing involved a larger group of people led by a ritual expert (*manang*). Derris root contains a poisonous substance that disorients the fish, making it easy to spear the fish or catch them with nets. Iban women used to be involved in much fishing activity, and a few still do some fishing, often together with their husbands, and sometimes alone from the riverbank with a pole and line. In the past, women used woven baskets (*pemansai*) made from rattan to catch smaller aquatic products, a 'scooping' technique where small holes in the basket serve as a filter. Women and men would regularly check and empty fish traps located near their longhouses. According to a 90-year-old Iban woman, these traps provided a good, stable supply of fish until the 1970s. Since then, the abundance of fish, particularly larger-sized ones, along the more populated sections of the Temburong river has declined, in the area below Nanga Belalong.[7]

It goes beyond the scope of this chapter to examine in detail how the shift from primary-sector work over the last fifty years has affected traditional knowledge and fishing practices. Brunei is a high-income, oil-rich country (Ananta, Hoon, and Hamdan 2023). Government revenues from the oil and gas sector have allowed for large investments in infrastructure, education, and healthcare. Modern education and service-sector jobs have since the 1970s become increasingly common and important for the people of Brunei, and also for Iban in Temburong. Today, transportation mainly takes place on roads, not so much on rivers. The promotion of ecotourism and the establishment of Ulu Temburong National Park have also contributed to changes in resource use (see below). While the knowledge and skills on how to build boats and catch fish and other edible aquatic products have not been wiped out by 'modernity', a smaller number of people are today in possession of the more advanced skills. At the same time, quite a few people occasionally use rivers to access hunting and fishing grounds, when they have time off from work, and after retirement. In addition, local knowledge of the environment and the skills to

7 The introduction of nylon nets and outboard engines in the 1950s, 60s, and 70s increased the capacity of fishers to catch fish and other aquatic products, contributing to the subsequent decline in catch levels. The gravel industry, which has transformed riverbanks and landscapes in some of the lower freshwater sections of the Temburong, has also had a negative impact on the fisheries.

navigate rivers with a canoe are increasingly used in tourism and research-related activities. The boatmen who bring tourists and researchers into the national park are all local Iban. In addition to sharing some of their knowledge of trees, plants, insects, birds, and land-based wildlife with tourists and researchers, Iban boatmen and guides may also share some of their knowledge of rivers, streams, and aquatic life with the visitors.

Given the restrictions of Ulu Temburong National Park on traditional resource use, a brief justification for our choice of framework is helpful. Despite the high local demand for freshwater river fish, and a somewhat limited supply of fish in the more densely populated areas of the Temburong river, further upriver, inside the park, there is far less fishing activity. We believe that this has less to do with the entry and use permit requirements than with other barriers to access, both socio-cultural and physical. The main social barrier to entry nowadays is that most people in Temburong, including Iban, have modern-sector jobs and other commitments (King and Knudsen 2021). Many do not have the flexibility of schedule for Wong Wan fishing trips, which may last two or three weeks, and many will not have the strength, stamina, and skills required to bring a boat through the tricky sections of the river. The physical barriers to entry are significant and require a separate section.

4 Knowledge of How to Bring a Boat to Wong Wan

The Iban in Temburong learn about rivers and river ecology through what Tim Ingold calls "the active, perceptual engagement of human beings with the constituents of their world" (2000, 60). From an early age, they learn to swim, dive, and be comfortable in freshwater currents. By walking and swimming in rivers, travelling on rivers in boats, and through fishing and gleaning, they learn about water flow, river landscape, and ecology. For our informants, fishing and using boats to reach fishing grounds are central to how they relate to and know the riverine environment.

Among Iban, the term *mudik* refers to travel upriver in a longboat, for the purposes of hunting or fishing, and it implies staying at least one night. Among the Temburong Iban participants in our research, *mudik* is exclusively used to refer to overnight fishing trips to *ulu*.[8] *Teleba* is their term for knowledge or skill. *Teleba mudik*, then, refers to the knowledge and skills of bringing a boat to *ulu*, knowledge possessed by seasoned travellers, for the purpose of overnight

8 As hunting activities are banned within the Ulu Temburong National Park, *mudik* have over the last thirty years come to imply overnight fishing trips to *ulu*.

fishing. In our estimate, there are today only four or five boatmen, each with two to three companions, who are capable of accessing the deep interior of Temburong river with a longboat. All of them have gone through various stages of learning.

Just upriver from Nanga Belalong, several shallow, stony rapids, called *rantau*, serve as entry barriers. In some areas, larger rocks are plentiful, in sections of the river called *grugu*. These can easily cause difficulties for boatmen and crew. Driftwood (*kayu agang*), too, poses a threat to canoe travellers and may block the ideal path of the boat, small channels that form in the gravel substrate of the river. The propeller easily gets damaged on rocks, and the canoe, while robust, may be severely damaged if it hits a larger rock or big piece of wood at high speed. If knocked out of course, the boat may be tossed around and capsize in strong currents. With a boat full of fishing and camping gear, it takes considerable navigation skills and place-specific knowledge to reach Wong Wan. To go further upriver, past Wong Wan, is even more challenging and is only attempted when the water level is low.

The longboats used for fishing expeditions to *ulu* have common characteristics, but they may have slightly different designs depending on the characteristics of each river and the owner's preferences. All canoes used for *mudik* expeditions in Temburong are called *perau panjay*. They are 30–32 feet long and narrow, and given the shallowness of the river, they have to float high in the water. Aki builds his *perau panjay* with a slightly higher bow than others in Temburong. This is to better handle strong currents, choppy water, and to get through Wong Wan, to reach the best *Semah* fishing spots. The preferred hardwood for making *perau panjay* is *emperawan* (meranti).[9] In addition to being robust, it has the capacity to 'glide' over rocks. The boat's length provides directional stability and good carrying capacity. The drawback is that a long boat is more difficult to turn than a shorter one. Punting poles (*suar*) and paddles (*sengayuh*) are used to help steer the boat and turn it around.

In addition to the technical material, and the physical and livelihood aspects of interacting with rivers, people learn about rivers through naming practices and storytelling. Iban participants shared with us how they develop cognitive maps of the river system and memorize critical sections. Initially, when we started work on the resource use map, we noticed that the participants had names for many of the smaller rivers and streams that drain into the Temburong. One common memorization technique was focused on getting the sequence of place names right; for example, starting from their own

9 Aki and Ombay expressed concern over how difficult it has become to get government approval to cut *emperawan* trees.

place of residence, they would name all the rivers and streams that they pass as they travel upriver. They also remember many other topographical features, such as very long stretches, large bends, deeper pools of water, challenging rapids, special rock formations, and human and spiritual beings associated with some of these features. These features "act as mnemonic devices, playing important roles in the way individual and collective memories are organised and retained" (Wagner 2018, 226, with reference to Harrison 2004). Table 11.1 provides some examples of place names, each with a brief explanation.

TABLE 11.1 Place names

Grugu Pantu	*Grugu* refers to an area of big stones in the river. *Pantu* is a palm tree. The shoots of the *pantu* are good to eat.
Sungai Babi	Wild boar river. Good place to set traps.
Lubok Rumput	*Lubok* is a deepwater area or 'pool.' *Rumput* is a type of grass.
Lubok Baya	Crocodile 'pool'
Sungai Sekurup Kassim	*Sungai* means river. *Sekurup* is a Murut name. My informants did not know the meaning of it. Kassim is the name of a Kedayan Malay who used to go *mudik*.
Sungai Sekurup Mit	Small Sekurup river
Batu Putar	'Stone turn around' refers to a place with a large rock formation in the middle of the river. When going downriver, the boatman has to turn the boat around before continuing.
Lubok Reba	A deepwater area (*lubok*) with many dead trees in the river.
Grugu Buluh	An area with many big stones in the river and bamboo trees (*buluh*) growing along the river. *Grugu Buluh* is located just below Wong Wan and is quite steep and very difficult to get up.
Wong Wan	'Waterfall wait'. The water level has to be low to get safely through *Wong Wan*. Stop, inspect the water flow and be respectful of the local spirit(s). When the water level is high, wait for it to drop.
Lubok Bungkang	Deepwater area (*lubok*) with *bungkang*, a type of tree, growing on the riverbank.
Lubok Apan	Deepwater area where the water is salty. Deer (*rusay*) like to drink water here.
Sungai Balan	Balan river. Balan is the name of a Murut man, a mythical figure who could jump across the river.
Rantau Padang	*Rantau* refers to stony shallow rapids. *Rantau Padang* is an area where the *rantau* is unusually wide.
Rantau Prin	A very long, straight *rantau* (about 2 km long).

On the trip up Temburong river, the first author noticed the importance of coordinated action between Aki, the 'captain' or driver (*ngamudi*) standing at the back of the boat and running the engine, and Ombay, the frontman (*neluan*) equipped with a punting pole (*suar*) and a paddle (*sengayuh*). The frontman looks out for rocks and driftwood that may damage the boat and indicates with hand signals the best route ahead. Up the more difficult shallow rapids (*rantau*), the frontman helps by using the punting pole. When it is too shallow to run the engine and punting no longer helps to move the boat forward, or if the boat starts moving backwards, the captain, frontman, and other crew quickly jump into the river to push or carry the boat up (*bebatak*). When going downriver, the frontman uses the paddle to help steer the canoe.

Mudik usually takes place during periods of little rain or when there are eight to ten days of no rain. While rain and heavy downpours can occur at any time of the year in the interior of Ulu Temburong, February to September is the main period for considering overnight fishing trips to Wong Wan, with February/March and July/August singled out as having been the most common periods for *mudik* expeditions. With the introduction of outboard engines (among Iban in Temburong from the 1960s and 70s and becoming more common in the 1980s), it became possible for those with the necessary skills to travel up to Wong Wan in a day or two when the water level was high. However, it is considered risky to travel to *ulu* in rainy periods, and unless there is a longer period of very little or no rain in *ulu*, going past Wong Wan is near impossible due to the strong and complex currents that form through the narrow, cliff-faced gorge area (see Figure 11.2).

When the water level is high, the best option is to wait. Similarly, those who successfully bring their boat through Wong Wan must be prepared to wait to get down again, for the water level to drop. Some of the participants in our research believed that the water level can suddenly rise in Wong Wan without there being rain in *ulu*, and they attributed this to the power of *Antu-ai*, a water spirit. If antagonized, the spirit can create a splash or eddy (*bebuli*) that may capsize your boat. *Antu-ai*, according to respondents who claim to have seen it, has what looks like big white deer horns, bursting out from the middle of the eddy. Aki said he did not support this explanation and was doubtful as to the existence of *Antu-ai*. "I've never seen it," he said. If the water level rises rapidly in Wong Wan, it is because of a heavy downpour higher up, Aki explained. Aki could, however, feel the presence of *Orang Nguan Menoa* in Wong Wan. In trying to describe *Nguan Menoa* to us, Aki said it is similar to what the Malays call *Penunggu*, a spirit that protects or guards a place. When you enter its domain, you should be quiet, not make noise or be boastful, and you should not urinate in the place. *Nguan Menoa* spirits are invisible to humans. They can be

FIGURE 11.2 The terrain near Wong Wan

protective, benign, or malevolent, but according to Aki, they do not mysteriously cause the water level to rise. Aki was puzzled by the weather lately, citing global warming as a possible contributing factor to an even more unpredictable rainfall pattern.

Even when the water level is low, it is challenging to get a boat through Wong Wan. Boats have sunk and been severely damaged here. If that happens, your only option is to swim downriver to Nanga Belalong, a distance of about 17 kilometres. To make the boat lighter, and to avoid losing gear if the boat capsizes, camping and fishing gear are commonly carried past Wong Wan, up a cliff and three hundred metres further upriver. Some also prefer to turn the engine off and use ropes to pull the boat through Wong Wan. We heard stories of men who had, in the past, carried their boat up the cliff to bypass Wong Wan. Even with four strong men, this is difficult. In addition to the challenge of bringing the boat up or down the steep cliff, it also requires cutting a path through a slippery jungle area.

Despite the many access barriers, the Iban in Temburong have a long history of exploring and exploiting the aquatic resources in the freshwater sections of the river system, including weeklong and longer fishing trips to catch Ikan Semah in Wong Wan. Aki and Ombay have gone through Wong Wan several times. More often than not, however, they reach the conclusion that it is too risky to attempt to go through. On our trip, we reached Wong Wan on the third day. Having experienced a major flooding event during the first night of the trip, the water level was still quite high, and we could not pass through Wong Wan. When the water level is high, a camp site a kilometre below Wong Wan serves as a good waiting area. This area can also be good for fishing. With only seven days allotted for our trip, the water level would not drop in time for us to go through Wong Wan. Instead, we focused our fishing activities in the area below Wong Wan.

Aki and Ombay were insistent about one thing: it is in some of the deeper pools (*lubok*) past Wong Wan where the truly large catches can happen. The first deep 'pool' (*lubok*) after Wong Wan is called Lubok Bungkang, where bamboo grows near the river. Next is Lubok Apan, a place where the water is salty (*ai garau*). Deer (*rusa*) drink water here. Rantau Padang and Rantau Prin are considered key spawning areas for Ikan Semah. It was in the area of Rantau Padang that another key informant hit the 'jackpot' in 2002. He went with four other men, all Iban from Kampung Amo. They travelled three days past Wong Wan and found a *lubok* full of Ikan Semah. With nets, they caught perhaps 200 kg of Semah. As they ran out of salt and the boat was full of fish, they returned home, with plenty of Semah to consume and share, and with a good story to tell. As this account affirms, there is much prestige in successful *mudik* expeditions.

Aki told us about his fishing trip to Rantau Prin in 1974. It took them about five days of travel from Wong Wan to reach the place. Here he caught a 4 kg *Semah*, his biggest so far. They continued higher up, to explore the terrain and look for new fishing grounds. After another week of travelling, they reached Sungai Tanawal,[10] a side stream. Here the fish were smaller in size. Since then, Aki has not been past Rantau Prin. This lends support to the argument that Ikan Semah help define the river area of interest to the Iban in Temburong. Semah are born, feed, and reproduce in the freshwater sections of the river. An important part of Iban knowledge of their environment is acquired through fishing, and the upper limit of their territory is defined by how high Ikan Semah swim, which is roughly to the spawning area of Rantau Prin.

In the next section, we document some of the knowledge that the fishers have of *Ikan Semah* and how to catch them under varying conditions.

10 Tanawal may be a Murut name, but its meaning is unknown. Aki suggested an alternative name, sometimes used among his group: Tamawai or Tembawai. Tembawai in Iban refers to the site of a former house which is planted over as a fruit grove.

5 Knowledge of *Ikan Semah* and How to Catch Them

Ikan Semah is an Iban name for *Tor Tambra*.[11] In Temburong, the habitat of Ikan Semah is mainly in Ulu Temburong river and Belalong river, where the water is fast flowing and clear with a gravel substrate. Being strong swimmers, mature Semah migrate ('*bejalai*') long distances upriver to reach the spawning grounds in *ulu*, including Rantau Padang and Rantau Prin. Being multiple-spawners, Semah will migrate up and down the river system several times to feed and reproduce. When reaching the spawning grounds in *ulu*, they wait for the right timing, staying in *lubok*. The exact timing of the spawning (*nyaru*[12]) is hard to predict in Temburong.[13] Aki explained that female Semah spawn at different intervals, and although there is no distinct spawning season, spawning is more likely to take place after several days of little or no rain. At the same time, Aki explained that the water should not be hot.

In Temburong, there are two types of Semah, one longer and darker with a hint of red colour which Iban call *Semah Panjai* (long-shaped Semah) and one shorter-bodied with silver-greyish fins and a more rounded body called *Semah Pandak* (short Semah). On our trip, we caught both types. To catch Semah, we used nets with 3 ½ inch mesh (*pukat jarang*) and two hook-and-line techniques, *ginti aris* and *ginti ambor*. The nets were set just before sunset and pulled in in the next morning. The longest nets were 100 feet in length. Nets are used when the water is clear (*ai chiru*) and the water level is medium to low. A good location to set the net is on the 'inside' just after a bend (*tanjung*) in the river, a location called "*ulak*" (see Figure 11. 3). At the *ulak*, the current is weak and it is the preferred route of fish swimming upriver. In addition, leaves and debris flow largely at the outer bend of the river. If you get plenty of leaves in your net, the fish will see it and swim around.

---

11 Biologists have limited knowledge of *T. Tambra*. It is classified as 'data deficient' in the International Union for the Conservation of Nature Red List (Kahar 2019). *Tor Tambra* and *Tor Tambroides* are found in intact freshwater river systems in the tropical regions of South, East and Southeast Asia (Kahare et al. 2014; Ambili et al. 2014; Pinder et al. 2019). Commonly referred to as "the king of the river" (Muchlisin Z.A., et al. 2022), Masheers (Tor sp.) are the most highly priced and valued river fish in many parts of Southeast Asia.

12 *Nyary* is the Iban name for the spawning of Ikan Semah and some other larger fish. *Ngaban* is their term for the spawning of Ikan Kulong and other smaller species of fish.

13 According to a recent biological study of freshwater fish in Temburong, female Ikan Semah display a prolonged period of spawning, and male Semah were found to be capable of spawning in nearly every month, meaning that they are ready to spawn whenever mature females are available (Kahar, Ahmad, and Arai 2023). The study attributes the lack of seasonality in spawning to the unstable hydrology of the Temburong basin.

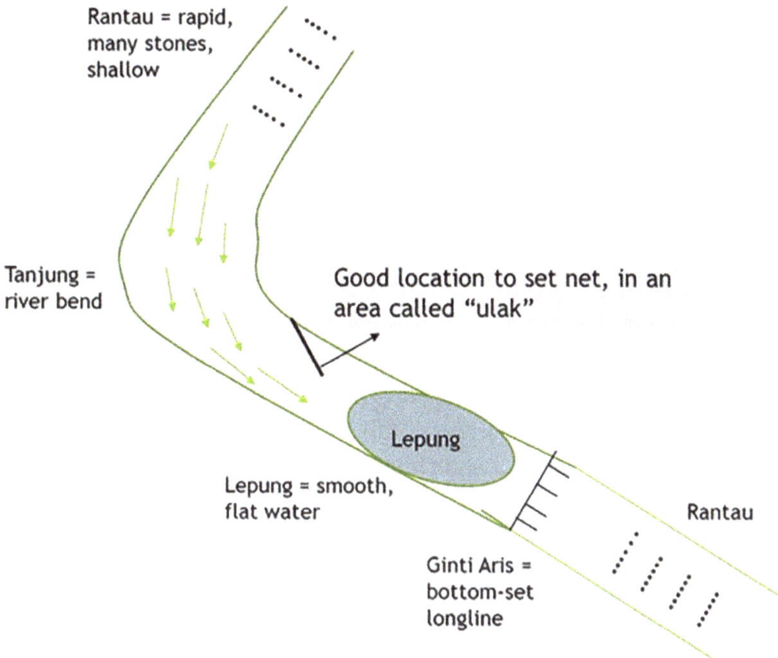

FIGURE 11.3 Classification of a river section

If the water is murky (*ai keruh*) or mixed (*ai kelabur*) and the water level is medium high, *ginti aris*—a bottom-set longline—is typically the better option. The line is set across the river just before sunset and checked or pulled in in the morning. Nine shorter lines with baited hooks are tied to the main line. A few rocks are placed on the main line to keep it in place. Ikan Kulong is the preferred baitfish, but Ikan Enseluai is also good.[14] Aki and Ombay used a cast net (*nyala*) and a small net (*pukat tilan*) to catch baitfish. When the water level is high, just after a flood (*ai bah*), *ginti ambor* is considered the better technique. It consists of a line with a baited hook tied to a branch along the river. According to our informants, Semah will only 'migrate' (*bejalai*) when the water is clear. When the water is murky or cloudy, Semah will stay more or less still, moving only in a small area. Aki explained: "Semah are afraid of the otters." Otters are active and clever hunters in cloudy water. Aki demonstrated considerable knowledge of inter-species relations and how the changing environmental conditions are changing the behaviour of Semah.

14 Ikan Enseluai in Temburong is *Rasbora dusonensis* (yellow tailed) and *Rasbora torneiri* (red tailed). Ikan Kulong is likely *Lobocheilos ovalis* (Siti Rafhiah binti Hj Abd Kahar, personal communication December 2018). The hooks can also be baited with palm fruit.

On our trip, Aki and Ombay caught about 22 kg of Ikan Semah. They also caught about 5 kg of Ikan Juak (*Hampala bimaculata*) and one 3 kg Ikan Bong (tropical anguillid eel, *Anguilla marmorata*). The Ikan Semah varied in size, from around 300 g to 1.5 kg, with an average of about 0.7 kg.[15] In the river sections where Aki and Ombay fished, below Wong Wan, this was considered a relatively good catch for the time spent fishing. Each time they caught a Semah, they were visibly satisfied, Ombay in particular putting on a big smile. To explain why they caught quite a lot of good-sized Semah here, Aki believed that the big flood that occurred on the evening of 19 September, on the first day of our trip, had 'flushed' more Semah downriver. Another factor affecting the catch level, according to Aki, is the moon's cycle. During a full moon, Ikan Semah fishing tends not to be good. Our trip took place in the third quarter, reaching a new moon on the last day of fishing, 26 September.

As they cleaned the fish, Aki and Ombay studied each one carefully, determining its sex, estimating the approximate time to spawning, and examining what the fish had been eating. All except for one of the Semah Aki and Ombay caught were female. Aki explained: the male Semah are generally smaller than the females, and in the area where we were fishing, too small to get stuck in the 3 ½ inch mesh nets and bite on the large hooks.

6 Why So Delicious?

We consumed about nine kilos of Ikan Semah on the trip. For lunch and dinner, Aki and Ombay made fish soup with vegetables and served rice on the side. We also ate leftover soup for breakfast. For lunch one day, Aki smoked a large Semah. It was delicious. The rest were salted (*kasam Ikan Semah*) and brought back home. We brought more than ten kilos of salt with us on the fishing trip. Ombay did the job of salting the fish. He cleaned the fish, chopped them into pieces, squeezed water out of the flesh, and worked salt into it. He packed the salted fish in plastic buckets and added a bit of cooked rice to each bucket, to reduce the smell of the fish and make it more delicious to eat.

Why is Semah so tasty? The answer has a lot to do with what Semah eat. Apart from consuming Ikan Kulong, Ikan Enseluai and some other small fish and aquatic species, Ikan Semah eat larvae, centipedes, and some other

15 In Temburong, based on sampling mostly around Nanga Belalong, Semah have been found to reach sexual maturity at a relatively small size, 377 mm for females and 255mm for males (Kahar 2019).

insects. Importantly, Semah are also fruit eaters. They eat more than five different types of fruits, including Ensurai (diptocarpus oblongifolius), two types of Kalampu (*Sandoricum koetjape* and *Sandoricum borneense*), Kasai (*Pouteria caimito*), and Engkabang (*Shorea spp.*). When trees growing along the river are fruiting, they attract Semah. If the trees are fruiting in *ulu*, Semah will swim to *ulu*. If trees are fruiting downriver, they will go there, or stay there. The fruits of the Engkabang tree, illipe nuts, are particularly high in fat content.[16] Although hard to open, the outer shell softens in water, and Ikan Semah have teeth in their throats (pharyngeal teeth) with a design that enables them to chew the nuts.

When we returned to Aki's house on 26 September 2022, plenty of family members were visiting. They knew that we were scheduled to return on that day and may have been keen to hear about the trip, and possibly to get access to some of the fish. Aki and Ombay divided the salted Semah among relatives. The first author was given a share to pass on to some of his Iban friends. Salted Ikan Semah is a popular dish that can be stored and consumed all year round. Fresh Semah is often cooked in water with vegetables and salt, but can also be fried or cooked in bamboo (*pansoh*). Each family will have their own recipes and preferred cooking styles, and the fish are today often consumed on special occasions.

When we asked what it is about the fish that they like so much, many answered that the fish are "sweet and fatty". The granddaughter of Aki said: "Semah are sweeter." Compared to most other freshwater fish in Temburong, many appreciate that the fish have larger chunks of meat and larger bones. When fermented, the texture is nice and "flaky." Finally, the fish are "not smelly" like many saltwater fish. Ombay said: "I don't like saltwater fish. They are smelly to me, except marlin."

Variants of a distinction between 'saltwater' and 'freshwater people' are common across the Asia–Pacific (Hviding 2018; Paredes 2016). In Temburong, a 'saltwater'/'freshwater' distinction is expressed in various ways, also in terms of food. Iban have a strong liking for freshwater fish, and many do not like the smell of saltwater fish. The Iban respondents also said that they do not like to swim in saltwater. In contrast, Kedayan and Brunei Malays tend to say that freshwater fish have a weak taste and are difficult to eat due to the many small and sharp bones in them.

16 Aki is aware of three different Engkabang tree species in Temburong.

7 Conclusion

Fish can be both "good to eat" and "good to think with" (White and Candea 2023, with reference to Lévi-Strauss 1963, 89). Iban, in the Borneo literature more commonly associated with swidden cultivation, inland hunting, and gathering and cash cropping (Freeman 1955; Appell 1976; Wadley and Mertz 2005), are very involved in river fishing in Brunei's Temburong District. This chapter documented some of the knowledge and skills that Temburong Iban have of fish, fisheries, and river ecology. Their knowledge of rivers has in important ways been shaped by river fishing. Ikan Semah, in particular, is an important food fish for them, and much of their knowledge of the upper reaches of the Temburong river stems from fishing trips to catch Semah.

A river-based cultural ecology framework that includes fish and fisheries looks promising for Borneo studies, providing fresh insights on livelihoods, ecological knowledge, food cultures, and more. Previous literature has largely overlooked the importance of river fisheries for the livelihood success and wellbeing of interior Borneo communities. By adding fish and fisheries to the menu, new questions arise, also about mobility and migration. While it is well-known that Iban traditionally branched off from their communities in search for new land for swidden cultivation (Freeman 1992), might they not have explored new rivers for their fisheries resources as well? Different rivers will have different aquatic life and different types of fisheries. From our fieldwork elsewhere in Brunei, we see that Iban are keen fishers also in the Tutong and Belait rivers, and although these do not have Ikan Semah, there are other valuable freshwater fish in these rivers and their tributaries. We also suggest that there may have been important inter-dependencies between fishing and farming, with salted Semah or other fatty fish providing nutritious food during labour peaks in the agricultural cycle, while rice was an important staple on long fishing expeditions. Such inter-dependencies undermine the idea of fishing being a small sideline activity among river-oriented Borneo groups. We also encourage scholars to examine more fully to what extent and how 'freshwater' and 'saltwater' figure as identity and boundary markers for the peoples and cultures of Borneo. Finally, we believe that further research along the lines suggested here will add important insights on the types of knowledge that indigenous rainforest communities have of their environment, and help demonstrate that, far from being untouched by humans, tropical forests and their rivers have long been valued, used, and shaped by them (Fletcher et al. 2021; Franco, Knudsen, and Hassan 2022).

References

Abas, Siti Noor Aqilah. 2021. "Social-ecological dynamics: Mangrove ecosystem and dependent communities in the Brunei Bay." Master diss., Universiti Brunei Darussalam.

Ananta, Aris, Chang-Yau Hoon, and Mahani Hamdan, eds. 2023. *Stability, Growth and Sustainability: Catalysts for Socio-Economic Development in Brunei Darussalam*. Yusof Ishak Institute, Singapore and Universiti Brunei Darussalam.

Anderson, A.M., Derek Headly, and J.A. Tubb. 1952. "Notes on the fisheries industry of Brunei Bay and Labuan Island." Accessed January 6, 2023. http://fao.org/apfic/publications/etail/en/c/420308/.

Anderson, Elizabeth. P., Sue Jackson, Rebecca. E. Tharme, Michael Douglas, Joseph. E. Flotemersch, Margreet Zwartebeen, Chicu Lokgariwar, Mariana Montoya, Alaka Wali, Gail T. Tipa, Timothy D. Jardine, Julian D. Olden, Lin Cheng, John Conallin, Barbara Cosens, Chris Dickens, Dustin Garrick, David Groenfeldt, Jane Kabogo, Dirk J. Roux, Albert Ruhi, and Angela H. Arhington 2019. "Understanding rivers and their social relations: A critical step to advance environmental water management." *WIREs Water* 6 (6): 1–32.

Appell, George N., ed. 1976. *Studies in Borneo Societies: Social Process and Anthropological Explanation*. Northern Illinois University, Center for Southeast Asian Studies Special Report.

Austin, Robert F. 1976. "Some Demographic Characteristics of the Iban Population of Brunei Part I: 1947–1960." *The Brunei Museum Journal* 3(4): 64–69.

Baumgartner, Lee J. and Arif Wibowo. 2018. "Addressing fish-passage issues at hydropower and irrigation infrastructure projects in Indonesia." *Marine and Freshwater Research* 69 (12): 1805–13.

Boomgaard, Peter. 1998. "The VOC trade in forest products in the seventeenth century." In *Nature and the Orient: The Environmental History of South and Southeast Asia*, edited by Richard H. Grove, Vinita Damodaran, and Satpal Sangwan. Oxford University Press.

Cinco, Elviro A., Lydia C. L. Teh, Kyrstn Zylich, and Daniel Pauly. 2015. "Reconstructing the marine and estuarine fisheries of Brunei Darussalam, 1950 to 2010." Working Paper No. 29, Fisheries Centre, University of British Columbia.

Choy, Satish C. 1996. "Distributional ecology of freshwater fishes in tropical rainforest streams of Borneo." In *Tropical Rainforest Research – Current Issues: Proceedings of the Conference held in Bandar Seri Begawan 1993*, edited by D. S. Edwards, W. E. Booth and S. C. Choy, S. C. Springer.

Cramb, Rob A. 1989. "Shifting cultivation and resource degradation in Sarawak: Perception and politics." *Borneo Research Bulletin* 21: 22–49.

Cramb, Rob A. 2007. *Land and Longhouse: Agrarian Transformation in the Uplands of Sarawak*. Nias Press.

Department of Economic Planning and Statistics. 2022. "Report of the Population and Housing Census (BPP) 2021: Demographic, Household and Housing Characteristics." Ministry of Finance and Economy, Brunei Darussalam.

Douglas, Ian, Tony Greer, Kawi Bidin, and Mike Spilsbury. 1993. "Impacts of rainforest logging on river systems and communities in Malaysia and Kalimantan." *Global Ecology and Biogeography Letters* 3: 245–52.

Eder, James F. 2003. "Of fishers and farmers: Ethnicity and resource use in coastal Palawan." *Philippine Quarterly of Culture and Society* 31: 207–25.

Eghenter, Cristina. 2001. "Towards a causal history of a trade scenario in the interior of East Kalimantan, Indonesia, 1900–1999." *Bijdragen tot de Taal-, Land- en Volkenkunde* 157 (4): 739–69.

Ellen, Roy F. and Katsuyoshi Fukuyi. 1996. *Re-defining Nature: Ecology, Culture and Domestication*. Routledge.

Fabinyi, Michael and Kate Barclay. 2022. *Asia-Pacific Fishing Livelihoods*. Palgrave Macmillan.

FAO FishStatJ (2019). FAO FishStatJ database: 2019 dataset. http://www.fao.org/fishery/statistics/software/fishstatj/en.

Fletcher, Michael-Shawn, Rebecca Hammilton, Wolfram Dressler, and Lisa Palmer. 2021. "Indigenous knowledge and the shackles of wilderness." *Proceedings of the National Academy of Sciences* 118 (40): e2022218118.

Firth, Raymond. 1966. *Malay Fishermen: Their Peasant Economy*, 2nd edition. Routledge and Kegan Paul.

Franco, Merlin, Magne Knudsen, and Noor Hasharina Hassan. 2022. *Case Studies in Biocultural Diversity from Southeast Asia: Traditional Ecological Calendars, Folk Medicine and Folk Names*. Springer.

Freeman, Derek. 1955. *Iban agriculture: A report on the shifting cultivation of hill rice by the Iban of Sarawak*. Her Majesty's Stationery Office, Colonial Office Research Study 19.

Freeman, Derek. 1992. *The Iban of Borneo*. Dewan Bahasa dan Pustaka.

Funge-Smith, Simon, and Abigail Bennett. 2019. "A fresh look at inland fisheries and their role in food security and livelihoods." *Fish and Fisheries* (20): 1176–95.

Hart, Don V. 1956. *Securing aquatic products in Siaton Municipality, Negros Oriental Province, Philippines*. Bureau of Printing, Institute of Science and Technology, Manila.

Harrison, Simon. 2004. "Forgetful and memorious landscapes." *Social Anthropology* 12 (2): 135–51.

Hviding, Edvard. 2006. Knowing and managing environmental knowledge in the Pacific Islands: Challenges of environmentalism in Marovo Lagoon. *International Social Science Journal* 58(1): 69–85.

Hviding, Edvard. 2018. "The River, the Water and the Crocodile in Marovo Lagoon." In *Island Rivers: Fresh Water and Place in Oceania*, edited by J. R. Wagner and J. K. Jacka. ANU Press.

Hviding, Edvard, and Tim Bayliss-Smith. 2000. *Islands of Rainforest: Agroforestry, Logging and Eco- tourism in Solomon Islands.* Routledge.

Iqbal, Iftekhar, Noor Hasharina Hassan, and Asiyah Kumpoh, A., eds. 2026. *Fluvial Phenomena: The River and Anthropocene Life-Worlds in the Asia-Pacific.* Brill.

Ingold, Tim. 2000. *The Perceptions of the Environment: Essays on Livelihood, Dwelling and Skill.* Routledge.

Johannes, Robert E. 1981. *Words of the Lagoon: Fishing and Marine Lore in the Palau District of Micronesia.* University of California Press, Berkeley.

Johannes, Robert E., and Edvard Hviding. 2000. "Traditional knowledge possessed by the fishers of Marovo Lagoon, Solomon Islands, concerning fish aggregating behaviour." *Traditional Marine Resource Management and Knowledge Bulletin* 12: 22–29.

Kahar, Rafhiah. 2019. "Mahseer fish (*Tor* sp.) of northwest Borneo: A first look on the reproductive ecology." Paper presented at the 29th International Congress for Conservation Biology (ICCB), 21–25th July, Kuala Lumpur, Malaysia. Accessed November 5, 2022. https://iber.ubd.edu.bn/freshwater-research-at-international-congress-for-conservation-biology/.

Kahar, Rafhiah, Norhayati Ahmad, and Takaomi Arai. 2023. "Year-round spawning of three tropical Cypriniformes fishes in Southeast Asia." *Scientific Reports* 13 (1): 8971. https://doi.org/10.1038/s41598-023-36065-9.

Kedit, Peter M. 1993. *Iban Bejalai.* Ampang Press for the Sarawak Literary Society.

Khare, Praveen, Vindhyia Mohindra, Anindya Sundar Barman, Rajeev Kumar Singh, and Kuldeep Kumar Lal. 2014. "Molecular Evidence to Reconcile Taxonomic Instability in Mahseer Species (Pisces: Cyprinidae) of India." *Organisms Diversity and Evolution* 14: 307–26.

King, Victor T. 1985. *The Maloh of West Kalimantan: An Ethnographic Study of Social Inequality and Social Change among an Indonesian Borneo People.* Foris Publications.

King, Victor T. 1993. *The peoples of Borneo.* Blackwell.

King, Victor T. and Knudsen, Magne. 2021. "The Iban of Temburong: Migration, adaptation and identity in Brunei Darussalam." Universiti Brunei Darussalam, Institute of Asian Studies, Working Paper No. 65.

Knudsen, Magne. 2022. "Changing tides: Temporal dimensions of low-cost, high-skill fisheries in the Central Visayas, Philippines." In *Case studies in Biocultural Diversity from Southeast Asia: Traditional Ecological Calendars, Folk Medicine and Folk Names*, edited by M. Franco, M. Knudsen and N. H. Hassan: 21–42. Springer.

Langub, Jayl, and Noboru Ishikawa. 2016. "Community, river and basin: Watersheds in northern Sarawak as social linkage." In *Borneo studies in history, society and culture*, edited by Victor. T. King, V.T., Ibrahim Zawawi, and Noor Hasharina Hassan Springer.

Leach, Edmund R. 1950. *Social science research in Sarawak: A report on the possibilities of a social economic survey of Sarawak presented to the Colonial Social Science Research Council.* His Majesty's Stationery Office.

Lévi-Strauss, Claude. 1963. *Totemism.* Beacon Press.

Lien, Marianne E. and Gísli Pálsson. 2021. "Ethnography beyond the human: The 'other-than-human' in ethnography." *Ethnos* 86 (1): 1–20.

Maxwell, Allen R. 1980. "Urang Darat: An Ethnographic Study of the Kadayan of Labu Valley, Brunei." PhD-diss., Yale University.

Muchlisin, Zainal A., Firman M. Nur, Siti Maulida, Luvi S. Handayani, and Sri Riska Rahayu. 2022. "Masheer, the history of the king of the river." E3S Web Conference 339, Article 03006, 10.1051e3sconf/202233903006.

Pálsson, Gísli. 1994. Enskilment at sea. *Man, New Series* 29 (4): 901–27.

Paredes, Oona. 2016. Rivers of memory and oceans of difference in the Lumad world of Mindanao. *TRaNS: Trans-Regional and -National Studies of Southeast Asia* 4 (2): 329–49.

Randall, Robert A. 1977. "Change and variation in Samal fishing: Making plans to 'make a living' in the Southern Philippines." PhD-diss., University of California, Berkeley.

Rousseau, Jerome. 1989. *Central Borneo: Ethnic Identity and Social Life in a Stratified Society.* Clarendon Press.

Sercombe, Peter G. 1999. "Adjacent cross-border Iban communities: A comparison with reference to language." *Bijdragen tot de Taal-, Land- en Volkenkunde* 155(4): 596–616.

Shirley, Rebekah G. and Jettie Word. 2018. "Rights, rivers and renewables: Lessons from hydropower conflict in Borneo on the role of cultural politics in energy planning for Small Island Developing States." *Utilities Policy* 55: 189–99.

Sovacool, Benjamin K., and L. C. Bulan. 2011. "Behind an ambitious megaproject in Asia: The history and Implications of the Bakun Hydroelectric Dam in Borneo." *Energy Policy* 39 (9): 4842–59.

Sutlive, Vinson H. 1992. *The Iban of Sarawak.* S.A. Majeed & Co.

Temburong District Office. 2011. Ethnic Composition of Temburong's Population 2010. Accessed March 10, 2021. http://www.information.gov.bn/PublishingImages/SitePages/Publication%20Lists/TemburungDistrict.pdf.

Wadley, Reed L., and Ole Mertz. 2005. "Pepper in a time of crisis: Smallholder buffering strategies in Sarawak, Malaysia and West Kalimantan, Indonesia." *Agricultural Systems* 85 (3): 289–305.

Wagner, John R. 2018. "Rivers of memory and forgetting." In *Island rivers: Freshwater and Place in Oceania,* edited by John R. Wagner and Jerry K. Jacka, J. K. ANU Press.

Wagner, John R. and Jerry K. Jacka. 2018. *Island Rivers: Freshwater and Place in Oceania.* ANU Press.

Wagner, John R., Jerry K. Jacka, Edvard Hviding, Alexander Mawyer, and Marama Muru-Lanning. 2018. "Introduction: River as Ethnographic Subject." In *Island rivers: Freshwater and place in Oceania*, edited by J. R. Wagner and J. K. Jacka. ANU-Press.
White, Thomas and Matei Candea. 2023 [2018]. "Animals." In *The Open Encyclopedia of Anthropology*, edited by Felix Stein. Facsimile of the first edition in *The Cambridge Encyclopedia of Anthropology*. Online: http://doi.org/10.29164/18animals.

CHAPTER 12

Ethno-Commercial Networks in Northwest Borneo: a View from the River

Md Mamunur Rashid

The economic vibrancy of colonial northwest Borneo—comprising present-day Brunei, Sarawak, and Sabah—was largely driven by the maritime trade networks of the South China Sea and its extensive river systems. For centuries, the South China Sea emerged as a busy trade corridor, fostering the growth of numerous ports along northwest Borneo's coastline. While historians have focused on the transoceanic networks and mobilities in understanding the dynamics of trade and commerce, what remains relatively understudied is the contributions of the rivers that connected the remote interior to the maritime ports. Archival evidence indicates that these rivers were vital to the region's economy, establishing trade links between upriver zones and sea routes. During the colonial period, rivers became essential arteries for socio-economic development, facilitating trade activities and transportation while shaping migration patterns, settlement dynamics, and environmental changes. By influencing the region's socio-economic trajectories, the rivers served as vital networks, akin to a trading belt that integrated interior marketplaces with major ports and urban centres. Forest products, including sago, timber, antimony, manganese, and many other trading goods were almost entirely transported via river, largely due to the absence of roads and railways. The increasing demand for global commodities prompted British colonial rulers to plant tobacco, rubber, and other cash crops along riverbanks and near the newly constructed railway lines, enabling both river and rail networks to support internal and export trade effectively.

While this connection between the river network and commercial buoyancy in northwestern Borneo is evident, this chapter contends that a more nuanced understanding of this economic dynamic can be achieved by examining the role of riverine ethnic groups who created a complex network between upriver inland ecology and the coastal trading centres. Some studies investigate how rivers function as ethnic networks shaped by the social and commercial movements of these groups. These studies also show that ethnic groups heavily use this network to funnel their trade and commerce from remote interiors

 | HTTPS://DOI.ORG/10.1163/9789004758100-014

to principal commercial spaces, managing their cultivation on the riverbanks, and serving their communication through this network (De la Torre 2018; Yü and Dean 2021; Lenggenhager et al. 2023).

The ethnographic study of rivers has recently been popular among academics, particularly anthropologists and river historians (Driessen 2016). The insights derived from these studies could be effectively applied to the river systems of northwest Borneo. Another promising avenue for research is the commercial ethnohistory of these rivers, which remains underexplored in the context of northwest Borneo. Since the latter half of the twentieth century, ethnohistory has been extensively examined by anthropologists and historians, particularly in relation to ethnic groups such as the Red Indians (DeMallie 2013; Strong 2015; Carlson, Lutz, and Schaepe 2018). Some American environmental historians have developed an approach to ethnohistory that emphasizes the changing modes of production, social relationships, and interactions with the environment. Their work has focused on the impacts of the fur trade, agriculture, mining, and other European intrusions on indigenous ecologies and economies, exploring the dependencies and innovations generated by these disruptions (White 1988; Strong 2015).

In colonial northwest Borneo, three major ethnic groups—the indigenous peoples, Malays, and Chinese—were integral to the river-based commercial network. Unlike in precolonial times, many indigenous groups shifted from exclusively dealing in natural products to becoming smallholding cultivators of newly introduced cash crops. This chapter argues that these ethnic communities played a vital role in both commodity production and mobility, with rivers serving as central conduits in this economic landscape.

1 Geography of River Networks and Settlements

The ecology, physical geography, and natural resources of northwest Borneo play a pivotal role in shaping its intricate water networks. Sabah's diverse topography is characterized by mountain ranges and extensive river systems. The river system includes three main groups. The first group consists of the long east coast rivers that run from the interior into the Sulu Sea. Among the most significant are the Kinabatangan, Segama, Sugut, Paitan, and Labuk rivers, along with their tributaries. The second group includes the short west coast rivers that flow into the South China Sea. Notable examples are the Papar river and the Padas river, both of which are navigable by small vessels. Finally, the third group consists of the short rivulets of the southeast coast that meet the

FIGURE 12.1 Major rivers in Sabah
SOURCE: KAUR 1994, 3

Celebes Sea. The major river in this category is the Kalabakang, which spans 560 kilometers from its origin in the mountains of the southwest to its estuary in the Sulu Sea, east of Sandakan (bin Awang Ali Bahar 2007).

The landscape of Sarawak can be broadly categorized into three main regions: (i) coastal plains and narrow deltaic and alluvial plains; (ii) an expansive hilly inland area; and (iii) the mountain highlands (Kaur 1995, 68). The region is traversed by two major river systems and their numerous tributaries. The larger rivers run from the mountain ranges along the Sarawak–Kalimantan border, while the shorter but mature coastal rivers flow through the mid-Sarawak plains (Kaur 1995, 68). The Sarawak river system is dominated by the Rajang river, the longest river in Sarawak. It begins in the Iran Mountain range and flows through Kapit, where it turns westward before completing its 760-kilometre journey to the South China Sea (bin Awang Ali Bahar 2007). Other significant rivers, including the Sarawak, Batang Lupar, Baram, and Limbang, connect the Rajang river system (Kaur 1995, 1998).

Brunei occupies a small section of northwest Borneo, featuring a diverse landscape that transitions from a flat coastal plain to mountainous terrain in the east and hilly, undulating lowlands in the west. The majority of Brunei is

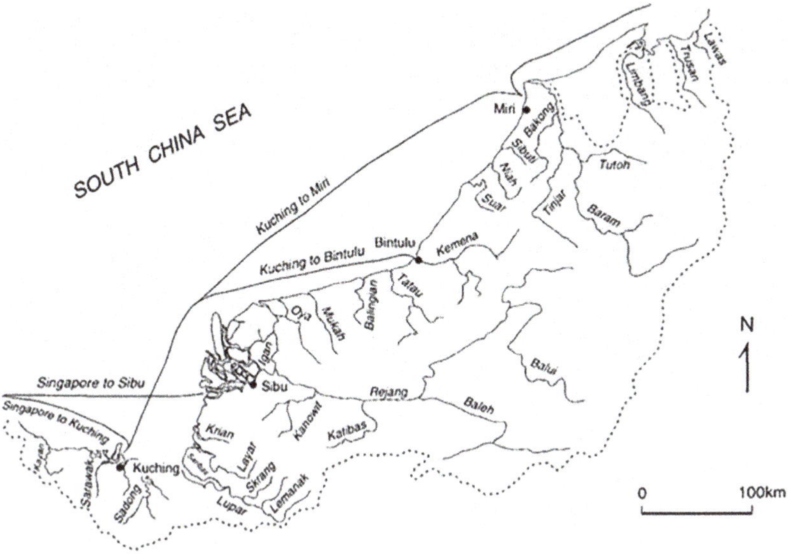

FIGURE 12.2 Sarawak major rivers and sea routes
SOURCE: KAUR 1993, 85

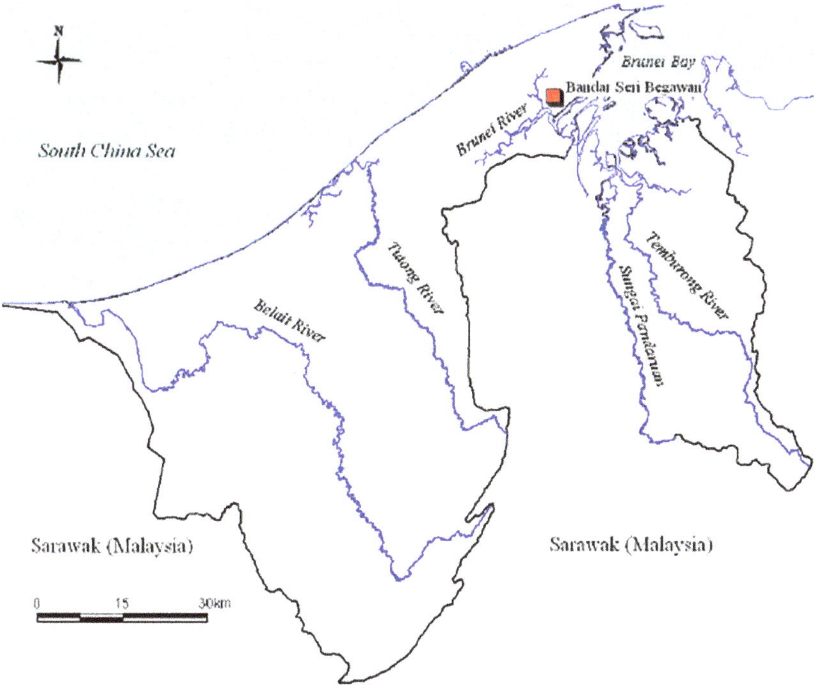

FIGURE 12.3 Major rivers of Brunei
SOURCE: BAC-BRONOWICZ, JOANNA AND BECEK, KAZIMIERZ. ENVIRONMENTALLY BASED PERCEPTION OF SPACE

covered by Borneo lowland rainforests, and 161 kilometres of coastline along the South China Sea. The region's river system is composed of four major rivers: the Belait river, Tutong river, Temburong river, and Brunei river (Bac-Bronowicz and Becek 2009).

The physical environment of northwest Borneo gave rise to three distinct ecological zones that served as key population settlements during the colonial period. The first zone, encompassing the coastal areas and estuaries, was the primary hub for habitation and commercial activities. This zone provided space for settlement, cultivation, and trading centres while facilitating a spatial connection between upstream and downstream areas. The second zone emerged along the riverbanks, where indigenous communities established their settlements. This zone also provided trading centres and farming activities. For this zone, the river network facilitated the commercial services of the adjoining forest for its inhabitants. Finally, the upland zones comprised forests and hilly regions. These areas were predominantly inhabited by nomadic and isolated ethnic groups who relied on a hunter–gatherer economy and practiced swidden cultivation (Kaur 1998, 13–14).

2 The Connectivity between the River and Ethnic Groups

From ancient times, rivers served as vital conduits of spatial connectivity, enabling trade between the remote upriver regions and the coastal or river estuary zones of northwest Borneo. As previously mentioned, three ecological zones of northwest Borneo constituted a crucial cultural division among various groups, influencing their habitation patterns, dialect, and religion (Kaur 1998, 15; Harrison 1949, 133). Ethnic groups in the remote upriver regions often formed trade relations with deltaic inhabitants to supply forest-based resources. On the other hand, downriver social groups provided household commodities and food to the upriver indigenous groups in exchange for jungle products (Burbidge 1989; Evans 1922; Cleary 1996a).

The ethnic groups of northwest Borneo were categorized based on their habitat into two main groups: those residing in the coastal areas or river estuaries, such as the Malays and Melanau, and those inhabiting the interior regions, collectively referred to as the Dayak (Kaur 1993, 97; Kaur 1995, 66). The Malays, being more exposed to external influences, predominantly settled along the coast and at the mouths of major rivers. This group included the Brunais, Melanau, Sulus, and Bajaus (Treacher 1891, 140). The indigenous groups were often named after the rivers that ran through their territories, such as the Kayan people, who were named after the Kayan river (Baring-Gould and Bampfylde 1909, 34). The term "Dayak" refers to determining the original peoples of Borneo. It originated from the Kenyan word "daya," which means inland or interior in the Melanau and Brunei Malay dialects (Baring-Gould and Bampfylde 1909, 33). The Dayak people comprised numerous subgroups, including the Bidayuh (Land Dayak), Iban (Sea Dayak), Kenyah, Kayan, Kedayan, Murut, Penan, Bisayah, Kelabit, and many others (Cleary 1996b; Kaur 1993, 97, 1998, 16). Another significant ethnic group in northwest Borneo was the Dusun, who are now known as the Kadazan and primarily reside in Sabah (Kaur 1998, 17).

When James Brooke expanded his control over Sarawak, he established small forts and administrative settlements along the riverbanks. He also divided Sarawak into three divisions based on various rivers for administrative purposes (Kaur 1993, 80–81, 1995, 70).[1] Similarly, in Sabah, the North Borneo Charter Company developed its business and administrative settlements along the riverbanks and coastal regions, leveraging the connectivity provided by numerous rivers (Kaur 1994). In Brunei, the Sultanate and early Malay settlements were primarily established along the Brunei river. During the British residential era, administrative and residential zones continued to develop along the riverbanks (Horton 1985; Cleary and Wong 1994, 11). Throughout the colonial period, rivers in northwest Borneo served as vital channels for communication and played a central role in resource extraction.

In the pre-colonial period, three social groups were mainly involved in commercial activities. The first group comprised indigenous communities,[2] such as the Kadyans, the Dayaks,[3] the Dusun, the Murut, the Punan, and others, who primarily acted as collectors of forest produce. They cultivated staple foods, such as wet rice and sago. The second group, the Malay people, resided along the coastal areas and strategic river estuaries, where they controlled the major rivers and their tributaries. They established trade relationships with indigenous communities in the interior, exchanging ecological products while exerting political influence through taxes, tolls, and tribute payments (Hornaday 1885, 344; Mundy 1848).

1 The First Division extended from Tanjung Datu to the Sadong River; the Second Division, from the Sadong River to the Rejang River; and the Third Division started from the Rejang River to Tanjung Kendurong.

2 Modern ethnologists divide the indigenous people of northwest Borneo into more than forty separate identities. For more detail, see Irwin, Graham. 1955. *Nineteenth-Century Borneo: A Study in Diplomatic Rivalry*. Brill.

3 The Dayaks are usually divided into "Land-Dayaks" and the "Sea-Dayaks" or "Iban", but during the nineteenth century, it was often applied to other ethnic groups as well.

The third group, the Chinese, joined these commercial spaces in the twelfth century and traded forest products with the Malay and indigenous people. Over time, they settled here and ventured upriver to interior villages, directly sourcing forest products from indigenous communities (Kaur 1993, 1998, 57; Gin 1995). During the colonial period, Europeans joined this commercial dynamic. Leveraging modern technology, they soon became dominant, consolidating both economic and political power in the region. Despite these changes, the role of indigenous people remained the same (Evans 1915, 24; Gin 1995; Cleary 1996a).

3 Trade and Commerce on the River Space

The significance of northwest Borneo, particularly Brunei, in the sixteenth century was primarily determined by the precious ecological products that contributed to the development of the region (Cleary 1996a, 305; Nicholl 1983; Hall 2011). The advent of colonial administration and the emergence of a modern capitalist economy ushered in a transformative shift in northwest Borneo's economic landscape. This period saw a substantial increase in the trade of forest products, facilitated by river networks. Colonial northwest Borneo also witnessed the rise of cash crop plantations established in river valleys. The rivers served as main arteries for transporting agricultural goods. Tobacco and rubber estates predominantly lined the riverbanks, while the establishment of numerous river and coastal ports fuelled a thriving export trade of these commodities. Cash crop production gradually became more significant as the colonial economy flourished profoundly in this region (Cleary 1996a, 1997; Kaur 1994, 1998). The involvement of ethnic groups in these broader economic changes further underscored their impact.

3.1 *Non-Malay Ethnic Groups*

From earlier times, the trade and commerce of northwest Borneo were dominated by locally sourced forest products, such as camphor, bird nests, rhinoceros' horns, feathers of birds of paradise, bezoar stones, hornbill ivory, rattans, benzoin, gutta-percha, aloe woods, rattan, dammar, firewood, beeswax, cutch, and nipah (Evan 1922, 36; Nicholl 1980, 1983; Cleary 1996a, 1997; Hall 2011). Ethnic groups ventured deep into the forests to collect these resources, exchanging them for beads, iron, cash, para crockery, pottery, or Ali Baba jars (Smythies 1961, 168). The traditional ethnic collector carried these goods via river networks to the local markets, where Chinese and Malay traders purchased these commodities and sold them to the regional and international market (Burbidge 1989, 68; Cleary 1997). The Dayaks, known for their forest-dwelling lifestyle, possessed exceptional skills for gathering wild rubber and other forest products Similarly, the Dusun, Murut, and other ethnic groups specialized in collecting forest resources and delivering them directly to local markets (Evans 1922, 42–43). Spenser St John (1862, 45) noted the Punan people's proficiency in gathering wax, edible bird nests, camphor, and rattans. Baring-Gould and Bampfylde (1909, 376) highlighted the reliance of the Ibans on forest products, which they sourced from the depths of the forest zones.

From the late nineteenth century onward, tobacco was widely planted as a "local crop" in northwest Borneo, reflecting the widespread smoking tradition among various ethnic groups. In Sabah, the Dusun (Kadazan) and Kwijaus primarily cultivated tobacco, transporting it via river routes to coastal areas where it was traded with Chinese merchants (Rutter 1922, 153). The local Bajau purchased this tobacco in exchange for goods such as salt, salted fish, iron goods, and cloth. During this period, tobacco cultivation among indigenous groups expanded notably when the British initiated experimental farming of Havana tobacco on their agricultural site near Darvel Bay (John 1973, 89–91). Earlier, tobacco had been grown in Sarawak by Dayaks for their consumption. In 1871, the Dayaks of the upper Rejang, near the government station at Balleh, successfully cultivated Java tobacco (Gin 1995, 171). However, tobacco plantations in Sarawak were not commercially successful, leaving tobacco as a small-scale crop limited to indigenous communities.

In Sarawak, ethnic groups played a crucial role in developing rubber cultivation. By 1908, rubber seeds were distributed to various locations and provided to indigenous smallholders. Over time, rubber cultivation among these smallholders experienced significant growth. By 1908, rubber seeds were sent to various places and distributed among indigenous smallholders. The coastal Melanau were among the first to transition to rubber farming, followed closely by the Dayak communities. The Ibans maintained more than half of the rubber smallholdings in Sarawak. By 1938, all ethnic smallholders collectively owned 76,940 rubber smallholdings, totalling 101,000 acres (Tremeer 1964; Kaur 1995). In Sabah, the Europeans owned most of the rubber industry. Gradually, indigenous farmers entered the sector as smallholders, employing family labour or engaging in share-cropping activities. Many of them often became short-term casual workers on other rubber estates (Kaur 1998, 42–43). In Brunei, indigenous smallholders cultivated rubber alongside European planters (Cleary and Wong 1994, 21).

3.2 *The Malays*

In the pre-colonial period, the Malay chieftaincies (Sabah) or Brunei nobles (Sarawak) along the Borneo coast practised their power primarily through their control of foreign trade, the imposition of tolls on river traffic, and the collection of tributes in forest products from dependent interior ethnic groups within their domains (Hornaday 1885; Mundy 1848; Kaur 1994, 1998). Malay traders or *nakhodahs*[4] navigated river valleys to gather goods from local markets operated by various ethnic groups along the riverbanks through barter trade (Gin 1995, 1997). Initially, these trade networks were managed solely by Malay traders, but later, Chinese traders became integral to this commercial chain. In addition to trading local products, their chief activities were wet rice cultivation, fishing, and commercial agriculture (Kaur 1994; Gin 1995; Cleary 1996a).

From the late nineteenth century, the supremacy of the Malay people in upriver trading networks began to diminish with the growing presence of Chinese traders (Evans 1915, 24; Cleary 1996a; Gin 1995). Consequently, the role of Malay traders changed into boat or itinerant traders. The colonial government's incorporation of the Malay elite into administrative roles further discouraged Malay nobles from participating in trade. However, the ordinary Malay individuals and ethnic groups remained active, cultivating staple grains and cash crops while operating small-scale river-based boat trades (Gin 1995, 85; Cleary 1996a). Among the Malay ethnic groups, the Melanau mainly cultivated sago in the upriver field, while the Malay traders supplied the sago to the processing factories. Kuching became a processing centre for sago processing, exporting the final product to Singapore and subsequently to the global market (Gin 1995, 82, 1997). In the timber sector, the Malays worked as labourers, focusing on wood cutting and log transportation along the rivers. When the North Borneo Trading Company established its business with reduced overhead costs, it relied on cheap workers from the Malay communities, employing 567 Malay workers out of a total workforce of 1,491 by 1919 (John 1974, 74–79).

3.3 *The Chinese*

Chinese traders in Borneo operated as independent agents, channelling the island's trade to South China (Tagliacozzo 2013, 70). By the 1870s, they had established clear supremacy and nearly monopolized internal trade. Initially, the Chinese dominated most of the country's coastal shipping. Chinese entrepreneurs and traders gradually ousted Malay traders from the river-based commercial network by making more reliable relationships with Europeans and

---

4 Malay traders were known as *nakhodahs*. It locally means "people who sail to the right". See The Sarawak Gazette, 31 July 1964, 165 for details.

indigenous groups (Evans 1915, 24; Gin 1995, 85, 1997). To further strengthen their ties with indigenous communities, Chinese traders often married Dayak women (Tagliacozzo 2013, 72). In the supply chain of forest products, the Chinese played a crucial role as intermediaries between indigenous groups and European traders. Chinese participation in the traditional trade was vital and persistent in Sarawak. Export statistics from 1902 highlight the dominance of Chinese traders in handling biota trade, such as gutta-percha and rattan. For example, in 1901, gutta-percha and rattan worth $65,000 were primarily collected by the Chinese and their agents from the Baram district for export (Tagliacozzo 2013, 73).

By 1886, the timber trade in Sandakan comprised five firms, including the Chinese companies Koh Heen Co. and Chop "Kong Watt" (John 1974, 74). In 1890, the China Borneo Company established many steam sawmills in Sandakan Bay to support growing export demands (Cook 1890, 69). By 1938, 14 sawmills owned by Chinese entrepreneurs had been established along the Rejang river, marking the community's near monopoly of the timber industry at the time (Gin 1995, 258, 1997; Tagliacozzo 2013). Chinese labour was integral to the industry, with workers primarily tasked with logging and hauling operations. When the North Borneo Trading Company started its trade in timber, the Chinese were engaged as contractors and low-cost labour. By 1919, 798 Chinese workers were employed in this industry (John 1974).

In Sabah, the Chinese Sabah Land Farming Company initially recruited Chinese and indigenous labourers into its workforce. Generally, the indigenous labourers were responsible for clearing and preparing fields for cultivation, while Chinese workers focused on planting. By the 1920s, rubber plantation acreage expanded significantly as more Chinese cultivators joined the industry in Sabah. In 1928, the total rubber cultivation area reached 27,114 ha, of which the Chinese and other smallholders planted 11,736 ha. At the end of 1928, Chinese planters had established ownership of several large estates in Sabah. Additionally, many Chinese joined many large rubber estates as labourers (Kaur 1994, 1998). In Sarawak, commercial tobacco plantations saw success during the 1930s. However, local demand for tobacco provided little employment and small profits to the Chinese cultivators throughout this decade (Gin 1995, 172, 1997). In 1908, rubber seeds were distributed to various locations and allocated to Chinese and indigenous smallholders, leading to significant growth in rubber cultivation among Chinese smallholders in Sarawak (Kaur 1995, 86). Aside from rubber, the Chinese planted pepper and gambier in west Sarawak. In the 1890s, gambier reached high prices in the world market, prompting the Borneo Company to develop a 20,000-acre concession known as the Poak concession (Kaur 1998, 46). The Chinese played a vital role in the sago trade. Following a rapid increase in sago cultivation after 1861, Chinese entrepreneurs established processing mills in several sago-producing areas (Gin 1995, 82, 1997). In Brunei, the Chinese cultivated pepper and worked as labourers on rubber estates (Horton 1985, 12–13).

In the mining sector of northwest Borneo, Chinese labour instrumental in the extraction of various minerals, including gold, precious stones, and coal. From the early nineteenth century, Chinese miners were actively engaged in gold mining within Sarawak. These mining sites were concentrated west of the Batang Lupar river, which was the most densely populated and developed part of Sarawak (Tagliacozzo 2013; Kaur 1998, 21–26). Many Chinese labourers were also employed in the Sadong and Brooketon Collieries owned by the government. By the 1920s, the Sarawak and Brunei Oil Company employed a substantial number of Chinese workers in its operations (Kaur 1995; Horton 1985).

4 Indigenous Livelihood

River-centric commercial networks played a pivotal role in shaping indigenous livelihoods in Northwest Borneo. Three primary ecosystems directly influenced the way of life for indigenous communities: first, harvesting sea products and fishing from the shoals and reefs along the coastal regions; second, cultivating wet rice in the fertile lowlands; and third, practicing swidden cultivation of dry or hill paddy, maize, and root crops in the interior uplands. These ecosystems—the coastal and estuarine belt, river valleys, and interior highlands—not only defined the region's broader ecological framework but also contributed to variations in habitat, language, and belief systems among indigenous groups (Kaur 1998). Initially, the indigenous livelihood practices of Northwest Borneo were centred around hunting, fishing, gathering, and swidden cultivation. These communities adopted diverse subsistence strategies, blending multiple approaches rather than relying on a single method (Ishikawa and Soda 2020). During the colonial era, livelihood patterns evolved with the introduction of cash crop cultivation, wage labour, and other economic shifts. Tobacco and timber emerged as significant commodities early in the colonial period, facilitating the expansion of cultivated areas due to the presence of extensive river systems and abundant natural resources (Kaur 1994).

The indigenous communities in Northwest Borneo showed remarkable adaptation to diverse environmental conditions, shaping their livelihoods and societies accordingly. Along the swampy coasts, the Melanau community sustained their livelihood through fishing and collecting swamp sago. In the interior, swidden cultivation of staple crops, along with forest sago and later hill paddy adoption, was predominant. Hill paddy cultivation was particularly common among the Murut and other upland communities. In the steep upland forest region, the indigenous communities followed their way of life based on shifting cultivation supplemented by hunting and gathering. Meanwhile, in the suitable areas for cultivation, permanent agriculture, including wet rice, flourished along the riverbanks. For example, the Malays settled along the coastal areas and played a significant role in developing wet rice cultivation along the lower reaches of the major rivers. In other areas, people like the Penan avoided agriculture, adopting a nomadic lifestyle based solely on hunting and gathering. Although largely independent and self-sufficient, these groups were not entirely isolated. Trade, war, and occasional intermarriage interconnected the various ethnic groups throughout the territory (Kaur 1995).

5 Ethnic Interaction with the Colonial Economy

The expansion of timber and forest products' trade and global crops' cultivation influenced the colonial ruler to stimulate the further exploration of the forests and prepare more lands on the riverbanks for the plantation of tobacco, rubber, sago, pepper, and other cash crops in northwest Borneo. Furthermore, the development of mining sectors from the late nineteenth century also played a pivotal role in bolstering the colonial economy. The ethnic collaboration with the colonial economy was evident in this context. It was also a pressing necessity for the colonial rulers to use the ethnic groups in the commercial sectors (see Table 12.1).

TABLE 12.1 Employment by race in Sarawak, 1935 (Number)

Employment sector	Chinese	Javanese	Indian	Malay & Melanau	Dayak, Kayan & Kenyah	Children	Total
Rubber Estate	423	786	22	769	400	15	2415
Gold Mining	2099	2	29	450	248	61	2589
Sago Factories	183		1				184
Jelutong Factories	26		2		16		44
Sawmills	53		1	281	82	23	440
Cutch Factory	24		2	70	81	5	182
Oilfield	1691		109	895	197	86	2978
Total	4499	788	166	2165	1024	190	8832

SOURCE: KAUR, 1998, 101

TABLE 12.2 Employment by Race in Sabah, 1952

Sector	No of concern	Chinese	Javanese	Indigenous people	Others	Total
Timber	49	1048	238	2386	152	3824
Rubber	59	1339	1260	6022	83	8704
Tobacco	1	243	155	214	401	1013
Building Construction	23	1060	30	223	7	1320
Public Works	N. A.	287	18	1149	16	1470

SOURCE: AN ECONOMIC SURVEY 1955, 124

Before the colonial period, forest product trade was funnelled by the indigenous people, Malay, and Chinese traders. Initially, groups such as the Kadyans, the Dayaks, the Dusun, the Murut, the Punan, and other collected forest products from the remote interior; then secondary traders sent such commodities downriver to the coastal ports. Malay or Chinese traders sold them to regional and international markets (Cleary 1997). The dynamics of this trade shifted dramatically with the growing European interest in forest products. Using the previous trading network, the colonial rulers soon carried these products into an extensive global market. Consequently, forest products remained a significant part of the export trade and revenues. In both Sarawak and Sabah, the exports of forest products more than doubled between the 1880s and 1930 (Cleary 1997).

TABLE 12.3 Export of forest produces in Sabah, 1938–53 (In Tonnes)

Year	Cutch	Damar	Firewood	Birds' nests
1938	2458	1684	39000	19.1
1938	4066	1751	65000	18.2
1948	2252	885	34202	13.4
1949	3815	639	74757	17.5
1950	4120	1936	67868	22.4
1951	4577	1798	58741	15
1952	4025	752	92190	21
1953	4941	829	54700 (est.)	17

SOURCE: AN ECONOMIC SURVEY 1955, 122

TABLE 12.4 Exports of Forest products in Sarawak, 1870–1960

Year	Gutta percha		Damar		Rattans		Cutch	
	Tonnes	£	Tonnes	£	Tonnes	£	Tonnes	£
1870	—	96470	—	210	—	90	—	—
1900	476	91970	298	1950	2458	3267	—	—
1910	101	21780	125	1720	1238	1083	—	—
1920	24	16320	679	42490	1054	1208	3640	56300
1930	1	110	470	8200	446	3100	2520	55840
1940	12	3770	1292	3700	190	1420	2510	55510
1950	42	25040	1506	58500	988	24220	1770	134550
1960	83	17150	2202	86780	—	—	120	6530

SOURCE: SARAWAK GAZETTE, SEPTEMBER 30, 1961, 174

In the last quarter of the nineteenth century, northwest Borneo became a significant source of global timber trade. During the colonial period, the Chinese, the Malays, the Dayaks, and others were actively employed in this sector. In 1883, the China Sabah Land Farming Company began to export timber (Tregonning 1965, 82). Gradually, many companies established steam sawmills in Sandakan Bay and exported rough and sawn timber to destinations such as China, Singapore, and England (Cook 1890). In 1885, Alex Cook, the Government officer of Sandakan, mentioned the role of the Chinese and Malays in timber industries: "The wood is cut and squared if necessary by Malays and Chinese, principally the former, and hauled by buffaloes or manual means to the river, where it is rafted to Sandakan Bay, into which run some seventeen timber-supplying rivers" (BNBCCo. Handbook, 68–75). As the twentieth century began, logging areas expanded around the ports of Sandakan and Tawau (John 1974). By 1930, timber trade volumes had grown significantly, reaching 3,525,452 cubic feet (Tregonning 1965, 83–84).

In Sarawak, Borneo Company Limited began timber exploitation at Rajang in 1886. The main concern was logs or rough-hewn logs, but the project proved unsuccessful due to teredo infection (Smythies 1961, 169). Until the First World War, the timber trade was trivial in Sarawak. In 1936, Borneo Company Limited turned its attention to the forests of the upper Rejang above Kapit. Much capital was invested in building the sawmill and installing an enormous boiler and new machinery (Smythies 1961, 170). The Chinese entrepreneurs and workers were involved mainly in the timber industry in Sarawak.

TABLE 12.5 Sabah timber exports, 1938–53

Year	Quantity ("000 cubic feet)	Value ($'000)
1938	4663	2177
1938	5420	2436
1948	2458	3018
1949	2553	5444
1950	3751	6534
1951	4427	10184
1952	3476	8177
1953	5697	12198

SOURCE: AN ECONOMIC SURVEY 1955, 122

TABLE 12.6 Sarawak timber exports, 1930–60

Year	Quantity in Hoppus tonnes	Value £
1930	?	12170
1940	5000	4960
1950	64400	348440
1960	527633	5087510

SOURCE: THE SARAWAK GAZETTE SEPTEMBER 30, 1961, 174

In the early twentieth century, massive forest logging resulted in immense river pollution in northwest Borneo. Forest logging led to empty trees on riverbanks, causing soil erosion and landslides that ultimately disrupted the river course. For instance, in 1903, Mr. Birch, the Governor of Sabah, mentioned that riverbanks were being exposed through massive deforestation and that government control was necessary. The Commissioner of Lands and Controller of Forests recommended that professional forest officers be employed to protect against unplanned logging in Sabah in 1909 (Tregonning 1965, 83).

By the last quarter of the nineteenth century, tobacco plantations began in northwest Borneo. The tobacco farms were strategically scattered and located in those areas where the positions and topographic conditions were suitable for cultivation, while rivers served as vital arteries for transportation (Rutter 1922; John and Jackson 1973). Indigenous groups like Dayaks and Chinese labourers mostly worked in this industry. In 1883, the Chinese Sabah Land

Farming Company pioneered the cultivation of Sumatran tobacco seeds with a trial plantation along the Suanlamba river in Sandakan Bay. The first sample bales were exported to London in February 1884 and were acknowledged by the experts as the best among others (Cook 1890; John and Jackson 1973; Tate 1996). Tobacco estates expanded rapidly, and by 1890, 61 tobacco estates had been established in Sabah, with huge areas along the East Coast rivers (John and Jackson 1973; Cleary 1992; Tate 1996). However, the tobacco industry began to decline with the onset of the rubber boom in the early twentieth century. By 1929, only one estate remained operational—the Darvel Bay Tobacco Plantation (John and Jackson 1973).

TABLE 12.7 Output of estate tobacco in Sabah, 1914–29

Year	Output (lbs.)	Year	Output (lbs.)
1914	2442738	1922	1208133
1915	1799880	1923	1250598
1916	1332424	1924	1343705
1917	1486150	1925	1428332
1918	1520130	1926	1134050
1919	1857380	1927	1202533
1920	1632960	1928	833200
1921	1157390	1929	641333

SOURCE: JOHN AND JACKSON 1973, 105

In Sarawak, Charles Brooke initiated the government tobacco estate at Lundu, which began cultivation in late 1887. However, the tobacco concern soon made huge losses as the trial planting of tobacco at Lundu faced a total deficit of $60,000 (Gin 1995, 193, 1997). As a result, commercial tobacco plantations achieved little success, and tobacco remained only as a smallholding indigenous crop in Sarawak (Gin 1995, 171–172, 1997).

The rubber plantation industry in northwest Borneo reached its peak during the economic boom of the 1920s. Rubber plantations in Sarawak and Brunei mainly depended on traditional river networks. Most rubber estates in both regions were strategically established along riverbanks or near rivers and ports to facilitate operations. The cultivation of rubber in Sarawak integrated the region into the global economy. In 1902, Ernest Hose first initiated experimental planting of rubber and established the Dahan Rubber Estate, spanning 2,500 acres (Tremeer 1964). By 1917, rubber cultivation had expanded significantly, with 2,136 hectares of land under production dominating the agricultural landscape (Tate 1996). Charles Brooke promoted the smallholding cultivation of rubber. However, the Chinese and indigenous smallholders held the lion's share of about 90 per cent of the area and production (Kaur 1995, 1998; Tate 1996).

Rubber plants were first imported to Sabah from the Singapore Botanical Gardens in 1882. However, commercial planting did not commence until 1892, when Mortgage Investment and Contract Corporation Ltd. began operations along the Labuk river (Rutter 1922, 45). The booming automobile industry, particularly in the United States, intensified the massive demand for rubber (Rutter 1922; Evan 1922; Tregonning 1965). By the end of 1928, thirty companies with four million pounds sterling capital were involved in the rubber industry. The Chinese, the Javanese, the Dusuns, the Bajaus, and other indigenous groups were deployed as labourers in this sector (An Economic Survey 1955).

TABLE 12.8 Rubber exports from Sabah, 1910–1931

Year	Tonnes	Value (£)
1910	24	34315
1915	1050	274422
1917	2444	635288
1919	3939	782037
1921	3121	289634
1923	4239	560770
1925	5425	1397562
1929	7381	590115
1931	6247	159932

SOURCE: CLEARY 1992, 176

In Brunei, the British residential administration initially sought to attract foreign capital for investment in a new plantation system, particularly in the rubber industry (Cleary 1996). Consequently, five British-owned estates began plantations during the residential era: three in Temburong district at Labu, Batu Apoi, and Biog river areas and two in Brunei district at Gadong and Kumbang Pasang (Horton 1985, 204). Additionally, the Chinese and indigenous smallholders widely cultivated rubber alongside their staple crops (Cleary and Wong 1994, 21–24).

Rivers served as vital channels of commercial connectivity for the mining industries in colonial northwest Borneo. Most mines were strategically located along riverbanks or near rivers, facilitating the transport of extracted minerals by river freight to coastal ports. Massive mining exploitation began to dot the northwest Bornean landscape in the late nineteenth century via European capital. Key minerals such as antimony, coal, and oil played a central role in generating significant prosperity for the colonial economy (Kaur 1995; Cleary 1996). The Chinese labourers were mostly used in the mining sector of northwest Borneo. The Land Dayaks were often forced to work in the mine in exchange for beads and brass. The Kayans and the Kenyahs generally worked in iron mines, while other subgroups of Dayaks were involved in the gold sector (St. John 1862; Roth 1896; Kaur 1995). The ecological effect of the mining was very significant; an immense amount of land was cleared for mining operations, and the chemical run-off from operations often contaminated river water (Tagliacozzo 2013, 75).

6 Conclusion

In northwest Borneo, the colonial era witnessed extensive agricultural expansion, particularly cash crop cultivation, which transformed the riverbanks into flourishing agroecological zones. European settlers, Chinese migrants, Malays, and indigenous ethnic communities collectively contributed to this agricultural expansion, reshaping their livelihoods in a competitive quest for growth and prosperity. The long and sustained relations between ethnic groups and the diverse river ecologies of northwest Borneo exemplify what Iqbal (2021) describes as the "ethno-morphology" of rivers. These river networks fostered the convergence of local ethnic groups, Malay communities, Chinese migrants, and adventurous Europeans, enabling them to seek opportunities and fortunes in the region.

During the colonial period, rivers played a crucial, multi-functional role in driving the economy of northwest Borneo. They served as vital conduits for transportation, commercial mobility, trading centres, plantations, industrial zones, new settlements, and numerous river and coastal ports. These functions brought economic prosperity to colonial rulers, indigenous communities, Malays, and Chinese alike.

Yet what needs to be emphasized is that deforestation and other forms of commercial exploitation of forests adversely impacted the ecology of the region. Most importantly, the very river space that was the catalyst of economic growth in colonial northwest Borneo became a victim of the same. Extensive logging throughout the colonial era caused soil erosion, landslides, and widespread river pollution (Langub 2020, 132). Even in postcolonial and recent times, the hydraulic interference of the river remained more catastrophic.

Many dam construction projects for hydraulic power in Sarawak led to the deterioration of rivers—a reflection of anthropogenic impact on rivers as exemplified in various cases studies presented in this book. Recently, contamination from chemical herbicides and fertilizers used in oil palm plantations has compounded the issue (Takeuchi et al. 2020, 450).

This ongoing decline highlights the need for further research into the interconnected role of rivers in economic development, environmental sustainability, and ethnic history—areas explored within this chapter. This book (Iqbal, Hassan, and Kumpoh 2026) illustrates the duality of the Anthropocene: it reveals both the perils of ecological decline and the ability and need for human reflection on our relationship with nature. This chapter has also highlighted how the interplay between ethnic composition and commercial activities initially operated in a sustainable manner but, under certain historical shifts, contributed to the degradation of water bodies in northwest Borneo.

References

Archival Documents

An Economic Survey of the Colonial Territories (1955). Vol. v, The Far Eastern Territories: Singapore, Hong Kong, Brunei, North Borneo and Sarawak. Her Majesty's Stationery Office.

Report by A. Cook to Colonial Secretary, 14 Nov. 1885, BNBCCo. Handbook, 1886.

Sarawak Gazette, 30 September 1961.

Sarawak Gazette, 31 March 1964.

Sarawak Gazette, 31 July 1964.

Views of North Borneo with a Brief History of the Colony, 1899. London: British North Borneo Company.

Secondary Literature

Bac-Bronowicz, Joanna, and Kazimierz Becek. 2009. "Environmentally based perception of space." In *International Cartographic Conference (ICC) 2009 Proceedings*.

Baring-Gould, Sabine, and Charles Agar Bampfylde. 1909. *A History of Sarawak under its two White Rajahs, 1839–1908*. H. Sotheran & Company.

bin Awang Ali Bahar, Awang Azfar. 2007. "Frequency Analysis of River Flow in Sabah and Sarawak." Project Dissertation. Universiti Teknolgi Petronss.

Burbidge, Frederick William. 1880. *The Gardens of the Sun: A Naturalist's Journal of Borneo and the Sulu Archipelago*. John Murray.

Carlson, Keith Thor, John Sutton Lutz, and David M. Schaepe, eds. 2018. *Towards a New Ethnohistory: Community-Engaged Scholarship among the People of the River*. University of Manitoba Press.

Cleary, Mark C. 1992. "Plantation agriculture and the formulation of native land rights in British North Borneo c. 1880–1930." *Geographical Journal*: 170–81.

Cleary, Mark C. 1996a. "Indigenous trade and European economic intervention in North-West Borneo c. 1860–1930." *Modern Asian Studies* 30 (2): 301–24.

Cleary, Mark C. 1996b. Devise and Rule: Colonial Constructions of the Borneo Dayak c. 1860–1920. *Singapore Journal of Tropical Geography 17*(1): 15–23.

Cleary, Mark C. 1997. From Hornbills to Oil? Patterns of Indigenous and European Trade in Colonial Borneo. *Journal of Historical Geography 23*(1): 29–45.

Cook, Alex. 1890. "Notes on the Recent Development, Explorations and Commercial Geography of British North Borneo." *The Journal of the Manchester Geographical Society*. 7: 63–75.

DeMallie, Raymond J. 2013. *Transforming Ethnohistories: Narrative, Meaning, and Community*. University of Oklahoma Press.

De la Torre, Oscar. 2018. *The people of the river: Nature and identity in Black Amazonia, 1835–1945*. University of North Carolina Press.

Driessen, Henk. 2016. "Towards an ethnography of rivers." In *Migration, networks, skills: Anthropological perspectives on mobility and transformation*, edited by Astrid Wonneberger, Mijal Gandelsman-Trier and Hauke Dorsch. transcript Verlag.

Evans, Ivor Hugh Norman. 1915. "North Borneo Markets." *Man* 12: 23–25.

Evans, Ivor Hugh Norman. 1922. *Among Primitive Peoples in Borneo: A Description of The Lives, Habits & Customs of The Piratical Head-Hunters of North Borneo, With an Account of Interesting Objects of Prehistoric Antiquity Discovered in The Island*. Seeley Service & Co. Limited.

Gin, Ooi Keat. 1995. "An Economic History of Sarawak during the Period of Brooke Rule, 1841–1946." PhD diss., University of Hull.

Gin, Ooi Keat. 1997. *Of Free Trade and Native Interests: the Brookes and the Economic Development of Sarawak, 1841–194*. Oxford University Press.

Harrisson, Tom. 1949. "Gold and Hindu Influences in West Borneo." *Journal of the Malayan Branch of the Royal Asiatic Society* 22 (4) (150): 33–110.

Hirth, Friedrich, and W. W. Rockhill. 1911. *Chau Ju-Kua: His Work on the Chinese and Arab Trade in the Twelfth and Thirteenth Centuries*. Entitled Chu-fan-chi (Translated from the Chinese and Annotated). Printing Office of the Imperial Academy of Sciences, St. Petersburg.

Hornaday, William.1885. *Two Years in the Jungle*. K. Paul Trench.

Horton, A. V. M. 1985. "The Development of Brunei during the British Residential Era, 1906–1959: A Sultanate Regenerated." PhD Diss. University of Hull.

Iqbal, Iftekhar. 2021. "Rivers of Mobility: Multi-ethnic Societies and Ecological Commons in a Fluvial Asia." In *Yunnan–Burma–Bengal Corridor Geographies: Protean Edging of Habitats and Empires*, edited by Dan Smyer Yü and Karin Dean. Routledge.

Iqbal, Iftekhar, Hasharina Hassan, and Asiyah Kumpoh, eds. 2026. *Fluid Phenomena: The River and Anthropocene Life-World in the Asia-Pacific*. Brill.

Irwin, Graham. 1955. *Nineteenth-Century Borneo: A Study in Diplomatic Rivalry*. Brill.

Ishikawa, Noboru, and Soda, Ryoji. 2020. "Commodification of Nature on the Plantation Frontier." In *Anthropogenic Tropical Forests: Human–Nature Interfaces on the Plantation Frontier*, edited by Noboru Ishikawa and Ryoji Soda. Springer.

John, David W., and James C. Jackson. 1973. "The Tobacco Industry of North Borneo: A Distinctive form of Plantation Agriculture." *Journal of Southeast Asian Studies* 4(1): 88–106.

John, David W. 1974. "The Timber Industry and Forest Administration in Sabah under Chartered Company rule." *Journal of Southeast Asian Studies* 5(1): 55–81.

John, Spenser St. 1862. *Life in the Forests of the Far East*. Smith Elder & Co.

Kaur, Amarjit. 1993. "Transport and the Sarawak Economy, 1841–1983." *Borneo Research Bulletin* 25: 76–100.

Kaur, Amarjit. 1994. "'Hantu' and Highway: Transport in Sabah 1881–1963." *Modern Asian Studies* 28 (1): 1–49.

Kaur, Amarjit. 1995. "The Babbling Brookes: Economic Change in Sarawak 1841–1941." *Modern Asian Studies* 29 (1): 65–109.

Kaur, Amarjit. 1998. *Economic Change in East Malaysia Sabah and Sarawak since 1850*. Palgrave Macmillan.

Langub, Jayl. 2020. "The Penan of Jelalong River, Sarawak: A Narrative of Migration and Adaptation." In *Anthropogenic Tropical Forests: Human–Nature Interfaces on the Plantation Frontier*, edited by Noboru Ishikawa and Ryoji Soda. Springer.

Large, Andy, David Gilvear, and Eleanor Starkey. 2017. "Ecosystem Service-based Approaches for Status Assessment of Anthropocene Riverscapes." In *Rivers of the Anthropocene*, edited by Jason M. Kelly, Philip Scarpino, Helen Berry, James Syvitski, and Michel Meybeck. University of California Press.

Lenggenhager, Luregn, Martha Akawa, Giorgio Miescher, Romie Nghitevelekwa, and Ndidzulafhi Innocent Sinthumule. 2023. The Lower! Garib-Orange River: Pasts and Presents of a Southern African Border Region. transcript Verlag.

Mundy, George Rodney. 1848. *Narrative of Events in Borneo and Celebes, down to the Occupation of Labuan: From the Journals of James Brooke Esq., Rajah of Sarāwak, and Governor of Labuan. Together with a Narrative of the Operations of* HMS *Iris*. John Murray.

Ng, Casey Keat-Chuan, Fatimah Abdullah, Helena Biun, Mohammad Khadafi Ibrahim, Shuhadah Mustapha, and Ahemad Sade. 2017. "A working checklist of the freshwater fish diversity for habitat management and conservation work in Sabah, Malaysia, North Borneo." *Biodiversitas Journal of Biological Diversity* 18(2): 560–74.

Nicholl, Robert. 1980. "Notes on Some Controversial Issues in Brunei History." *Archipel* 19 (1): 25–42.

Nicholl, Robert. 1983. "Brunei Rediscovered: A Survey of Early Times." *Journal of Southeast Asian Studies* 14 (1): 32–45.

Nicholl, Robert. 1989. "Some Problems of Brunei Chronology." *Journal of Southeast Asian Studies* 20(2): 175–95.

Pachuau, Joyc L.K., and Willem van Schendel. 2022. *Entangled Lives: Human-Animal-Plant Histories of the Eastern Himalayan Triangle*. Cambridge University Press.

Tremeer, R. E. 1964. "The Early History of Rubber Planting in Sarawak, 1880–1910." *Sarawak Gazette* 31 March.

Roth, Henry Ling. 1896. *The natives of Sarawak and British North Borneo*. Vol. 2. Truslove & Hanson.

Rutter, Owen. 1922. *British North Borneo: An Account of its History, Resources, and Native Tribes*. Constable & Company Limited.

Smythies, B. E. 1961. "History of Forestry in Sarawak." *Sarawak Gazette* 30 September.

Strong, Pauline Turner. 2015. "Ethnohistory." In *International Encyclopedia of the Social and Behavioral Sciences* (2nd ed., Vol, 8), edited by James D. Wright. Elsevier.

Takeuchi, Yayoi, Ryoji Soda, Hiromitsu Samejima, and Bibian Diway. 2020. "Current Status and Distribution of Communally Reserved Forests in a Human-modified Landscape in Bintulu, Sarawak." In *Anthropogenic Tropical Forests: Human–Nature Interfaces on the Plantation Frontier*, edited by Noboru Ishikawa and Ryoji Soda. Springer.

Tagliacozzo, Eric. 2013. "Borneo in fragments: Geology, biota, and contraband in trans-national circuits." *TRaNS: Trans-Regional and-National Studies of Southeast Asia* 1(1): 63–85.

Tate, D. J. M. 1996. "Planting in Nineteenth Century Sabah and Sarawak." *Journal of the Malaysian Branch of the Royal Asiatic Society* 69(1) (270): 37–63.

Treacher, William Hood. 1891. *British Borneo: Sketches of Brunei, Sarawak, Labuan, and North Borneo*. Government printing Department.

Tregonning, Kennedy Gordon. 1965. *A History of Modern Sabah (North Borneo 1881–1963)*. University of Malaya Press.

White, Richard. 1988. *The Roots of Dependency: Subsistence, Environment, and Social Change among the Choctaws, Pawnees, and Navajos*. University of Nebraska Press.

Yü, Dan Smyer, and Karin Dean, eds. 2021. *Yunnan–Burma–Bengal corridor geographies: Protean Edging of Habitats and Empires*. Routledge.

PART 3

Development and Governance

CHAPTER 13

Dam Trouble: Transdisciplinarity and the Case of a Riverine Livelihood Study in Northeastern Cambodia

Ta-Wei Chu and Paul J. Carnegie

1 Introduction

Transdisciplinarity gained purchase as a research paradigm in the early 1990s in response to the restrictive views on what constituted acceptable and authoritative 'knowledge' sources for problem solving—universities being widely regarded as the most privileged sites of empirically relevant scientific knowledge (Bernstein 2015; Nicolescu 2014). In theory and often in practice, this scientific knowledge production—which is acquired through positivist objective methods—is considered the 'legitimate' evidentiary base from which to address complex issues. Consequently, the value of normative thinking and knowledge acquired through subjective experience is relegated to a distinctly subordinate, if not wholly discounted role. This is despite its ability to enhance our understanding of the human and social dimensions of the problems under consideration (Simpson et al. 2015, 352; Bernstein 2015). In an attempt to retrieve the utility of more subjective forms of knowledge acquisition for grasping and responding to contemporary challenges and problems, transdisciplinarity places an emphasis on the 'knowledge' contributions of non-academic stakeholders. These people possess awareness of facts, information, and skills "built from experience of living in a place over an extended period of time … [or knowledge] accumulated by working in a particular location, or on a particular issue, for an extended period" (Bracken et al. 2015, 1293). As such, the scope of non-academic knowledge can include indigenous, local, or situated site-specific political and ecological awareness, familiarity, and understanding.

Although this distinguishing feature of transdisciplinarity can raise sceptical eyebrows in more scientifically minded academic circles, an unintended consequence of such doubt is that diverse local stakeholders end up discounted from the formulation and synthesis of the solutions to the issues that they face in the first place. By acknowledging the relevance and value of these stakeholders, transdisciplinarity commits to promoting the integration

 | HTTPS://DOI.ORG/10.1163/9789004758100-015

of their heterogeneous knowledge for problem-solving (Gibbon et al. 1994). As Godemann (2008, 627–28) notes, for these types of projects, research questions and objectives are geared not toward academia but toward real life. Referring to participants as 'stakeholders' emphasizes the view that different forms of knowledge accumulation can enjoy a degree of parity and that contributors can, to an extent, "have a mutual understanding and shared vision concerning some activity or interest" (Simpson et al. 2015, 353). According to Pohl (2005, 1161), these sorts of cooperative interactions provide an auxiliary problem-solving paradigm and can enhance mutual learning. For Schmidt and Pröpper (2017, 365), the process holds the potential to generate what they refer to as "socially robust knowledge". In short, the process entails a thinking through of issues and problems, that is context specific, inclusive of a plurality of viewpoints, locally informed, and ultimately accepted.

This emphasis of transdisciplinarity does differentiate it slightly from both interdisciplinarity and multidisciplinarity: in the former, academic cooperation across disciplines is largely "integrating, interacting, linking, and focusing", while in the latter, the cooperation is often "encyclopaedic, additive juxtaposition or, at most, some kind of coordination" (Alvargonzalez 2011, 388). Having said that, Schmidt and Pröpper (2017, 366) do strike a note of caution for transdisciplinarity, "the realisation of this ideal [knowledge co-production] proves more difficult than expected-and often, more difficult than is admitted in the literature."

With these considerations in mind, the following chapter considers the impact of the state-capital-development nexus on rivers and transdisciplinary projects in the Global South. It underscores how maintaining the fidelity of transdisciplinary projects is far from straightforward. The chapter details the difficulties of eliciting and integrating credible knowledge co-production about riverine concerns from relevant stakeholders arising from the Lower Sesan II Dam construction in Northeast Cambodia. By adopting a social conflict-cum-politics of scale lens, it reveals the impact of inimical and asymmetrical relationships and provides a microcosm of how the state's assertive and functional use of the Sesan river has degraded the riverine system.

2 Transdisciplinary Projects in the Global South

Recently, various scholars have conducted transdisciplinary projects in the Global South (TPGS). The collaborations usually dealt with problems such as environmental degradation, natural resource management, and climate change. In conducting these projects, identifiable challenges arose such as language barriers, internet access, and unexpected funding expenses (Siew et al. 2016, 821, 825; Schmidt and Pröpper 2017, 371, 376; Djenontin and Meadow 2018, 890, 893). Tuck Fatt Siew et al. (2016) reviewed four projects in China, the Philippines, and Vietnam. The research teams encountered language barriers: foreign academic stakeholders and local non-academic stakeholders had little or no command of the local language and English, respectively, and compounding this problem was a perceived 'unprofessionalism' of hired translators (Siew et al. 2016, 821, 825). Moreover, the teams found that local Vietnamese and Chinese stakeholders needed financial compensation for their participation in the projects. These unanticipated expenses, coupled with an equally unanticipated turnover of academic personnel, exacerbated the already arduous projects (Siew et al. 2016, 820). If we consider the project undertaken by Laura Schmidt and Michael Pröpper (2017, 376), who examined sustainable land management in southern Africa, they found that interactions between German and African stakeholders "reproduced North–South power asymmetries and dependencies" involving funding allocation and knowledge contribution. In their review of nine transdisciplinary projects, Djenontin and Meadow (2018, 890) identified various challenges, with poor internet access singled out as particularly aggravating.

Whilst not wanting to trivialize such tangible difficulties, they are not insurmountable. Overcoming them can make our scholarly endeavours laborious and frustrating, but as this chapter shows, we should exercise caution in assigning primacy to them. There is little empirical evidence to suggest that they are fatal to a project. Indeed, the experiences described above constitute good lessons for transdisciplinarians conducting TPGS. Lesson 1: Take care to hire professional interpreters who are familiar with the relevant issues. Lesson 2: Take care to conduct preliminary fieldwork to better appreciate local political cultures. Lesson 3: Take care in estimating approximate expenses. Of course, these lessons are subsidiary to the importance of establishing trust with local stakeholders "through extended periods or recurrent visits" (Siew et al. 2016, 826).

These concerns are not unique to transdisciplinarity but the unintentional blurring of the boundaries between what constitutes interdisciplinary, multidisciplinary, and transdisciplinary projects promotes a tendency in the literature to downplay the latter's accent on the value of non-academic knowledge. There is limited attention paid to the attendant difficulties of incorporating that into projects. As our chapter details, the overarching challenges that a state's political-economic dynamics and the agency its associated actors bring to bear on the conduct of a TPGS is significant.

For instance, the target state in the specific project under consideration is Cambodia, where former Prime Minister Hun Sen,[1] his inner governmental circle, members of his extended family, and business oligarchs form a powerful ruling coalition. The Hun Sen government has granted sundry land and forest concessions to such figures. This has led to the expropriation of large tracts of customary land and deforestation that plunge rural subsistence communities into precarious situations. Ruling coalition interests have subsequently clashed with emerging opposition groups over various contentious resource-and-development projects, no more so than dam-building projects (Baird 2011, 2016). Hun Sen's ruling coalition and opposition groups constitute the local non-academic stakeholders of the project under consideration, whose positionality and mutual antagonism are largely a product of the country's political-economic dynamics. As our chapter will show, it is a dialectic that produces contestation and a struggle for legitimacy over who has the authority to speak and act on such matters. We reflect on and detail how antagonism, in turn, triggers contestation over what was and is deemed relevant and/or 'authoritative' contributions to the project and the implications that this had for its intent and aims.

The chapter is divided into three further sections. The first section sets out the conceptual utility of a social conflict-cum-politics-of-scale lens to decipher the ways in which political-economic dynamics can affect projects. Establishing the conceptual import of such a lens allows us to gain insight into a state's political-economy and why its endogenous characteristics present major challenges for TPGS. Our focus is specifically on Cambodia as it was the site of the first author's project: a *Study of Sesan Riverine Communities' Livelihood* (SSRCL). The study investigated the livelihood difficulties stemming from the Lower Sesan II Dam (LS2) project (see Figure 13.1).

In the second and third sections, we reflect on this case study—specifically, the ways in which Cambodia's political-economic dynamics played a significant role in conditioning antagonistic relations that emerged between non-academic stakeholders. The third section recounts in detail the contestation over authoritative knowledge contributions that arose between the local non-academic stakeholders involved in the project. It also highlights the limited efficacy of the attempts to minimize such contestation. Although such contestation receives slim attention in published transdisciplinary research, our contention is that it significantly influences project aims and outcomes. This

---

1 Hun Sen recently stepped down as PM and his son, Hun Manet, was confirmed as the new PM by all 123 members of Cambodia's lower house of parliament on 22 August, 2023, but the father remains at the heart of power and politics in Cambodia.

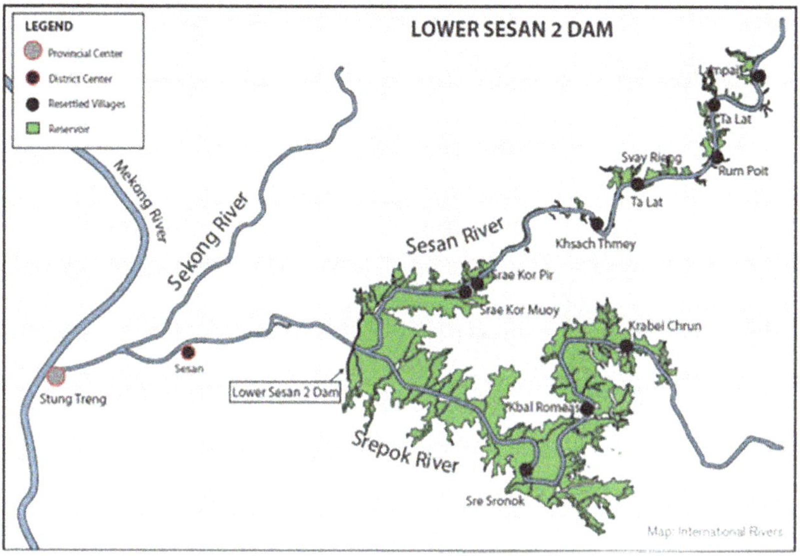

FIGURE 13.1 Map showing the reservoir area flooded by the construction of the LS2 and the resettled villages.
SOURCE: COURTESY OF INTERNATIONAL RIVERS

is especially the case when functionaries representing ruling coalition interests engage *mala fides* in a project with the purpose of co-opting and subverting the process of knowledge co-production unduly in favour of the partisan agendas and policies that they are there to represent. Our chapter's primary contribution is to give greater clarity on the underlying reasons behind the often-antagonistic relationships that form among non-academic stakeholders during projects and the implications this has for the aims of transdisciplinarity.

3 Methodology

The primary data for this chapter came from three sources. The first source was the preparatory fieldwork that the first author (TWC) conducted at the LS2's resettlement sites in June and December 2017 and July, October, and November 2018. The purpose of the fieldwork was to get a sense of the dam-affected villagers' resettled lives and invite the related stakeholders to join the SSRCL. The second primary-data source was on-site paperwork from the project. This consisted of observational notes from TWC, and his research assistants taken while documenting the observations of stakeholder interactions in the two preparatory meetings and four one-day workshops. Additional paperwork consisted of the flipchart papers contributed by the workshop stakeholders. The third primary-data source was the preparatory meetings with the NGOs involved, which took place in October and November 2018. The purpose of the meetings was to discuss the project's research design and potential with village and government stakeholders. Due to limited proficiency in the Khmer language, TWC relied on research assistants to act as interpreters in the workshops and fieldwork. To preserve anonymity, pseudonyms were used for all stakeholders.

In order to grasp the overarching challenges to TPGS, namely the influence that political-economic dynamics exert on local non-academic stakeholders and project efficacy, the following section sets out the utility of a social-conflict-cum-politics-of-scale lens. It details how such dynamics can influence a project's aims and outcomes before making specific references to Cambodia and the LS2 in the sections after that.

4 Social Conflict and the Politics of Scale

Social conflict is a key explanatory concept prevalent in the work of the 'Murdoch School' whose primary focus on the comparative politics and political economy in Southeast Asia emerged out of the Asia Research Centre, Murdoch University in the 1990s (Hameiri and Jones 2014, 3). In brief, underlying a social conflict perspective is that "different social forces naturally struggle against one another to establish state forms that will privilege their interests at the expense of others" (Jones 2012, 25). The forces can be "classes, class fractions, [and] ethnic and religious groups" (Jones 2010b, 551). The contestation between such social forces reflects what Bob Jessop (2008) terms as something akin to a zero-sum game. As Jones (2011, 411) notes, "states are never neutral apparatuses … rather, they exhibit an inherent 'strategic selectivity', marginalising some interests while advancing others". Furthermore, due to the fact that the contestation between social forces involves a competition over power and resources, the resultant struggles can be diverse and brutal (Jones 2013; Carnegie et al. 2021). Given this context, "the state is best understood not as an institutional ensemble but as a 'field of power' or a 'social relation'" (Jones 2010b, 551). This field of forces and interests significantly underpins and conditions the aspects of societal configurations.

Employing the concept of social conflict, Jones (2010a, 485–87, 2012, 39–91, 2013) identified various mixtures of sovereignty, intervention, and social order in Cold War Southeast Asia. During that period, capitalist-oriented socio-political forces ruled the founding members of the Association of Southeast

Asian Nations (ASEAN): Indonesia, Malaysia, the Philippines, Singapore, and Thailand. These members adopted the principle of non-interference embodied in three agreements: the Bangkok Declaration (1967); the Zone of Peace, Freedom, and Neutrality (1971); and the Treaty of Amity and Cooperation in Southeast Asia (1976) (Jones 2010a, 485). Internally, the principle helped the members set aside intra-ASEAN conflicts such as the Philippines–Malaysia dispute over Sabah "to foster the stability necessary for the economic growth required to undercut the appeal of communism" (Jones 2010a, 485). As Jones (2010a, 485) notes, "externally, non-interference sought to insulate ASEAN societies from 'subversive' external influences to help stabilise capitalist social order". However, the principle did not prevent ASEAN members from cooperating with each other against left-wing movements. For example, when, in 1975, Indonesia annexed East Timor, where the left-wing movement *Frente Revolucionária de Timor-Leste Independente* (FRETILIN) agitated "for a radical post-independence state", Malaysia backed the Indonesian government, supplying it with military personnel and arms (Jones 2010b, 553, 2012, 68). The fighting led to heavy casualties and a severe refugee crisis (Lawless 1976).

In the mid-2010s, Lee Jones in collaboration with Shahid Hameiri (2013, 2014, 2015a, 2015b, 2017) and drawing on the work of political geographers such as Swyngedouw (1997) and Brenner (2001), further conceptualized social conflict according to the politics of scale. According to Hameiri and Jones (2017, 60), "scales may reflect existing political tiers within a state—a village, a province, or the 'nation'—or cut across them, like 'bio-regions', 'transgovernmental networks' or 'the global'". They argue that the question of "whether a political issue is defined as urban, local, provincial, national, regional, global and so on is not neutral but, because each scale involves different configurations of actors, resources and political opportunity structures, always privileges certain societal interests and values over others" (Hameiri and Jones 2015b, 450). In short, this pattern of societal contestation involves an expanding or contracting back-and-forth struggle between antagonistic forces: in the struggle, one coalition of socio-political forces favours itself by treating a particular issue as an international, national, or sub-national matter; in response, other actors resist the coalition's actions by 're-scaling' the issue in the opposite direction, be it in a national, sub-national, or international direction. It is a phenomenon known as 'scale jumping'—"shifting political contestation to a different scale to bring in new actors and resources" (Hameiri and Jones 2013, 466). As such, the politics of scale extends our theoretical thinking on social conflict beyond national boundaries, by linking and attributing to international actors a prominent role in localized social conflicts (Hameiri and Jones 2013, 2015a, 2015b, 2017).

Hameiri and Jones adopted their 'social conflict 2.0' to investigate three non-traditional security issues pertaining to governance in Southeast Asia: money laundering in Myanmar and the H5N1 avian flu and haze in Indonesia (2013, 2015a, 2015b). In these cases, international actors, rather than treat issues on a national or sub-national scale, treated them on international scales by helping relevant state agencies align their domestic political governance with "international-standard regulatory practices" (Hameiri and Jones 2013, 2015a, 154–55, 2015b, 454–56). However, these international actors clashed with domestic regime-affiliated actors: the Myanmar military junta (money laundering), the powerful poultry industry in Indonesia (the avian flu), and political and agro-business interests in Indonesia (haze). In these three cases, the countering moves came not from international actors but from national and sub-national actors who struggled to rescale 'hot-button' issues to national and sub-national levels, where the actors could avoid external pressures to meet international standards. In all three cases, the measures designed by international actors to address the problems were ineffective, precisely because the domestic actors, fronted by oligarchic capitalists, selectively and strategically embraced measures largely in the protection of their own self-interests rather than for their overall effectiveness (Hameiri and Jones 2013, 2015a, 156, 2015b, 456). For Hameiri and Jones, the politics of scale is a dynamic process that can play out in various directions, pitting several types of actors against one another. What their work reveals is the way that domestic struggles are shaped by politics, economics, and "broader social power relations" (2015, 446).

5 Cambodia: a Political-Economic Context

Applying the aforementioned social conflict-cum-politics-of-scale lens to Cambodia assists in gaining purchase on underlying political-economic dynamics and their implications for TPGS. Both Hun Sen and his Cambodian People's Party (CPP) gained predominance in Cambodia's politics after his successful coup d'état against First Prime Minister Prince Norodom Ranariddh in 1997. After the coup, Hun Sen and his followers took control of the state apparatus and its ability to exert influence across Cambodia via patron–client relations. To elaborate, the CPP government was able to issue land and logging concessions to tycoons and associated business 'cronies', who then reciprocated by donating considerable sums of money to government-run rural development projects, which in turn garnered votes from poor rural populations (Un 2005; Hughes 2006). Significantly, the extended Hun Sen family and its wealthy allies have dominated lucrative business sectors including ones associated with special economic zones (SEZ), infrastructure (Hughes 2011), urban renewal (Willem 2012), media, and international trade (Global Witness, 2016). Domestically, these powerful actors have been driving forces behind the 'development' of natural resources such as timber (Global Witness 2007) and sand (Global Witness 2010). In this way, Hun Sen, his extended family, the CPP, and privileged acolytes have been able to consolidate an immensely influential ruling coalition; a state–capital–development nexus.

The above-mentioned business activities have frequently operated under the guise of state-led development projects (SDPs), the legitimating pretext of national economic development, and the raising of people's living standards. However, it is fairly incontrovertible that a host of these projects have actually harmed communities, caused environmental degradation, and triggered protests and clashes with project-affected villagers who, in turn, receive support from domestic and international NGOs (Sokphea 2016; Chu 2017). Together, these protestors and their supporters have formed a series of loosely knit collectives of opposition groups. Opposition activities have 'scaled' the problem of SDPs to the international level by using methods such as social-media reporting and official filings with the European Union and other regional organizations (Sokphea 2016, 602–3; Chu 2016, 1107). The CPP government responded to this opposition strategy by passing the Law on Associations and Non-Governmental Organizations No. 0415/010 (LANGO 2015), which has severely curtailed NGO operations in Cambodia (Curley 2018). Moreover, arrests and even extra-judicial killings of democratic and environmental activists have been on the rise in Cambodia, with the CPP government widely suspected of orchestrating the abuses (Blake 2019, 72). Struggles between the Hun Sen-led ruling coalition and opposition groups over development projects reflects a 'politics of scale' in the shaping of these adversarial relationships.

This brings us to a set of important questions. Given that a core principle in any transdisciplinary project is "everyone who has something to say about a particular problem and is willing to participate can play a role" (Augsburg 2014, 263), how and in what ways do political-economic dynamics play out in TPGS? And how do those dynamics affect the rationale of *bona fides* knowledge co-production? What sort of stress does this core inclusionary principle come under in Global South contexts? These are pertinent considerations due to the fact that many of the issues engaged are either directly or indirectly linked to SDPs (Siew et al. 2016; Bréthaut et al. 2019). If that pits local non-academic anti-SDP stakeholders against entrenched oligarchic pro-SDP forces, then it takes little stretch of the imagination to appreciate that the resulting tensions and, in some instances, outright hostilities can easily extend to a given project irrespective of whether it is focused directly or indirectly on the

SDP in question. Given such contexts, it is fairly unsurprising that knowledge contestation can arise quickly in TPGS, between the perceptions of local non-academic stakeholders and those of entrenched pro-establishment stakeholders. The subsequent antagonism can express itself through mutual distrust, visible dislike, and fractious disagreement. In other words, a project can run a high risk of becoming embroiled in the situational political-economic context of its locale. This can lead to a tendency in which the project intent [of facilitating *bona fide* knowledge co-production for potential solutions to specific problems] is arrogated and politicized. If that unfolds in an untrammelled manner, a project then faces significant hurdles in trying to achieve its aims and intent.

For example, researchers who conduct fieldwork in an authoritarian state context on sensitive political and economic issues can expect to come under surveillance of varying degrees by government officials and experience obstacles or certain restrictions on their ability to collect data. In order to manage such challenges, Morgenbesser and Weiss (2018) suggest that researchers thoroughly familiarize themselves with a target state's local political-economic dynamics before heading on site and seek to establish contact and relations with local partners such as NGO workers and academic institutes. In this sense, it is the political-economic dynamics of a target state that constitute the main exogenous challenge not only to data collection but to a scholar's safety. Again, because stakeholders represent a range of actors, the agents of oligarchic capitalism end up becoming intrinsic participants in TPGS. Although Bréthaut et al. (2019, 160), claim that their "participatory and transdisciplinary" research "led to some disruption of established power dynamics," researchers should not assume that this means that entrenched state-affiliated actors will refrain from trying to dominate the exercise in knowledge co-production.

The ubiquity of these actors and their tendency to dominate proceedings is evident in the project under consideration and those of other researchers (e.g., Siew et al. 2016). Moreover, matters are regularly compounded by the timeframes of transdisciplinary projects. Projects sometimes extend for three to four years, during which time the researchers struggle to accomplish their research objectives, let alone resolve the tensions of antagonistic local relationships. In short, the priority that transdisciplinarity places on inclusivity means that political-economic dynamics are the endogenous challenge that they cannot avoid but which invariably get downplayed. The scholarly concern for methodological challenges actually diverts our attention away from the significance of this contextual factor. Having said that, a growing body of work has begun to map the ways in which politico-economic forces and power dynamics shape people's exposure to and understandings of such circumstance

(Baird 2016; van Voorst 2016; Rosario and Rigg 2019; Carnegie et al. 2016 and 2019; Carnegie and King 2020).

6 Local Non-Academic Stakeholders and the LS2: Deciphering the Interactions

Before moving on to consider the effect of Cambodia's political-economic dynamics on the case study, it is important to understand how they condition the relationships among local Cambodian non-academic stakeholders. Doing so, helps highlight the links between micro-subjective experiences and wider state-capital-development practices and how they play out at the level of individuals and communities.

In the 1990s, Cambodia lacked sufficient electricity production which necessitated importing it from Laos, Thailand, and Vietnam. As a result, Cambodian electricity prices rose to some of the highest in the world (Heng 2015, 406). From the CPP government's perspective, "the high cost of electricity affects all productive sectors and hinders industrial investments and competitiveness" (RGC 2006, 24). Meanwhile, the Asian Development Bank (ADB) initiated the Greater Mekong Sub-region (GMS), focusing on infrastructure creation, with an emphasis on the energy industry (Dosch and Hensengerth 2005). This set of circumstances eventually occasioned the CPP government's decision to generate electricity by damming the Mekong river. From the government's perspective, it is viewed as an economic resource capable of facilitating the industrialization and modernization of the state.

The LS2 stems from the convergence of two prominent factors: Cambodia's state-led development policy and the emergence of sub-regionalism in the country. As Baird (2016, 263) notes, the origins of the LS2 trace back to "a 1999 study of hydropower dam potential in the Nam Theun, Sekong and Sesan river basins in Laos"; however, the engineering consultancy company, Halcrow and Partners, did not take further action owing to the proposed dams' "marginal financial viability" and the environmental and social harm of the dams. In 2007, the state-owned company, Electricity of Vietnam (EVN), became the LS2's main investor, with the EVN subsidiary—EVN International Joint Stock Company (EVNI)—taking 51 per cent of the LS2's shares. At the same time, the Cambodian Royal Group (CRG), under the directorship of Kith Meng, known for his close relations to Hun Sen, took 49 per cent of the shares (Blake 2019, 80). EVNI and CRG created a joint venture, known as Hydropower Lower Sesan 2 Company (henceforth, LS2 Company). The EVN shareholder, Power Engineering Consulting Joint-Stock Company No. 1 (PECC1), oversaw the dam's feasibility study and conducted the dam's environmental impact assessment (EIA) alongside the Cambodian national consultancy company, Key Consultants Cambodia (KCC). The EIA was completed in June 2010 without any input from dam-affected villagers (Hensengerth 2017, 96; Grimsditch 2012, 26).

The LS2 is located in the Sesan river Basin, which is one of the largest tributaries of the Mekong river. The source of the Sesan river is in Vietnam's Central Highlands and flows to Ratanakiri and converges with the Sekong and Srepok Rivers in Stung Treng Province, west of Ratanakiri. Many people living along the Sesan river in Stung Treng are Khmer, Laotian, or members of indigenous groups. Prior to the dam's completion, these people eked out a living by fishing and collecting non-timber forest products (NTFPs). The indigenous groups, in particular, have a deep spiritual connection to the land and believe that the forests, mountains, rivers, and graveyards harbour powerful spirits. It was recognized early on, that construction of the LS2 would result in the relocation of about 1,000 households, many of which hailed from four villages: Sre Kor, Kbal Romeas, Sre Sronouk, and Krabei Chrun (Grimsditch 2012, 28). Moreover, according to the EIA report, the LS2's reservoir would destroy 30,000 hectares of forest (Grimsditch 2012, 29). According to Ziv et al. (2012, 5609), the LS2 would reduce fish stocks by an estimated 9.3 per cent throughout the Lower Mekong river Basin. Concerns over worsening food security for locals and the erosion of traditional indigenous culture from forced relocation to new villages were also raised. Despite the warnings, it did little to deter Cambodia's ruling coalition from pressing forward. From the perspective of their partisan development agenda, the river has but one central function: economic.

Resistance from LS2-affected villagers to the CPP government's forceful river governance has been visible in local protests and communal elections in 2012 and 2017. In February 2012, with assistance from the domestic NGO, the 3S Rivers Protection Network (3SPN), dam-affected villagers planned to hold a public protest near the LS2's construction site. The plan raised "concerns" among district police that led to the community leadership cancelling the protest (Baird 2016, 269). Nevertheless, the villagers managed to hold a protest, at which one banner read, "We must preserve the river, which is the livelihood of the people!" (Baird 2016, 269). Villagers who were critical of the dam emphasized the river's multiple functions. It is interesting to note that rural areas, including Stung Treng, have been strongholds for Hun Sen's CPP (Baird 2016, 261). Yet, in the 2012 communal elections, the Cambodia National Rescue Party (CNRP)—dissolved by the CPP government in 2017—won a majority of the seats in the Sre Kor commune. This commune had been strongly opposed to the LS2. Although the CPP won the commune election in Kbal Romeas, the opposition party's vote dramatically increased. According to Mak Sithirith

(2016, 70), widespread opposition to the LS2 spurred the high voter support for the CNRP. In 2017's commune elections, the CNRP won again in the Sre Kor commune. The CPP government encountered a significant local backlash.

Despite the resistance from locals, the National Assembly of Cambodia approved the LS2 in November 2012 without hesitation. The most interesting issue in the approval process was the decision by EVNI to partially withdraw from the LS2. Taking the place of EVNI was Hydrolancang International Energy Co. (HIE), a subsidiary of the China Huaneng Group—one of the top five state-owned electric-utility enterprises in China. Now, HIE held the largest share in LS2's stocks, namely 51 per cent, whereas CRG's share was 39 per cent and EVNI's was 10 per cent. In February 2013, the National Assembly passed the Cambodian Government Guarantee of Payments to the LS2 Company Law. It established two basic conditions: (1) the government would cover the costs of power purchases in the event of non-payment by *Electricité du Cambodge*[2] and (2) the government would financially oversee the project if a crisis prevented the company from performing its duties (Kimkong et al. 2013, 50). The CPP government had essentially extended significant economic advantage to the HIE by helping it "avoid taking any significant risk when investing in the LS2 project" (Chu 2017, 1105).

In January 2014, the LS2 Company released "The Compensation and Solution Policy". Many dam-affected villagers disagreed with its provisos and submitted petitions articulating this disagreement to the relevant government agencies and the Chinese embassy in Phnom Penh (Chu 2017, 1107). The petitions stressed the signatories' rejection of the compensation policy and called for a cancellation of the dam project (Chu 2017, 1107). Furthermore, in October 2014, eighteen NGOs including domestic and international ones issued a public statement calling for the government to "release details of proposed changes to the project's design, mitigation measures, and their operation, to the public and affected communities" and conduct a new EIA (Earth Rights, 2014). These developments indicated that dam-affected villagers and domestic and international NGOs were in the process of forging an anti-LS2 alliance in contradistinction to the plans of the ruling coalition.

The anti-LS2 alliance's politics-of-scale strategy did not stop the CPP government from advancing "The Compensation and Solution Policy". According to a report conducted by Mekong Watch, provincial government officials employed various methods to elicit support from the villagers for the policy. The Mekong Watch report asserted, for example, that a government official had asked an

2 Electricité du Cambodge is Cambodia's state power company.

adult male villager for his thumbprint, claiming that it was part of a survey for villager compensation; later, however, the official used the thumbprint as proof that the villager had agreed to the compensation policy (Mekong Watch 2015, 2). A female adult villager relayed a similar experience: "[Officials] said that this development [i.e., the LS2 project] would make poor people like us rich. They also said that if [my mother] protested, the government would not assume compensatory responsibility for my family" (Mekong Watch 2015, 2). By using the term 'development,' 'poor people', and 'rich,' the officials again emphasized the river's economic function. The LS2's proponents—Hun Sen's ruling coalition—seemed intent on rejecting the possibility of the river's multiple functions.

The displaced villagers started moving into the government-designated relocation sites in about mid-2016. Two resettlement villages—Kbal Romeas and Sre Sronok—are located along National Road No. 78, which links Stung Treng and Ratanakiri Provinces (see Figure 13.2). The villagers' new surroundings are no improvement over their pre-dam settlements. For example, the relocation sites have tainted wells, poorly constructed and vermin-infested houses, missing land titles, a spotty home-ownership system, teacherless schools, and insufficient cultivatable land.[3] These resettlement problems go far in explaining why some villagers from Sre Kor and Kbal Romeas refused to set up homes in the new sites. The dam's operations, slated to begin in December 2017, would flood the original Sre Kor and Kbal Romeas villages, so these residents had to move about three to five kilometres away. The new locations lacked adequate educational provision and sufficient cultivatable land.[4]

After the June 2017 commune elections, the Hun Sen regime kept the LS2's resettlement sites under surveillance. During preliminary fieldwork in 2017, a local NGO worker suggested that TWC not visit the resettlement sites owing to the regular patrols by local police.[5] The 3SPN's director conveyed a similar concern, informing him that the 3SPN might be on the government's blacklist.[6] During the inauguration of the LS2, which was held in September 2017, Hun Sen made no reference to surveillance and intimidation. Instead, he continued emphasizing the river's economic function: "The Lower Sesan II dam will impact the development of the whole country. It will supply [electricity to]

3 Observations and interviews in the preliminary fieldwork, conducted in resettled Kbal Romeas and Sre Srenok villages, June and December 2017.

4 Observations and interviews in the post-workshop fieldwork, conducted in new Sre Kor Village, 22 June, 2019. As of writing, the provincial government is renovating the school.

5 Preliminary fieldwork, conducted in Stung Treng Town, 20 December, 2017.

6 Interview in the preliminary fieldwork, conducted in Banlung, 20 June, 2017.

FIGURE 13.2 A newly built but understaffed school in the resettled Sre Srenok village.
SOURCE: COURTESY OF TA-WEI CHU

households, as well as firms in the agricultural, industrial, and service sectors" (Khmer Times 2018). Nevertheless, the on-the-ground impression is that surveillance and blacklisting were part of the CPP government's response to the anti-LS2 alliance's politics-of-scale strategy.

7 The SSRCL: Preparation and Challenges

In April 2018, TWC set up the SSRCL on the basis of transdisciplinary principles. As mentioned earlier, both the generally tense political atmosphere in Cambodia and the specific sensitive nature of the LS2 prompted him to invite and consult with NGO stakeholders. In July 2018, he went to Stung Treng, Ratanakiri, and Phnom Penh, where he invited four NGOs to join the project: the 3SPN, the Cambodia Indigenous Youth Association (CIYA), the Fisheries Action Coalition Team (FACT), and My Village (MVi). All of them agreed to join the project. MVi suggested that participation in the project extend to the NGO Forum on Cambodia (NGOF), which had published two reports related to the LS2's resettlement and compensation (NGOF 2012, 2015). One of the key criteria for choosing NGOs rests on their experience working on protection, conservation, fisheries, and riparian livelihood, all in relation to the Sesan river. The suggestion was approved, and the NGOF became the fifth NGO to join.

In October 2018, the first preparatory meeting was held to discuss the project's research topic, research question, research objective, and potential village stakeholders. As Bergmann (2012) notes, workshops and seminars are often the most common places where stakeholders engage in knowledge sharing. For this reason, they agreed to hold at least four one-day workshops. In addition, a few NGO stakeholders suggested expanding the scope of the research site from the specific LS2 resettlement sites to the wider Sesan Riverine Basin in Stung Treng Province. Two concerns prompted this suggestion. First, because the LS2 has been controversial, the expansion of the research sites could lessen any governmental suspicions of the project. Second, some Sesan Riverine villagers have been negatively impacted by two dams in Laos: Don Sahong (Champasak Province) and Xe-Pian Xe Namony (XPXN, La Attapu Province). The former has created food insecurities for villagers by reducing local fish populations (Baird 2011); the latter collapsed in July 2018 (South China Morning Post 2018), spreading concerns among dam-affected Cambodian villagers about the LS2's safety. The NGO stakeholders suggested that the project could include these villagers and benefit from their input. Finally, it was agreed that the project should invite villager-representatives from three locations: (1) the LS2-affected villages: Kbal Romeas, Sre Kor, and Sre Srenok; (2) the Don Sahong-affected village: Leu; and (3) the XPXN-affected village: O'Chay.

Bearing in mind the lessons drawn from previously mentioned projects, TWC decided to conduct preparatory work. First, he undertook preliminary fieldwork and visited the resettled Kbal Romeas and Sre Sronok villages and interviewed villagers and NGO workers. Second, to avoid potential interpreter problems, he hired a Cambodian graduate student, Sen, who had transdisciplinary training and high English proficiency, as a research assistant. He was to serve as the project interpreter and to organize its first preparatory meeting. Third, he secured a research grant for the project's expenses. However, despite solid preparations, methodological challenges arose. An NGO worker told him that in Cambodia, hosts of workshops, seminars, and the like are expected to financially reimburse participants for their related daily expenses, including those pertaining to travel and accommodations. However, the project's research grant was not enough to cover these expenses. In addition, Sen was unable to continue working in the project after the first preparatory meeting. As documented in many of the previously mentioned projects, it was also encountering its fair share of problems.

Fortunately, NGO stakeholders and TWC were able to minimize the problems that these challenges posed. First, he elected to partly self-fund the project.

Meanwhile, some NGO stakeholders, aware that the project's grant was limited, expressed to him their willingness to participate in the workshops without reimbursement. Particularly helpful in this regard was the NGOF, which voluntarily sponsored the workshops. Second, he hired a new MA research assistant, Kem, who quickly integrated into the project owing to his previous experience in many other local research projects. However, Kem was unfamiliar with the LS2 issue, so, he hired a local researcher, Ben, who had been an NGO worker involved in the LS2 issue and who is, as of this writing, an instructor at a local university. Both Kem and Ben were workshop facilitators and assistants in the post-workshop fieldwork. Kem, Ben, and TWC were the project's core academic team. It is worth pointing out that the above-mentioned challenges required patience and fairly laborious 'fixes', but at no point did they pose an existential threat to the project.

8 Dealing with an Intransigent Reality

Somewhat differently, the challenges posed by Cambodia's political-economic dynamics were to prove far more daunting than those outlined in the previous section. In November 2018, NGO stakeholders and TWC held the second preparatory meeting. An NGO stakeholder suggested that because male villagers had dominated daily life in their communities, the research project should give more voice to marginalized females. This suggestion met with agreement from the meeting's participants. Subsequently, six of the nine village stakeholders were female. Moreover, TWC suggested recruiting local government officials who had been involved in the LS2 issue. He believed that the SSRCL needed a diverse range of stakeholders, which would have to include government officials, whose extensive knowledge of the Cambodian political bureaucracy was highly relevant for the aim of knowledge co-production. Without realizing the extent of the underlying 'politics' that his suggestion would manifest, he somewhat naïvely assumed that collectively and collaboratively the stakeholders would generate and synthesize socially robust knowledge. Eventually, three government officials were invited to participate in the SSRCL. They were from provincial, district, and commune levels, respectively.

By insisting that the project open itself to the participation of government officials, he had inadvertently fostered a volatile context in which the village and government participants would easily slip into antagonistic patterns of interaction. On 25 January, 2019, the MVi held the kick-off workshop. Its aim was to identify the Sesan Riverine villagers' livelihood difficulties (see Figure 13.3). The academic team designed a discussion-and-debate model

FIGURE 13.3 A villager's written knowledge contribution in the first workshop.
SOURCE: COURTESY OF TA-WEI CHU

for group discussions. According to the model, same-attribute stakeholders should hold "inner discussions" before proceeding to "outer debates" involving all the different-attribute stakeholders. The academic team divided stakeholders into three attribute-based groups: (1) villagers, (2) government officials, and (3) NGOs. Each group had one to two facilitators who would guide discussions. Inner discussions, as noted above, were followed by outer debates, and the stakeholder groups were required to record their thoughts on the offered flipchart papers and to present their ideas to one another. After the presentation, other stakeholder groups could ask questions and give feedback. Stakeholder groups presenting their ideas would also respond to questions and feedback. The purpose of the discussion-and-debate model was to avoid direct antagonism between the village and government stakeholders.

Despite the efforts to minimize conflict, intense knowledge contestation marred the first workshop. The village stakeholders identified four main livelihood difficulties: (1) Crises—dams' risk of collapse and flooding and dam-caused fish decline; (2) Basic needs—resettlement sites' lack of potable water, healthcare services, and teachers; (3) Land issues—missing property titles, uncultivable lands, and land provisions smaller than those stipulated in resettlement contracts; and (4) Illegal activities (e.g., illegal fishing and logging along the Sesan river).[7] After a village stakeholder voiced personal experiences of

7 Note-taking and flipchart papers created by the village, government, and NGO stakeholders, Stung Treng Town, 25 January, 2019.

livelihood difficulties, the provincial-government stakeholder, Aiso, declared, "You should not bring these issues to the table."[8] The village stakeholder, Dang, responded to Aiso by raising the issue of thumbprints. Dang said that, from her perspective, the government had deceived her.[9]

Furthermore, the academic team noticed the reluctance of government stakeholders to discuss the LS2 and its compensation and resettlement problems. On their flipchart papers, the government stakeholders stated that villagers faced two main types of livelihood difficulties: illegal logging and illegal fishing by lawbreakers in neighbouring provinces.[10] In other words, the government stakeholders sidestepped the issue of the LS2 by refusing to attribute any of the villagers' livelihood difficulties to the dam. From the outset, contestation over what were the issues, the relevance of perspectives, and the meaning ascribed ('authoritative knowledge') between village and government stakeholders manifested itself in direct verbal exchanges and in personal written notes. It was clearly observable that the discussion-and-debate model would have to deal with substantial antagonism between the stakeholders. The intransigent reality of Cambodia's political-economic dynamics was conditioning their relative interactions with one another.

On 21 June, 2019, the 3SPN held the second workshop. Similarly, it followed the discussion-and-debate model. At the beginning of the workshop, it was emphasized that all the views expressed by stakeholders contribute to knowledge co-production and thus merit respect. Regrettably, this exhortation did little to ease the stakeholders' antagonism toward one another, as detailed below.

The second workshop's objective was to identify the possible solutions to the livelihood difficulties identified in the first workshop. During the workshop, the village stakeholders mainly requested that the Cambodian government take responsibility for their LS2-based livelihood difficulties. The government stakeholders denied the existence of some difficulties. For example, Aiso said that resettled villagers' land-shortage problems had geographical causes: "when the water level goes up, the size of [villagers'] lands shrinks."[11] Her comment subtly but undeniably decoupled the villagers' livelihood difficulties from the LS2. Knowledge contestation was emerging again.

8 Research assistant's note-taking, Stung Treng Town, 25 January, 2019.

9 Note-taking, Stung Treng Town, 25 January, 2019.

10 The flipchart papers created by the government stakeholders, Stung Treng Town, 25 January, 2019.

11 Research assistant's note-taking, Banlung, 21 June, 2019.

When responding to village stakeholders' concerns, the government stakeholders couched their comments in Cambodia's deeply entrenched hierarchical political culture and state–society relations. For example, regarding illegal fishing, poaching, and logging, the district-government stakeholder, Diser, asserted that the government is "educating local people about forest laws and land laws" and that the government tries "to meet the local people's needs, because it [the relationship between the government and the villagers] is like [the relationship between a father and his children]."[12] The paternalistic tone of Diser's comments reinforced and reasserted the asymmetrical relations between the two parties by using phrases like "a father and his children" and "educating local people".

Although the discussion-and-debate model, serving as the basis for these interactions, gave village stakeholders a chance to voice their concerns, Diser's statement, coupled with Aiso's denial of the existence of villagers' livelihood difficulties, gave an indication of the attempts by government stakeholders to dominate the process of knowledge co-production. Contra Bréthaut et al. (2019), it also brings into question the extent to which such exercises can disrupt the established power dynamics.

FACT held the third workshop on 21 November, 2019. Due to the lack of knowledge co-production in the second workshop, the third workshop's objective remained unchanged: identify the possible solutions to villagers' livelihood difficulties. This time, however, the stakeholders were asked to discuss possible solutions in the context of two specific questions: (1) What can I do? (2) What support do I need? The third workshop did not give rise to strong knowledge contestation because some of the more combative village and government stakeholders were not in attendance. Aiso, for example, was not present but had assigned a colleague to join the proceedings. The surrogate government stakeholder, who happened to specialize in provincial finances, was almost completely unfamiliar with the project's details. Despite this setback, NGO and village stakeholders offered possible solutions to the livelihood difficulties. One solution was village stakeholders' proposal that local villagers, police officers, and NGO workers jointly patrol nearby forests and riverbanks in search of illegal fishing, poaching, and logging.[13] But the absence of substantial government stakeholder's contributions to the project's knowledge co-production meant that the proposal was left weak and fragmented.

12 Note-taking, Banlung, 21 June, 2019.

13 Flipchart papers created by the village and NGO stakeholders, Stung Treng Town, 21 November, 2019.

After the third workshop, the SSRCL encountered a methodological challenge of unexpected proportions. The COVID-19 pandemic prevented most participants, including TWC, from travelling and meeting. The NGOF, therefore, postponed the fourth workshop until late September 2020. At the eventual holding of the workshop, in which he participated via videocall, participants were required to wear masks and use sanitizer. Adherence to health-and-safety protocols and the use of virtual communication were adopted to help minimize the COVID-19 impact.

As before, Cambodia's political-economic dynamics remained a formidable reality in stifling credible knowledge co-production during the fourth workshop. One objective in this workshop was to generate policy suggestions for the government's handling of livelihood difficulties. To this end, the NGOF invited additional government officials from provincial and national levels to participate in the workshop. One week before the event, the NGOF distributed to each of these individuals the initial SSRCL report documenting the progress made from the first to three workshops. In the fourth workshop, a village stakeholder complained about many unresolved livelihood difficulties and suggested that the government had ulterior motives for collecting the thumbprints of villagers on the resettlement-and-compensation contracts.[14] A government stakeholder responded to this concern by trying to dispel any suspicions: "please be aware that, at the national level, the government is always present to solve problems in cooperation with the people, and the thumbprints are merely standard information on official documents."[15] Moreover, the government stakeholder added that the SSRCL report should mention the positive sides of the LS2,[16] which in the context was an indirect allusion to Hun Sen's attempts to use hydrodevelopment as an engine of modernization in the Cambodian state and a way to seek legitimacy for the LS2. Such assertions again demonstrate the divergence of political-economic perspectives conditioning village and government stakeholder relations. Overall, in the four one-day workshops, NGO stakeholders and the academic team managed to overcome many challenges, but the political-economic reality remained a stubborn obstacle.

Overall, there was little common ground between the government stakeholders and village stakeholders in terms of generating credible knowledge coproduction. The village stakeholders bravely contributed their knowledge and experience of the post-relocation living conditions, which were observably inferior to the pre-relocation conditions. The government stakeholders viewed

14 The final workshop's minutes created by NGOF, Phnom Penh, 30 September, 2020.

15 The final workshop's minutes created by NGOF, Phnom Penh, 30 September, 2020.

16 The final workshop's minutes created by NGOF, Phnom Penh, 30 September, 2020.

this contribution as a challenge to and criticism of the state-led hydropower project and vigorously sought to reject, deny, or impede its articulation and acceptance as part of the co-production knowledge base. This sort of factual contestation characterized the interaction between the village stakeholders and the government stakeholders.

9 Conclusion

Somewhat differently to interdisciplinary and/or multidisciplinary studies, transdisciplinary studies consistently place strong emphasis on the value of non-academic knowledge in thinking through and contributing to possible solutions for site-specific socio-political, economic, or environmental problems. However, as this chapter has shown, eliciting and integrating such awareness and understanding into project aims and outcomes is especially vulnerable to a state's political-economic dynamics. Although other methodological challenges can disrupt projects, they are usually not insurmountable, and they will rarely sink them. In contrast, the undue influence of asymmetrical political-economic relations on a transdisciplinary project can be severe.

Employing a social-conflict-cum-politics-of-scale lens brought into focus and helped decipher the often-antagonistic relationships that form among various non-academic stakeholders. It foregrounded the debilitating effects of entrenched patronage politics and state-led, commercially collusive developments. As the data findings showed, when dominant ruling coalition actors gain ascendancy, it can severely constrain *bona fide* attempts at knowledge co-production. In sum, a fuller understanding of the overarching challenges facing TPGS comes from a grasp of the target state's political-economic dynamics and the influence exerted by its dominant actors.

As the observations from the SSRCL indicated, the positionalities and struggles that emerged among local non-academic stakeholders over the Sesan River's functions are a microcosm of Cambodia's political-economic dynamics and their deleterious environmental impact. The chapter demonstrated how the forces and interests of the country's ruling coalition play out at the level of individuals and communities. The undercurrents detailed in the four workshops gave a palpable sense of power asymmetries, interest incompatibilities, and knowledge contestation at work. Dominant actors gained ascendancy over what was deemed relevant and/or authoritative contributory or factual knowledge on issues and unduly influenced the raison d'être and efficacy of the project.

In its own way, the motifs of the SSRCL serve as a telling reflection of the Anthropocene and its impact on rivers. We are living through an era in which human activity has a profound often detrimental effect on the environment. The concept of the Anthropocene outlined in the introductory chapter of this volume highlights how this has become the dominant force shaping river ecology. Dam construction in Southeast Asia is a key example of such activity. Large-scale dams—driven by development agendas, energy demands, and regional geopolitics—exemplify modes of human intervention that alter river ecosystems, displace communities, disrupt sediment and migratory fish flows and impact livelihoods. These infrastructural developments raise critical questions about sustainability, ecological resilience, and the balance between economic growth and environmental justice.

Along with other contributions in this volume, our chapter challenges us to reconsider the long-term consequences of such interventions and explore alternative, less disruptive solutions. It revealed that the Cambodian state's assertive and functional use of the Sesan river to drive 'modernization' and 'development' has degraded the riverine system faster than at any time in the past. Dam-affected villagers are now living with heightened uncertainty and precariousness. While the villagers remain determined to protect their river, as the data from the workshops indicated, the tensions between the idea of state-led modernization and its environmental impacts on the one hand, and the preservation of ecologically sustaining ways of life on the other (that indigenous knowledge represents) are not easily resolved.

References

Augsburg, Tanya. 2014. "Becoming transdisciplinary: the emergence of the transdisciplinary individual." *World Futures* 70(3–4): 233–47.

Alvargonzalez, David. 2011. "Multidisciplinarity, interdisciplinarity, transdisciplinarity, and the sciences." *International Studies in the Philosophy of Science* 25(4): 387–403.

Baird, Ian G. 2011. "The Don Sahong Dam: potential impacts on regional fish migrations, livelihoods, and human health." *Critical Asian Studies* 43(2): 211–35.

Baird, Ian G. 2016. "Non-government organizations, villagers, political culture and the Lower Sesan 2 Dam in Northeastern Cambodia." *Critical Asian Studies* 48(2): 257–77.

Bergmann, Matthias, Thomas Jahn, Tobias Knobloch, Wolfgang Krohn, Christian Pohl, and Engelbert Schramm. 2012. *Methods for transdisciplinary research: A primer for practice.* Campus Verlag GmbH.

Bernstein, Jay H. 2015. "Transdisciplinarity: a review of its origins, development, and current issues." *Journal of Research Practice* 11(1): 1–15.

Blake, David. 2019. "Recalling hydraulic despotism: Hun Sen's Cambodia and the return of strict authoritarianism." *Austrian Journal of South-East Asian Studies* 12(1): 69–89.

Bracken, Louise J., Harriet A. Bulkeley, and Geoff Whitman. 2015. "Transdisciplinary research: understanding the stakeholder perspective." *Journal of Environmental Planning and Management* 58(7): 1291–308.

Bréthaut Christian, Louise Gallagher, James Dalton, and Jeremy Allouche. 2019. "Power dynamics and integration in the water-energy-food Nexus." *Environmental Science and Policy* 94: 153–62.

Brenner, Neil. 2001. "The limits to scale? methodological reflections on scalar structuration." *Progress in Human Geography* 25(4): 591–614.

Carnegie, Paul J., Victor T. King, and Zawawi Ibrahim, eds. (2016). *Human insecurities in Southeast Asia*. Springer.

Carnegie, Paul J. and Victor T. King. 2020. "Mapping Circumstances in Oceania: Reconsidering Human Security in an Age of Globalisation." In *Mapping Security in the Pacific*, edited by S. Amin, D. Watson and C. Girard. Routledge: 15–29.

Carnegie, Paul J., Victor T. King, and Magne Knudsen. 2021. "Human security, marginality, and precariousness in Southeast Asia." *International Journal of Asia Pacific Studies*, 17(1): 1–28.

Chu, Ta-Wei. 2017. "Riparians versus the state in Southeast Asia: human security and hydropower struggles along the Mekong's Sesan tributary." *Asian Survey* 57(6): 1086–109.

Curley, Melissa. 2018. "Governing civil Society in Cambodia: implications of the NGO Law for the 'rule of law'." *Asian Studies Review* 42(2): 247–67.

Djenontin, Ida Nadia S., and Alison M. Meadow. 2018. "The art of co-production of knowledge in environmental sciences and management." *Environmental Management* 61(6): 885–903.

Dosch, Jörn, and Oliver Hensengerth. 2005. "Sub-regional cooperation in Southeast Asia." *European Journal of East Asian Studies* 4(2): 263–85.

Earth Rights International. 2014. "Thousands of lives to be devastated by LS2." https://earthrights.org/media/thousands-of-lives-to-be-devastated-by-lower-sesan-2-dam-international-groups-demand-new-eia/.

Gibbons Michael, Camille Limoges, Helga Nowotny, Simon Schwartzman, Peter Seot, and Martin Trow. 1994. *The new production of knowledge: the dynamics of science and research in contemporary societies*. SAGE Publications.

Global Witness. 2007. *Cambodia's family trees: illegal logging and the stripping of public assets by Cambodia's elite*. Global Witness.

Global Witness. 2010. *Shifting sand: how Singapore's demand for Cambodian sand threatens ecosystems and undermines good governance*. Global Witness.

Global Witness. 2016. *Hostile takeover: the empire of Cambodia's ruling family*. Global Witness.

Godemann, Jasmin. 2008. "Knowledge integration: a key challenge for transdisciplinary cooperation." *Environmental Education Research* 14(6): 625–41.

Grimsditch, Mark. 2012. *3S Rivers under threat*. Banlung Town, Cambodia and Berkeley, USA: 3SPN and International Rivers.

Heng, Pheakdey. 2015. "China's role in the Cambodian energy sector." *South East Asia Research* 23(3): 405–22.

Hensengerth, Oliver. 2017. Regionalism, identity, and hydropower dams: the Chinese-built Lower Sesan 2 Dam in Cambodia. *Journal of Current Chinese Affairs* 46(3): 85–118.

Hameiri, Shahar, and Lee Jones. 2013. "The politics and governance of non-traditional security." *International Studies Quarterly* 57(3): 462–73.

Hameiri, Shahar, and Lee Jones. 2014. "Murdoch international: the 'Murdoch School' in international relations." *Asia Research Centre Working Paper No.* 178. Murdoch University.

Hameiri, Shahar, and Lee Jones. 2015a. "Regulatory regionalism and anti-money-laundering governance in Asia." *Australian Journal of International Affairs* 69(2): 155–63.

Hameiri, Shahar, and Lee Jones. 2015b. "The political economy of non-traditional security: explaining the governance of Avian Influenza in Indonesia." *International Politics* 52(4): 445–65.

Hameiri, Shahar, and Lee Jones. 2017. "Beyond hybridity to the politics of scale: international intervention and 'local' politics." *Development and Change* 48(1): 54–77.

Hughes, Caroline. 2006. "The politics of gifts: tradition and regimentation in contemporary Cambodia." *Journal of Southeast Asian Studies* 37(3): 469–89.

Hughes, Caroline. 2011. "Soldiers, monks, borders: violence and contestation in the Greater Mekong Sub-Region." *Journal of Contemporary Asia* 41(2): 18--205.

Jessop, Bob. 2008. *State power: a strategic-relational approach*. Polity Press.

Jones, Lee. 2009. "Democratization and foreign policy in Southeast Asia." *Cambridge Review of International Affairs* 22(3): 387–406.

Jones, Lee. 2010a. "ASEAN's unchanged melody? The theory and practice of non-interference in Southeast Asia." *The Pacific Review* 23(4): 479–502.

Jones, Lee. 2010b. "(Post-)colonial state-building and state failure in East Timor." *Conflict, Security & Development* 10(4): 547–75.

Jones, Lee. 2011. "Beyond securitization: explaining the scope of security policy in Southeast Asia." *International Relations of the Asia-Pacific* 11(3): 403–32.

Jones, Lee. 2012. *ASEAN, sovereignty, and intervention in Southeast Asia*. Palgrave Macmillan.

Jones, Lee. 2013. "Sovereignty, intervention, and social order in revolutionary times." *Review of International Studies* 39(5): 1149–67.

Jones, Lee. 2016. "Explaining the failure of the ASEAN Economic Community." *The Pacific Review* 29(5): 647–70.

Khmer Times. 2018. "Dam in Sesan begins operation." Accessed April 1, 2021 https://www.khmertimeskh.com/560364/dam-in-sesan-begins-operations/.

Kimkong Ham, Samchan Hay, Thea Sok, Vichet Sim, and Lor Rasmey. 2013. *Improving hydropower projects' decision-making processes in the Mekong Basin: case studies of the Lower Sesan 2 and Kamchay Hydropower Projects, Cambodia.* Royal University of Phnom Penh.

King, Victor T., and Paul J. Carnegie. 2018. "Towards a social science understanding of human security." *Journal of Human Security Studies*, 7(1): 1–17.

Lawless, Robert. 1976. "The Indonesian takeover of East Timor." *Asian Survey* 16(10): 948–64.

Mekong Watch. 2015. *Lower Sesan 2 Dam compensation and resettlement program implementation.* Mekong Watch.

Morgenbesser, Lee, and Meredith Weiss. 2018. "Survive and thrive: Field research in authoritarian Southeast Asia." *Asian Studies Review* 42(3): 1–17.

NGOF. 2012. *Lower Sesan 2 Hydropower Dam: current livelihoods of local communities.* NGOF.

NGOF. 2015. *Survey on the compensation policies and market property price of the Lower Sesan 2 Dam Development Project.* NGOF.

Paling, Willem. (2012). Planning a future for Phnom Penh. *Urban Studies* 49(13): 2889–912.

Pohl, Christian. 2005. "Transdisciplinary collaboration in environmental research." *Futures* 37(10): 1159–78.

Rosario, Teresita Cruz-Del, and Jonathan Rigg. 2019. "Living in an age of precarity in 21st Century Asia." *Journal of Contemporary Asia*, 49(4): 517–27.

Rosendahl, J., M.A. Zanella, S. Rist, and Weigelt J (2015). Scientists' situated knowledge: strong objectivity in transdisciplinarity. *Futures* 17(27): 17–27.

Royal Government of Cambodia. 2006. *National Strategic Development Plan 2006–2010.* Royal Government of Cambodia.

Schmidt, Laura, and Michael Pröpper. 2017. "Transdisciplinarity as a real-world challenge." *Sustainability Science* 12(3): 365–79.

Simpson, Hugh, Rob de Loë, and Jean Andrey. 2015. "Vernacular knowledge and water management – towards the integration of expert science and local knowledge in Ontario, Canada." *Water Alternatives* 8(3): 352–72.

Siew, Tuck Fatt, Thomas Aenis, Joachim H. Spangenberg, Alexandra Nauditt, Petra Döll, Sina K. Frank, Lars Ribbe, Beatriz Rodriguez-Labajos, Christian Rumbaur, Josef Settele, and Jue Wang. 2016. "Transdisciplinary research in support of land and water management in China and Southeast Asia." *Sustainability Science* 11(5): 813–29.

Sithirith, Mak. 2016. "Dams and State Security: Damming the 3S Rivers as a Threat to Cambodian State Security." *Asia Pacific Viewpoint* 57(1): 60–75.

Sokphea, Young. 2016. "Popular Resistance in Cambodia: The Rationale behind Government." *Asian Politics & Policy* 8(4): 593–613.

South China Morning Post. 2018. "Laos dam Collapse." Accessed May 18, 2021, from https://www.scmp.com/news/asia/southeast-asia/article/2160635/laos-dam-collapse-mekong-river-projects-move-forward.

Swyngedouw, Erik A. 1997. "Neither Global nor Local: 'Glocalization' and the Politics of Scale." In *Spaces of Globalization: Reasserting the Power of the Local*, edited by Kevin Cox. Guildford: 137–66.

Un, Kheang. 2005. "Patronage Politics and Hybrid Democracy: Political Change in Cambodia, 1993–2003." *Asian Perspective* 29(2): 203–30.

van Voorst, Roanne. 2016. *Natural Hazards, Risk and Vulnerability: Floods and Slum Life in Indonesia*. Routledge.

Ziv, Guy, Eric Baran, So Sam, Ignacio Rodriguez-Iturbe, and Simon A. Levin. 2012. "Trading-off Fish Biodiversity, Food Security and Hydropower in the Mekong River Basin." *Proceedings of the National Academy of Sciences* 109(15): 5609–14.

CHAPTER 14

Ecological Governance of the Padma and the Teesta Rivers for Sustainable Development in Bangladesh

Sk. Tawfique M. Haque and Md. Parvez Hasan Yousuf

The rivers of tropical Asia are of immense importance in the sustainable development of South Asia. The rivers play an essential role in shaping the lives and livelihoods of the people of this region. The Ganges, the Brahmaputra, and the Meghna/Barak (GBM), three of the world's largest rivers, merge into the Bay of Bengal, creating a network of rivers across South Asia. This river system flows across several nations, including Bangladesh, China, India, and Nepal. China is the only one of these to contribute to the Brahmaputra's flow, and Nepal is responsible for the Ganges' flow. China and Nepal are also upper riparian nations, and the tributaries that flow into the GBM systems have yet to be fully used.

The Padma and the Teesta are two major rivers in Bangladesh. The Padma is the principal distributary of the Ganges. It flows primarily southeast for 356 kilometres (221 mi) until it meets the Meghna river near the Bay of Bengal (Allison and Mead 1998). The Ganges is a river that flows south of China, Nepal, and India into the Indo-Gangetic Plain. It has several distributaries and tributaries that pour into it. The Ganges has its source in the highlands of the Himalayas and the Tibetan Plateau. The Ganges and Brahmaputra merge in a large delta area in Bangladesh and India before flowing into the Bay of Bengal. The basin has a diverse spectrum of topography, ecosystems, and biodiversity, ranging from the alpine desert rain-shadow regions of the Tibetan Plateau to the steep mountain ranges of the highest peaks in the world, including Mt. Everest (8848 m) to the wide-open plains (Rahaman and Mamun 2020).

The Teesta river has a length of 414 kilometres (257 mi) and originates in the eastern Himalayan Pauhunri Mountain. It then passes through the Indian states of Sikkim and West Bengal through Rangpur before flowing into the Bay of Bengal (Mullick, Babel, and Perret 2010). It drains 12,540 km^2 of land (4,840 sq mi). It merges with the Brahmaputra river in Bangladesh's Phulchhari Upazila. The river's length is 305 kilometres (190 mi) in India and 109 kilometres (68 mi) in Bangladesh (Rahaman and Mamun 2020).

The Ganges is the largest river in West Bengal, whereas the Teesta is the major river in Sikkim, India. The Ganges takes the name of the Padma after it

 | HTTPS://DOI.ORG/10.1163/9789004758100-016

enters Bangladesh. The Padma and the Teesta are two important transboundary rivers causing water-related disputes between Bangladesh and India.

In the plain regions of both Bangladesh and India, floods occur frequently during the monsoon season. Still, during the lean season, the river transforms into a various-channel river with sand bars and streams switching back and forth between the mainstream banks (Rahaman and Mamun 2020).

The impact forecasts on hydrology and water availability have significant policy repercussions and provide unambiguous indicators of trends that must be considered in future planning. The potential difficulties varied between basins, as well as between the upstream and downstream regions and various catchments among the basins (Nepal and Shrestha 2015).

The rivers in Bangladesh are becoming unstable and ecologically vulnerable due to the sharp decline in the dry season flow, which is followed by enormous volumes in the wet season, which harms the morphology of the rivers. Riverbeds desiccate and harden during the dry season, which lessens the previous rainy season's bed scouring. Rivers erode their banks to handle the high volume during the rainy season, which has disastrous impacts on the lives and property of those who reside in the river basins. Bangladesh's rivers are consequently growing broader and shallower. Freshwater species like fish, dolphins, crabs, and other aquatic animals and plants are also being affected due to a lack of proper river governance. Excessive fishing in the estuaries and mainstream is causing river resource depletion. The fishermen are catching fewer fish, which is changing their livelihood options. The aquatic plants are being destroyed due to pollution and less water during the dry spells.

The term *Anthropocene* refers to the current geological epoch in which human activity has taken centre stage in shaping the Earth's environment (Chua and Fair 2019). The unprecedented scope and intensity of human-induced changes to the Earth—such as climate change and its effects, soil, water, and air pollution, species extinction, and biodiversity loss—define the Anthropocene. Addressing the environmental problems caused by human activity, and ensuring a habitable planet for future generations, underscores the critical need for sustainable practices and collective efforts.

There remains an "epistemic dualism" within the concept of the Anthropocene: devastation and healing are equally important aspects of this epoch. In recent decades, anthropogenic interventions have increasingly been implemented under the supervision of postcolonial states (Iqbal, Hassan, and Kumpoh 2026).

Rivers are significantly affected by human activities such as pollution, resource overuse, sand extraction, encroachment, the construction of unplanned dams, embankments, bridges, and oil spillage. Transboundary rivers are particularly vulnerable to anthropogenic pressures as they cross multiple international borders and are often subject to limited monitoring. Human interventions—such as collective ecological water governance, reducing river pollution, and implementing planned water management—can help maintain the health and ecological well-being of rivers. As rivers serve as the lifeblood of a country's economy, they need to be navigable, ecologically healthy, and jointly governed by all riparian states.

In the neoliberal protocol and the Anthropocene, many policies are devised from the business point of view to make more profit, neglecting the concerns of exploiting the natural resources. There are many laws, acts, rules, and regulations to govern Bangladesh's environment, rivers, and water bodies. Though some human interests are considered, there is little consideration for non-human beings like fish, crabs, dolphins, and other aquatic animals and plants inhabiting the rivers while formulating such environmental laws and implementing the river governance mechanisms. This chapter will delve into the river governance of the Padma and the Teesta from an ecological point of view, where healthy rivers will be resourceful and safe havens for both human and non-human beings, thus ensuring comprehensive sustainable development.

1 Ecological Governance of the River

A river is considered "ecologically healthy" when the ecosystem's requirements are met, allowing various species to thrive and achieve a favourable ecological status. Therefore, the creation of a healthy river is facilitated by the disciplines of social-economic, legal, ecological, and hydrological studies. Effective water quality regulation largely depends on the relationships between these disciplines (Wuijts, Driessen, and Van Rijswick 2018). Thus, managing water quality requires addressing the connections between how humans use ecosystems—ecosystem services—and the regulatory framework needed to consider ecosystems' inherent worth when making societal decisions (Watson and Zakri 2003). The decisions should also consider the well-being of aquatic life by regulating pollution and monitoring the water quality regularly. There should be no industries on the banks of the rivers. Even if some are established, they must set up Effluent Treatment Plants (ETPs) before releasing wastewater into the rivers.

Moreover, solving the transboundary river issues is essential because this will determine the water flow in the dry and monsoon seasons. The people and aquatic life suffer due to less water. The fish cannot thrive much as they die due to water scarcity. The farmers cannot irrigate their lands. However, during the monsoon season, as the rivers flood their banks, the riverine people suffer the most from losing property, crops, and livelihood options. The aquatic life is washed away, and the biodiversity is hampered, too. So, equitable water sharing is a prerequisite for collaborative ecological transboundary river governance between neighbouring countries.

Governance is more than simply government; it involves institutions and communities that control the use and distribution of resources (like water) through political channels. It also covers official and informal institutions, such as markets, laws, rules, and social norms. Management is typically considered as making things happen immediately on the ground, operationalizing governance processes, and encouraging transparency in decision-making (Chan, Roy, and Chaffin 2016).

The World Water Assessment Programme has underlined the importance of putting water at the forefront of environmental decision-making at all scales, including local, regional, and global (Rahaman 2009). Goal 6 of the Sustainable Development Goals (SDGs), "Ensure availability and sustainable water and sanitation management for all", was created in response and includes many targets. In addition to measurable objectives for gaining access to clean drinking water and sanitation, these goals explicitly contain water governance components, updating the procedures that the Millennium Development Goals (MDGs) initiated.

Collaborative governance has become the norm in environmental management, particularly river basin management (Baudoin and Gittins 2021). Overuse and disagreements are prevalent as each actor has a distinct concept of effective water management (Baudoin and Gittins 2021). Urban populace, modern economies, and climate change are putting unprecedented strains on rivers. River basins, or watersheds, are essential to moving freshwater between sources and sinks in the landscape. In the process, they drain all land related to the water cycle. However, river basins can be vast and contain social components like actor groups and organizations as well as ecological and hydrological factors. River basins are Social-Ecological Systems (SESs) in this sense (Ostrom 2009). Like those in a sizable river basin, a shared water system's actors frequently have little in common outside their reliance on the same essential resource. The management of river basins presents issues related to hydrogeomorphology, biology, and water quality and quantity, among other interconnected themes. Water resources are considered common-pool resources since they are rivalrous and non-excludable. As a result, they may be overused and vulnerable to unfavourable externalities like pollution, which could cause the resources to run out (Ostrom 1990).

The world's nations are challenged by the lofty goals set forth by the United Nations' Sustainable Development Goal (SDG) No. 6, which calls for the full implementation of Integrated Water Resources Management at all levels by 2030. Climate change and socioeconomic advancements compound this difficulty and make it longer than the SDGs, necessitating a comprehensive strategy to maintain healthy ecosystems (Wuijts et al. 2019). Thus, proper river governance is required to achieve the following SDG goals: 6. Clean water and sanitation, 13. Climate Action, 14. Life below water, and 15. Life on land.

2 River Governance in Bangladesh: Policies and Challenges

The demands of a river frequently have to be weighed against those of society, including flood control, agriculture, urban and industrial pollutants, fishing, and shipping. Politicians must decide which issues to prioritize and how to strike a balance between the requirements of the river and other societal concerns to maximize efficacy. The river could be given legal rights to address this matter within the existing institutional and legal framework. This transfer may present a chance to discuss the significance of healthy rivers for present and future generations, but it needs to be supported by legally binding regulations (Wuijts et al. 2019).

In Bangladesh, river governance comprises various laws, acts, and policies. There are 56 Acts or Ordinances about water in Bangladesh. Of them, eight were passed during the British era (before 1947), fourteen during the Pakistani era (from 1948 to 1971), and thirty-four during the Bangladeshi era (1972 to present). Some vital policy frameworks are discussed below.

3 Rivers' Living Being Status

An extraordinary judicial decision was made in Bangladesh, a nation with hundreds of rivers that are becoming more and more polluted, at the start of July 2019. The Supreme Court accorded living being status to all of the nation's rivers to safeguard them better. It is a crucial step for the rights of the environment. The Court ruled that protecting rivers should be done "whatever the cost" and stated, "Water is certainly the most pressing environmental subject of the next century". Its goal is to safeguard the largest delta in the world from further deterioration brought on by pollution, unlawful dredging, and other human activities (IFGR 2019). Granting rights to non-human entities may be an effective strategy to combat the deterioration or even irreversible destruction of nature and protect everyone interacting with and depending on it for survival.

Nonetheless, we need to reconsider how we interact with the natural world and uphold the principles of the law. Human rights advocacy groups in Bangladesh are alerting people to the increasing danger of displacing impoverished fishermen and farmers who rely upon the rivers for their livelihoods and live in shacks. Therefore, establishing regulatory frameworks that consider the entire ecosystem is the goal (IFGR 2019).

Ecological river governance has gained ground in Bangladesh following this supreme court ruling. The living status of rivers will help them be ecologically healthy if the laws are appropriately applied. Anyone polluting and grabbing the rivers can be held accountable as they are trying or threatening to destroy a living being. Now, people need to be aware of the rights of the rivers because, like a person, a river also needs to be healthy and pollution-free to thrive. If a river is properly governed using nature-based solutions, our ecosystem and weather would benefit us. This will also help us to tackle adverse climatic conditions and ensure sustainable development.

4 National River Conservation Commission (NRCC)

Following the directive from the Honorable High Court Division in Writ Petition No. 3503 of 2009, the National River Conservation Commission (NRCC) was established by the National River Conservation Commission Act 2013. The National River Conservation Commission (NRCC) is the official custodian of the nation's waterways. Fighting to recognize rivers as legal entities may restrict some highly polluting industrial activity. Therefore, the goal is to create regulatory frameworks that consider the entire ecosystem. The river commission is made up of a lawyer, an expert in the environment, a hydrologist, and a chairperson. The panel will guide the government and oversee the work of the ministries in charge of river and water management. The Act needs more precise directives and signatures about the instruments needed to put the Act into effect (Arifuzzaman, Hannan, Rahman, and Rahman 2019). Specific vague arrangements can be opened to reveal deceptive information. The Commission lacks formal authority and functions only as a recommendation body. Should its recommendations be ignored, it cannot take any further action. Three major rivers—the Ganges, the Brahmaputra, and the Meghna—that originate in the Himalayas and meet in Bangladesh before flowing into the Bay of Bengal present another challenge. The government of Bangladesh cannot compel its neighbours, particularly India, to uphold the legislation (IFGR 2019).

The NRCC has limited power to govern the rivers. If the power is enhanced and they fully implement the river laws, then the rivers of Bangladesh will be more protected. The Joint River Commission (JRC) is under the Ministry of Water Resources, while the NRCC is under the Ministry of Shipping. Moreover, the River Research Institute (RRI), the Bangladesh Water Development Board (BWDB), and the Department of Bangladesh Haor and Wetlands Development are under the Ministry of Water Resources, while the Bangladesh Inland Water Transport Authority (BIWTA) is under the Ministry of Shipping. It hampers our country's river governance as the actions between various government organizations need to be coordinated. Inter-ministerial coordination is required to govern the rivers properly. The river grabbers and polluters must face the music. Their punishments should become exemplary so that others become warned of the consequences of polluting and grabbing the rivers. As the Padma and the Teesta are transboundary rivers, proper communication and collaboration between the NRCC and JRC is required to properly share our river waters for ecological river governance and sustainable development.

5 Bangladesh Environment Conservation Act (BECA), 1995

The Bangladeshi government passed the Bangladesh Environment Conservation Act (BECA) in 1995 as a collection of regulations to protect the country's environment. Its primary objectives were to "provide for conservation of the environment, improvement of environmental standards and control and mitigation of environmental pollution" (Bangladesh Environmental Conservation Act, 1995). The Act establishes operational terminology such as ecosystem, pollution, waste, and dangerous chemicals that were not used previously. The National Environment Management Action Plan (NEMAP) in 1992, the Forest Policy in 1994, and the Forestry Master Plan (1993–2012) in 1993 were all preceded by the Act. The Ministry of Environment and Forest was established in 1989. The Act became operative on 30 May, 1995.

This law designates seven sites in Bangladesh as Ecologically Critical sites. Bangladesh's environmental degradation persists despite the Act and the rules and policies that support it, primarily due to the enormous population and low enactment of laws. The Department of Environment designated the four rivers surrounding the nation's capital, Dhaka, as Ecologically Critical Area (ECA) in September 2009. These rivers are the Buriganga river, Shitalakshya river, Turag river, and Balu river.

Numerous areas for improvement and gaps exist in the Bangladesh Environment Conservation Act, 1995, and Rules, 1997. Despite the Act's establishment in 1995, Bangladesh's environmental conditions have not significantly improved. The BECA of 1995 is a cure-oriented law that only addresses therapeutic measures rather than preventive ones, as evidenced by the terms "improvement of environmental standards" and "mitigation of environmental pollution" used in its preamble.

Many rivers die from pollution, grabbing, siltation, and poor governance. The BECA can be utilized and integrated to ensure proper ecological river governance because protecting the rivers' ecosystem and conserving biodiversity are essential to keeping the rivers healthy and alive.

6 Conservation of Fisheries and Aquatic Resources

The Bangladeshi government has established a multitude of acts, ordinances, and rules to establish a framework for the management, development, and conservation of the country's aquatic resources and fisheries industry. Additionally, several actions have been taken by the Bangladeshi fisheries authorities to preserve and oversee artisanal Hilsa fishing. The Marine Fisheries Ordinance (1983) is the primary statute governing marine fisheries, whereas the Protection and Conservation of Fish Act (1950), as revised by the Protection and Conservation (Amendment) Ordinance (1982), is the fundamental act governing Hilsa fisheries. Based on these two acts and other laws, there are several conservation and protection measures for Hilsa. These consist of prohibiting fishing in specific locations, limiting the type of gear allowed, limiting the season, and establishing rules for fishing vessels (Islam, Mohammed, and Ali 2016).

Additionally, four Hilsa spawning areas have been marked off. After an 11-day fishing moratorium each year in October, fishing is permitted to support the successful spawning of gravid Hilsa, or female Hilsa-carrying eggs. Any fish captured during the prohibition period in sanctuaries or spawning zones using any gear may be seized and forfeited, according to the 1985 modification to the Protection and Conservation of Fish Rules. The government has implemented some management measures to preserve the Hilsa fishery based on these policies and other fisheries rules. Any government agencies that are engaged in fishing, as well as any bodies of water used for fishing, are covered by these policies. Numerous other pertinent topics, such as land use, industrial pollution, and quality control, are also covered by these rules (Islam, Mohammed, and Ali 2016).

The Bangladeshi government implemented the Hilsa Fisheries Management Action Plan (HFMAP) in 2005. It comprised several measures, such as the establishment of fish sanctuaries to facilitate the continuous rearing and feeding of young fish, the enforcement of patrols against the possession and use of monofilament nylon nets (*Current Jal*), the introduction of modest food grain compensation to underprivileged fishermen (currently 40 kg of rice/fisherman for nearly seven months) during fishing prohibition periods, and the facilitation of Alternative Income Generating Activities (AIGAs). These programmes support Bangladesh's more consistent use of Hilsa resources.

Protecting fisheries is impacted by specific regulations that originate from outside the industry. The 1992 Environment Policy, for example, strongly emphasizes the development and conservation of fisheries and the assessment of any initiatives that may influence these resources. Additionally, a magistrate may run a mobile court to handle offences "on the spot" following the Mobile Court Ordinance 2007. This mobile court is in use during the fishing prohibition time and has the authority to sentence offenders following Act V of the 1898 Code of Criminal Procedure. It may also request assistance from law enforcement or other organizations (Islam, Mohammed, and Ali 2016).

These laws and regulations related to fisheries can help us understand the importance of ecological river governance, as it largely depends on properly utilizing resources and conserving their natural habitat by saving the rivers from pollution.

7 Bangladesh Water Act (BWA) 2013

A significant piece of water legislation in Bangladesh for water governance is the Bangladesh Water Act (BWA) 2013. The BWA is a generally effective law for the National Water Policy. This law serves as a foundation for unifying and organizing the administration of the country's water resources. According to the Water Act, the National Water Resources Council's (NWRC) Executive Committee is the primary water regulator. The ECNWRC's secretariat is the Water Resources Planning Organization (WARPO). The NWRC is responsible for setting policy and implementing the Bangladesh Water Act, National Water Policy, and National Water Resource Plan. The Executive Committee (EC) is responsible for maintaining the normal flow of water, protecting water zones, tackling the drinking water crisis, and issuing Clearance Certificates for any development project relating to water resources. Without the EC's approval, no person or organization may extract, distribute, use, develop, protect, or conserve water resources. They also may demolish structures, obstruct the natural flow of rivers and creeks, and conserve biodiversity.

The Water Act aspires to develop a new, comprehensive plan for the protection, improvement, and sustainable utilization of the country's rivers, reservoirs, estuaries, marine waters, and groundwater. The BWA commonly agrees that a comprehensive legal framework is necessary to address the expanding issues around water rights, resource preservation, usage, and management (Bangladesh Water Act 2013). Under the Act, any area facing a critical situation can be declared a 'Water Stress Area' and demarcated into the water zones for efficient use of water resources. According to the BWA, there should be less abuse of water resources, and they should be shielded from contamination and harm of any type. The provisions of this Act permit restrictions or outright bans on textile firms' ability to pollute water. Companies that violate this Act may be removed (Mim 2022). Thus, industrial water pollution can be controlled by fully enacting this act. However, the act must still be fully implemented, as water pollution is rampant in Bangladesh.

8 Shortcomings of the Bangladesh Water Act 2013

The Act may be abused because of the Executive Committee's unrestricted authority and lack of liability. The people, ultimately the owners of water resources, are not permitted to bring legal action or complaints against other people or government organizations that violate the Act. Previous water policies mandated that the government create a code to ensure that water is used rationally for all purposes, rights, and administration. Still, the matter must be included in this Water Act 2013. The creation of Effluent Treatment Plants (ETPs) and the determination of discharge standards are undefined. The Act does not sufficiently address the punishment for the deterioration of water quality brought on by discharges from factories and other causes of pollution.

Although the Bangladesh Water Act (2013) and the National Water Policy (1999) are regarded as the country's critical pieces of legislation, they both fall short in terms of offering practical guidance for authorization, extraction limits, oversight, quality protection, water harvesting techniques, and recharging mechanisms. Current organizations are plagued by persistent crises, politics, corruption, a lack of public involvement, a lack of inter-institutional coordination, poor management, and empirical assessment (Bhattacharjee et al. 2019). As a result, ecological river governance is also being hampered due to lax implementation of the BWA 2013.

9 Challenges in Water Governance: the Case of Bangladesh–India Transboundary Rivers

According to Chan, Roy, and Chaffin (2016), the key obstacles to effective water governance in Bangladesh include (i) a rising national demand for water and increased strain on groundwater resources, (ii) adverse effects of climate change, such as droughts, floods, and other natural calamities that significantly affect the availability of reliable water and water for crop yields, and (iii) an alarming increase in pollution as a result of unplanned industrial expansion and subpar sanitation conditions. The first two challenges are simultaneously related to national river governance as well as transboundary water management. The following discussion deals with Bangladesh's transboundary river governance issue.

9.1 *Ganges Water Treaty of 1996*

Both of the agreements, which were signed in 1977 and 1996, regarding the sharing of Ganges water during the dry season, were developed by the Joint River Commission (JRC). Bangladesh and India agreed to share the Ganges' flow downstream of Farakka for 30 years in 1996. The need for more accurate and trustworthy data on the flows of common rivers is one of the fundamental issues that both Bangladesh and India encounter when discussing water-related issues. Data on flow is necessary for flood forecasting and warning in addition to serving the crucial purpose of establishing the baseline. This is especially important for flashy rivers like the Monu and the Muhuri on Bangladesh's eastern border (Faisal and Nishat 2000).

The only significant difference between this Treaty and the 1977 Agreement is that the water-sharing scheme was unrelated to the flow augmentation issue. During the dry seasons, separated into 10-day intervals between 1 January and 31 May, water is distributed according to a formula. In other words, both countries receive an equal portion (50%) if the flow is less than or equal to 70,000 cusecs. Bangladesh receives 35,000 cusecs of flow if it is more significant than 70,000 but less than 75,000 cusecs, with the remaining cusec going to India. India receives 40,000 cusecs, and Bangladesh receives the remaining 40,000 cusecs if the flow is 75,000 cusecs or above. Each nation is given a guaranteed 35,000 cusecs in alternate 10-day intervals during the most crucial times between 11 March and 10 May (Rahman, Islam, Navera, and Ludwig 2019).

Notably, this agreement states that "the upper riparian will take every effort" to protect the flows of water at Farakka, as in the 40-year average availability, due to the worry about more significant water diversion upstream of Farakka.

This is not a promise, just an official acknowledgment that flows to Farakka have been continuously dropping over the past three decades (Faisal and Nishat 2000).

The 30-year Ganges Water Sharing Treaty of 1996 is frequently touted as a success. Unfortunately, after the Farakka Barrage was put into operation, the Ganges flow in Bangladesh reduced from more than 2,000 cusecs during the dry season to less than 500 cusecs, and this flow essentially stayed unaltered after the 1996 deal. The absence of any assurance clause in this pact complicates Bangladesh's situation. India can now allow the Ganges flow into Bangladesh to fall far below the minimum environmental flow. Significant evidence of the two nations' desire to resolve their impasse and find a solution acceptable to both was the signing of the Ganges Water Treaty (GWT) in December 1996 (Brichieri-Colombi and Bradnock 2003). Bangladesh can join the growing chorus of Indians calling for the destruction of the Farakka Barrage. Farakka fails to remove sediment from the Hooghly river, leading to flooding and erosive conditions in Murshidabad and Bihar. The moment has arrived to repair the harm that Farakka has caused to both Bangladesh and India (Islam 2022).

Rahman et al. (2019) stated that the Ganges Water Sharing Treaty, signed in 1996, was seen as an opportunity for regional collaboration between India and Bangladesh. However, after the treaty was signed, its effectiveness was questioned multiple times because of the limited flow availability at Farakka. According to Rahman et al. (2019), the main obstacles to the Treaty's successful implementation are (i) an inaccurate estimate of the future obtainable flows at Farakka, (ii) an inappropriate guarantee of flow during crucial dry spells, (iii) insufficient protection for flows at Hardinge Bridge, (iv) the absence of a guarantee clause for Bangladesh, and (v) a failure to take environmental and economic drivers into account.

In their study, Rahman, Islam, Navera, and Ludwig (2019) discovered that in 94 out of 300 instances from 1997 to 2016, Bangladesh received less water at the Hardinge Bridge than what was discharged through the Farakka barrage in India. The authors note that, if they account for the essential dry spells, Bangladesh failed to receive its guaranteed supply 39 out of 60 times throughout the same timeframe. These failures were common during the dry spells from 2008 to 2011.

According to the treaty's terms, India has to give Bangladesh a certain amount of water every ten days during the dry season. When both countries' water demands are at their highest—between 11 March and 10 May—India often denies Bangladesh its fair share of water. Furthermore, India can declare itself in compliance with the agreement almost any year by discharging additional water than needed throughout its remaining water-sharing seasons (Thomas 2017).

A Joint Technical Committee was established "to conduct a study for optimum utilization of water received by Bangladesh under the provisions of the Ganges Water Sharing Treaty, 1996", which was also welcomed by both countries. The Joint Rivers Commission included additional transboundary rivers for interim water sharing. It was well-received by both Bangladesh and India, who saw it as a means of "prioritizing data exchange and formulating the framework of the interim water sharing agreements" (Ranjan 2022).

Discussing the 1996 Ganges Water Treaty's future during the last Golden Chapter of India–Bangladesh relations is appropriate. Article XII states that the pact will be in effect for 30 years, or until 2026, and may be extended by mutual agreement. Disagreements have previously arisen over it between Bangladesh and India. For example, many people in Bangladesh think that during the dry season, the nation does not receive as much water as promised by the treaty (Ranjan 2022). Upon reviewing the 1996 treaty, Bangladesh and India are expected to face significant obstacles. First, the disparity between the supply of and demand for water is growing in the Ganges Basin region. The Ganges Basin's severe water shortage has been worsened by population growth, industrialization, urbanization, agriculture, and climate change. In addition, the groundwater table is falling. Second, both nations are concerned about the Farakka barrage. Bangladesh accuses Farakka of producing silt and depriving the Sundarban delta of necessary waters, thereby endangering it. Even the Indian state of Bihar's chief minister, Nitish Kumar, requested the Farakka barrage's decommissioning in 2017. According to him, the barrage is useless and annually causes floods in Bihar. Third, and perhaps most importantly, is how the Indian states through which the Ganges river flows will significantly impact the deal's political outcome (Ranjan 2022).

The Ganges Treaty of 1996 is set to expire in 2026. It further stipulated that talks based on equality and non-harm principles should be conducted to share the water of other shared rivers. If it is not extended, there will not be a negotiated agreement for Ganges water sharing or a foundation for talks on other rivers in the years beyond 2026. Before the Ganges Treaty expires in 2026, a revised agreement should be reached involving Bangladesh, India, and Nepal, i.e., the states that make up its basin. It is time to go forward with future negotiations by applying the lessons learned from the past and considering the latest advancements in international law that support integrated river basin development (Islam 2023).

The Ganges water treaty has yet to be successful, and the Farakka Barrage has negative consequences, as is evident from the ongoing water scarcity in both countries. Humans need water for their lives and livelihood, and aquatic living and non-living objects also need water to survive. So, both countries need to settle the issues from the environmental and anthropogenic perspectives to usher in ecological governance while renewing the treaty.

9.2 *Uncertainty Over the Teesta Treaty*

Living along the banks of the Teesta, the people of northwest Bangladesh experience flooding and erosion during the monsoon season and water scarcity during the dry season. When the water flow is nearly entirely blocked by the Gajoldoba barrage of India during the dry season, the Teesta runs out of water. However, the Teesta swells and widens to up to 10 kilometres in certain spots during the monsoon (Roy and Roy 2022). According to one estimate, the Teesta river affects the livelihoods of 20 million people. Even while there is a draft agreement, if we ignore their concerns for too long and go on to other rivers that do not affect as many people, it will undoubtedly turn into a political and party issue. The Teesta river in India is home to 30–32 small and large dams for hydroelectric power that are estimated to impact each dam's water flow by about 5%, resulting in a seasonal flow deficit that drains into Bangladesh (Islam 2022). This is the main focus of Mamata Banerjee's (the current Chief Minister of West Bengal, India) argument over the Teesta. Bangladesh can fix this by collecting rain/flood water from the rainy season and letting it flow during the winter, as India will not demolish the dams. There are financial and environmental repercussions; one estimate places the total at about USD 1 billion. China expressed some interest in funding and implementing the Teesta river Management project, but India immediately opposed it. One option to address the issue is to form a consortium where many stakeholders can contribute resources and expertise to this infrastructure. India must understand that Bangladesh will only put off finding a solution for a while.

Bangladesh is dealing with the additional issue of the Indian barrages being now a second source of significant river flow changes, leading to terrible floods at unforeseen periods of the year. This has been especially true for the Teesta Basin, where a sudden onslaught of water that Gajoldoba operators released by activating all of the barrage's vents caused the people of the basin much suffering by ruining their crops and sweeping away their household belongings. There were six such incidents in 2022 alone (Tasneem 2022). However, West Bengal government's stance on the Teesta water-sharing problem in recent years made it abundantly evident how powerless the Union government might be if a riparian state were to decline to collaborate on the transboundary water agreement. The Union government agrees to confer with the state administration and involve the West Bengal Chief Minister, Ms. Mamata Banerjee, before taking further action on the Teesta waters issue. Nevertheless, when talks about renewing the treaty are underway, India–Bangladesh relations will be tested (Ranjan 2022).

There have been no notable developments in the negotiations on sharing other transboundary rivers, including the Teesta because they have yet to learn from the mistakes made with the Ganges. The Ganges pattern has been replicated in most of these negotiations: other basin states have been excluded, sharing of data and dialogue have been restricted to water availability at the river's tail end, the need to maintain environmental flows has been disregarded, and a solid and independent river commission has not been established (Islam 2023).

The Teesta treaty for equitable water sharing is inevitable for protecting the people and ecology of northern Bangladesh, especially in the Rangpur division. The river dries in the summer and floods during the monsoon season, causing enormous losses for agriculture, fishers, transport, biodiversity, and ecosystems in the region. So, governing Teesta by ensuring its natural wellbeing is also crucial for Bangladesh to achieve its sustainable development goals related to water, climate action, sustainable agriculture, life below water, and life on Earth.

10 Policy Gaps in Transboundary River Governance

Transboundary water issues have become complex and securitized due to rising geopolitical and geo-economic conflicts, maritime border conflicts, and claims to sovereignty for islands and coral reefs. In order to provide food security, minimize poverty, optimize the economic benefits of water use, and avoid and monitor water pollution, cooperative transboundary river governance is crucial. The fundamental issue with transboundary river water sharing is that there needs to be more water available to meet each water claim in its entirety. Water disputes arise from countries withdrawing more water to enhance its utility (Yuan et al. 2020).

Bangladesh acceded to the Convention on the Protection and Use of Transboundary Watercourses and International Lakes (UN Water Convention) in June 2025, becoming the first South Asian country to join this legally binding framework on sustainable transboundary water governance. The Convention provides an institutional and normative basis for managing shared river systems—such as the Ganges–Brahmaputra–Meghna basin—by promoting equitable and reasonable use, cooperation, and the peaceful resolution of water-related disputes among riparian states. However, Bangladesh has yet to sign the 1997 United Nations Convention on the Law of the Non-Navigational Uses of International Watercourses, which explicitly safeguards several legal entitlements of downstream riparian countries. Accession to this Convention would strengthen Bangladesh's legal position and could enable it to encourage India and other South Asian states to adopt a common, widely recognized normative framework for the governance of international rivers. Ratification may also facilitate Bangladesh's effective engagement with international legal and diplomatic forums in advancing its claims over shared transboundary rivers (Islam 2022).

To help India expedite the growth of its seven northeast provinces, Bangladesh might link the problem of river sharing with the subject of transit, transshipment, ports, and other amenities that it has previously granted and promised to grant to India. The people of Bangladesh and India must have favourable feelings toward one another so that their countries can work together in a long-lasting and comprehensive way. The resumption of the rivers' total natural flows would improve the relationship between the two countries. India should, consequently, reassess its river governance policies vis-à-vis Bangladesh, particularly in light of the political developments following the July–August 2024 mass uprising in Bangladesh.

The 1996 Ganges Treaty's objectives of equity and non-harm have yet to be entirely met. The state parties have yet to demonstrate an interest in fixing the 1996 treaty's shortcomings, and the joint river commission tasked with overseeing its implementation meets irregularly. Both sides have frequently voiced concerns about the treaty's subpar implementation (Islam 2023).

The governance of transboundary water resources is based on various ideas. International treaties, legal rulings, and conventions support managing transboundary water resources. These include the idea of "limited territorial sovereignty", "equitable and reasonable use", "the duty not to materially injure", "cooperation", "information exchange", "notice and consultation", and "the peaceful resolution of conflicts". The 1966 Helsinki Rules on the Uses of the Waters of International Rivers (hence referred to as the Helsinki Rules) and the 1997 UN Convention on Non-Navigational Uses of International Watercourses (herein referred to as the UN Watercourses Convention) are both based on these concepts (Giordano and Wolf 2003, 167; Rahaman 2009). The need to avoid causing "substantial harm" in terms of cooperation, information sharing, notification, consultation, and the peaceful resolution of conflicts were all integrated into this treaty (Rahaman 2009).

The four riparian nations of the Ganges river (India, Nepal, Bangladesh, and China) have never joined in a water-management treaty for the Ganges river, nor have any joint river basin management forums for these riparian states been established, despite the river's enormous importance to the livelihood of most riparian countries. This may result from the political and ideological divisions among the riparian nations and their various and potentially competing interests (Salman and Uprety 1999).

All of the South Asian basin states, particularly Bangladesh and India, ought to have a broader perspective in order to realize that the best way to meet the diverse needs of each international river basin state and guarantee its proper protection would be through integrated, multilateral, and basin-wide water resource management. The regulations reflected in the 1997 Watercourse Convention and other pertinent documents should be respected and upheld. They should also take into account the growing emphasis on human rights issues, related environmental obligations, and the necessity of an adequate institutional framework, as expounded in the 2004 ILA Berlin Rules and the 1992 UNECE Watercourse Convention (global participation opened in 2015) (Islam 2023).

Although Bangladesh has acceded to the 1992 UN Water Convention, it should take the initiative to sign the 1997 UN Watercourse Convention, given its position as a downstream riparian state highly vulnerable to the impacts of approximately 54 transboundary rivers. In addition, it ought to require adherence to current international environmental accords, such as the 2015 Paris Climate Change Agreement and the 1992 Convention on Biodiversity, which have relevant clauses about the usage of global watercourses. Thus, Bangladesh could persuade other co-basin nations that watercourses are valuable natural resources that must be developed, utilized, and maintained via an effective and all-encompassing plan that includes all relevant states and addresses all purposes. The long-term viability of the area and the sustainability of the reliant ecosystems depend on this realization (Islam 2023).

Despite sharing 54 transboundary rivers, India and Bangladesh have signed only the Ganges Water Sharing Agreement thus far. The Bangladesh–India Joint River Commission (JRC) ministerial-level meeting was conducted in Delhi in August 2022, following a lengthy 12-year break. The meeting ended without any progress regarding signing a Teesta water-sharing agreement and negotiations on six other transboundary rivers—Monu, Muhuri, Khowai, Gumti, Dharla, and Dudhkumar. There is a lack of joint dialogue between the countries regarding the deterioration of the health of the transboundary rivers from both ecological perspectives (human lives and non-human organisms). The JRC needs to hold regular meetings to resolve the transboundary river issues.

11 Conclusion

Bangladesh has more or less sufficient policy frameworks and institutional setups for ecological river governance, but these still need to be fully operational. In South Asia, in many cases, 'policies are made to be forgotten'. River-related policies in Bangladesh, unfortunately, are examples of such 'forgotten' policies.

The country needs to benefit from these policies and implementing agencies. Authorities, like the Bangladesh National River Conservation Commission (NRCC), require more power to enforce laws that protect rivers. The Padma and the Teesta rivers need collaborative efforts from the neighbouring countries, especially India, for proper ecological governance.

The two countries, India and Bangladesh, heavily rely on the resources from the GBM system. The rapidly growing population and increased demand for domestic, commercial, and farming water have escalated the water-related problems that these two neighbours have to cope with. Although the region generally receives much more water than is necessary throughout the year, the access to water availability in terms of time and space could be more balanced. Thus, one of the main issues between India and Bangladesh is the decreasing water supply during the summer months. Bangladesh's predicament could be better because it shares 54 rivers with India or approximately 80% of its annual freshwater supply.

In addition, due to bank erosion, the use of flood control techniques, and deteriorating water quality, Bangladesh and India share comparable water-related problems. Given that China and India oppose third-party involvement in the more extensive GBM system framework, concerned countries may first try to participate in multilateral agreements to settle a troublesome issue. If everything else fails, only third-party mediation may be considered. One concern that will be essential in the future is using similar ecosystems in the coastal area. The Joint River Commission needs to be more active in solving the transboundary river issues. Thus, governance could be faster-paced as the conditions of the rivers are deteriorating. A partnership in ecological river governance is necessary to meet the SDG goals. The rivers must be governed properly and protected for a sustainable and resilient future. Safeguarding the freshwater resources and river ecosystem is also essential for sustainable development. The governance needs to focus on the ecological aspects of the rivers to save the aquatic lives and livelihood of people living on the banks of the Padma and the Teesta.

References

Allison, Mead A. 1998. "Geologic Framework and Environmental Status of the Ganges-Brahmaputra Delta." *Journal of Coastal Research. Coastal Education & Research Foundation, Inc.* 13 (3): 826–36. JSTOR 4298836.

Arifuzzaman, Md, Mohammad Abdul Hannan, Md Redwanur Rahman, and Md Atiqur Rahman. 2019. "Laws Regulating Water Pollution in Bangladesh." *Journal of Sociology and Anthropology* 3 (1): 15–24. https://doi.org/10.12691/jsa-3-1-3.

Bangladesh Environment Conservation Act, 1995. Act No. 1 of 1995. Accessed 26 April, 2024. https://faolex.fao.org/docs/pdf/bgd42272.pdf.

Baudoin, Lucie, and Joshua R. Gittins. 2021. "The Ecological Outcomes of Collaborative Governance in Large River Basins: Who Is in the Room and Does It Matter?" *Journal of Environmental Management* 281 (March): 111836. https://doi.org/10.1016/j.jenvman.2020.111836.

Bhattacharjee, Shubhra, Bijeta Saha, Bohnni Saha, Md. Sadid Uddin, Chowdhury Hasanul Panna, Prosun Bhattacharya, and Ratnajit Saha. 2019. "Groundwater Governance in Bangladesh: Established Practices and Recent Trends." *Groundwater for Sustainable Development* 8 (April): 69–81. https://doi.org/10.1016/j.gsd.2018.02.006.

Brichieri-Colombi, Stephen, and Robert W. Bradnock. 2003. "Geopolitics, Water and Development in South Asia: Cooperative Development in the Ganges-Brahmaputra Delta." *The Geographical Journal* 169 (1): 43–64. https://doi.org/10.1111/1475-4959.t01-1-00002.

Chan, Ngai, Ranjan Roy, and Brian Chaffin. 2016. "Water Governance in Bangladesh: An Evaluation of Institutional and Political Context." *Water* 8 (9): 403. https://doi.org/10.3390/w8090403.

Chua, Liana, and Hannah Fair. 2023. "Anthropocene." In *The Open Encyclopedia of Anthropology*, edited by Felix Stein. Facsimile of the first edition in *The Cambridge Encyclopedia of Anthropology*. https://doi.org/10.29164/19anthro.

Faisal and Nishat. 2000. "An Assessment of the Institutional Mechanisms for Water Negotiations in the Ganges-Brahmaputra-Meghna System." *International Negotiation* 5, 2 (2000): 289–310, doi: https://doi.org/10.1163/15718060020848776.

Giordano, Meredith A., and Aaron T. Wolf. 2003. "Sharing Waters: Post-Rio International Water Management." *Natural Resources Forum* 27 (2): 163–71. https://doi.org/10.1111/1477-8947.00051.

IFGR. 2019. Initiatives for the Future of Great Rivers [IFGR] "Rights for All the Rivers and Watercourses of Bangladesh." *Initiatives Pour l'Avenir Des Grands Fleuves*. 27 August, 2019. https://www.initiativesrivers.org/actualites/rights-for-all-the-rivers-and-watercourses-of-bangladesh/.

Iqbal, Iftekhar, Hasharina Hassan, and Asiyah Kumpoh. 2026. "Introduction." In Fluid Phenomena: The River and Anthropocene Life-World in the Asia-Pacific, edited by Iftekhar Iqbal, Hasharina Hassan, and Asiyah Kumpoh. Brill.

Islam, S. Nazrul. 2022. "The Mirage of Teesta Treaty." The Daily Star. September 6, 2022. https://www.thedailystar.net/opinion/views/news/the-mirage-teesta-treaty-3112541.

Islam, Md Nazrul. 2023. *Sharing Ganges Water: Indo-Bangladesh Treaties and International Law*. University Press Limited.

Islam, Md Monirul, Essam Yassin Mohammed, and Liaquat Ali. 2016. "Economic Incentives for Sustainable Hilsa Fishing in Bangladesh: An Analysis of the Legal and

Institutional Framework." *Marine Policy* 68 (June): 8–22. https://doi.org/10.1016/j.marpol.2016.02.005.

Mim, Protyasha Ahmed. 2022. "Water Pollution by Textile Industries." The Daily Star. 25 November, 2022. https://www.thedailystar.net/law-our-rights/news/water-pollution-textile-industries-3178966.

Mullick, Md. Reaz Akter, Mukand S. Babel, and Sylvain R Perret. 2010. "Discharge-Based Economic Valuation of Irrigation Water: Evidence from the Teesta River, Bangladesh." *Irrigation and Drainage* 60 (4): 481–92. https://doi.org/10.1002/ird.597.

Nepal, Santosh, and Arun Bhakta Shrestha. 2015. "Impact of Climate Change on the Hydrological Regime of the Indus, Ganges and Brahmaputra River Basins: A Review of the Literature." *International Journal of Water Resources Development* 31 (2): 201–18. https://doi.org/10.1080/07900627.2015.1030494.

Ostrom, Elinor. 2009. "A general framework for analyzing sustainability of social-ecological systems." Science 325, no. 5939: 419–22.

Rahaman, Muhammad Mizanur. 2009. "Principles of International Water Law: Creating Effective Transboundary Water Resources Management." *International Journal of Sustainable Society* 1 (3): 207. https://doi.org/10.1504/ijssoc.2009.027620.

Rahaman, Muhammad Mizanur, and Abdullah – Al – Mamun. 2020. "Hydropower Development along Teesta River Basin: Opportunities for Cooperation." *Water Policy*, June. https://doi.org/10.2166/wp.2020.136.

Rahman, Kazi Saidur, Zahidul Islam, Umme Kulsum Navera, and Fulco Ludwig. 2019. "A Critical Review of the Ganges Water Sharing Arrangement." *Water Policy* 21 (2): 259–76. https://doi.org/10.2166/wp.2019.164.

Ranjan, Amit. 2022. "The Ganges Waters Treaty: What will happen after 2026?" Institute of South Asian Studies (ISAS). National University of Singapore. https://www.isas.nus.edu.sg/papers/the-ganges-waters-treaty-what-will-happen-after2026/#:~:text=However%2C%20the%20real%20challenge%20is,5%20to%208%20September%202022.

Roy, Pinaki, and S Dilip Roy. 2022. Teesta sorrows everflowing. The Daily Star. 5 September, 2022. https://www.thedailystar.net/news/bangladesh/news/teesta-sorrows-everflowing-3111036.

Salman, Salman MA, and Kishor Uprety. 1999. "Hydro-politics in South Asia: a comparative analysis of the Mahakali and the Ganges Treaties." *Natural Resources Journal*: 295–343.

Tasneem, Shuprova. 2022. "Bangladesh Can't Wait Forever to Resolve the Teesta Issue." The Daily Star. September 5, 2022. https://www.thedailystar.net/opinion/views/news/bangladesh-cant-wait-forever-resolve-the-teesta-issue-3111741.

Thomas, Kimberley Anh. 2017. "The Ganges Water Treaty: 20 Years of Cooperation, on India's Terms." *Water Policy* 19 (4): 724–40. https://doi.org/10.2166/wp.2017.109.

Watson, Robert T., and A. H. Zakri. 2003. "Millennium Ecosystem Assessment-Ecosystems and human wellbeing." *A framework for assessment.*

Wuijts, Susanne, Jappe Beekman, Bas van der Wal, Cathy Suykens, Peter P. J. Driessen, and Helena F. M. W. Van Rijswick. 2019. "An Ecological Perspective on a River's Rights: A Recipe for More Effective Water Quality Governance?" *Water International* 44 (6–7): 647–66. https://doi.org/10.1080/02508060.2019.1615773.

Wuijts, Susanne, Peter Driessen, and Helena Van Rijswick. 2018. "Towards More Effective Water Quality Governance: A Review of Social-Economic, Legal and Ecological Perspectives and Their Interactions." *Sustainability* 10 (4): 914. https://doi.org/10.3390/su10040914.

Yuan, Liang, Weijun He, Dagmawi Mulugeta Degefu, Zaiyi Liao, Xia Wu, Min An, Zhaofang Zhang, and Thomas Stephen Ramsey. 2020. "Transboundary Water Sharing Problem; a Theoretical Analysis Using Evolutionary Game and System Dynamics." *Journal of Hydrology* 582 (March): 124521–21. https://doi.org/10.1016/j.jhydrol.2019.124521.

CHAPTER 15

Islands of the Mekong, or the Existential Flood: Co-Creation and Fragmentation in the Wake of Hydropower

Andrew Alan Johnson

Mon stood in the Mekong, a river next to which he was born and had lived for most of his life. This particular river—*nam khong,* to Mon—was simply "the river", as its etymology suggests (*khong* being an older word for "river", although this link had long since fallen out of modern Thai and Lao). No other body of water in the region could compete with its significance. But for Mon, it was a part of his home. He swam across its red-brown flow as a child, across the international border between Thailand and Laos—one which, at that time, was fraught with Cold War tensions. As an adult, he fished in its depths, visiting his gill-nets in the cool of the morning or evening, throwing a net in public shallows in the heat of the day, sitting on its bank and drinking at night. Even here, the river was the primary topic of conversation (if occasionally replaced with food, women, or ghosts). It was, in other words, a part of him, and he of it.

Mon had been many things in his forty or so years—a migrant construction worker in Bangkok, a fisherman, an entrepreneur, and a father. His family had allegedly traced their lineage to displaced Laotian nobles fleeing the destruction of Vientiane at the hands of Siamese forces in 1828, but it was the fact that he was wealthy enough to own a few fishing plots in the river, and not to this possibly apocryphal story, that he had what status he had in the village, a place I call Ban Beuk.

As Mon's family and personal stories both indicate, the river is a key part of his orientation towards the world. One navigates on the river by reference to this or that rock (*hin*), or this or that rapid (*kaeng*), names that remain even when the rock is submerged by the river's monsoon pulse. The road follows the river, the hills follow the river, the spirits originate in the waters, the market stocks fish. Ban Beuk—a small village of about 300 households—is a river town.

I speak here of Mon, the individual, not Ban Beuk or to Lao people in general. In anthropological and other discussions of community, it is easy to forget that culture and community are a collection of practices undertaken in time and space by individuals—a confluence of imaginaries, processes, and

 | HTTPS://DOI.ORG/10.1163/9789004758100-017

activities. While some aspects of his experience are shared by his neighbours, he alone observes the flow passing by. He alone seeks to make sense of the world. And from time to time, from instance to instance (as the Buddhist philosopher Nagarjuna—see Nagarjuna (1995)—teaches us), he himself changes. Just as the water changes second by second, we ascribe an essence to what exists at one moment in time, and people and their communities likewise exist in a continual process of becoming. Generalities and commonalities can often be drawn—but such an attempt becomes difficult in the context of change. And the Mekong is changing.

This chapter is an attempt to place Mon in the context of a wider literature on more-than-human belonging. What we may term "local knowledge" (Yos 2012; Jakkrit 2007), a deep understanding of ecologies and hydrologies reflected in everyday practices (see Ingold 2005), can often become simply an indigenous-inflected[1] validation of ecological knowledge—consider a statement like "local communities possess local knowledge that helps them care for the river", a phrasing that would subsume Mon into the larger abstraction of the community, a diverse set of ways of being into one "local knowledge" (Thai: pumipanya thong-thin), and "care" into a set of practices that ecologists might deem wise. Regarding the latter point, De la Cadena and Blaser (2017) instead call for a "pluriverse": a bringing together of different, incommensurable ways of being united in a common purpose—rather than assuming that "local knowledge" is always already commensurable with ecology, their approach foregrounds a shared, often short-term goal. But here, too, I wish to focus on the pluriverse *within* Mon himself: by focusing on Mon as a "Thai-Lao fisherman", we lose sight of his own subjectivity, his own existential being, as someone whose aspirations and memories may be inconsistent, multivocal, and subject to change. And in so doing, I place Mon's story in parallel with that of the Mekong—likewise a constantly changing thing, torn between the desires of nation-states and spirits, and not always in sync with those of his fellow villagers.

The struggle for a fleeting continuity in the face of change is the nature of all riparian communities (see Khan 2023). But, on that day when I stood with Mon in the river's flow, that change was manifest. The water moved past clear and blue instead of red-brown, and thick mats of algae had begun to grow. This algae could be eaten, but more often than not it was a nuisance, clogging nets and making them too heavy to pull. Floods in the dry season, in addition

1 Mon is not "indigenous" to the region in the sense of belonging to a community who has faced settler colonialism. I use this term here to draw links to work by anthropologists such as Nadasdy (1999) and De la Cadena (2015) whose work on similar issues is tied with such groups.

to lower than normal water levels, made the river unpredictable. An unexpected flood a year ago had destroyed nearly everyone's nets. These gill nets had been strung up in particular places in the river, and the surge of water tore them from their locations and swept them downstream—they were often able to be recovered, but were hopelessly tangled and ripped.

Further, these floods disrupt the world of the fish. Fish, and the fishermen who follow them, have a particular rhythm to their lives—when a flood comes in egg-laying season, fish will swim up narrow channels to deposit eggs in safe places, and when the dry season comes, vegetation exposed by the water's retreat bears leaves and fruits. But a flood out of season comes when eggs are not ready, and the submersion of vegetation that has not yet had the opportunity to bear leaves chokes it off.

It was only those fishermen who—like Mon, who had a sister working for the army, as well as a position as an Assistant Village Head—had the capital to replace them who recovered. Many were too cautious to lay them again and just left fishing altogether.

These were the downstream effects of a number of hydropower projects across the mainstream of the Mekong—one at Xayabury, in Laos (since 2019), and others in China, at Jinghong (2008), Manwan (1998), and Nuozhadu (2012). In total, there are 11 major dams in China on the Mekong, and 86 outside of China along the river's course. The Mekong River Commission (MRC) estimates that hydropower will be a USD 160 billion business by 2040—albeit a rosy image. But observe the language here—dam estimates cite abstractions: megawatts of production and dollars of profit, but these are material infrastructures spanning the river, altering its ecology and hydrology (see Biba 2018; Middleton and Allouche 2016). While monitoring services at the MRC and independent observers have noted the alterations in the river's flow, from the standpoint of fishermen on the river, the change was obvious—and more intimate. Fish catches have cratered, with fishermen estimating that they caught about one-third of their usual yield in a year, and many have turned to blasting or electrocution of fish—harmful practices that devastate fish stocks even in normal years.

But that day, for Mon, he was watching the water. "See it move? It rises and inch or two, then falls back. In a few hours, it can rise a metre, even flood houses built above where it normally goes." He mentioned elsewhere how they had found bodies floating in the water, washed down from somewhere upstream, likely in Laos.

The crisis was not only one that affected humans. Mon had detailed to me his theory on the behaviour of naga—great water serpents—within the river. These powerful beings, for Mon, had tunnels under the red jagged hills in which Ban Beuk was set where catfish could hide in the absolute low dry season, but the uncertain flood affected these, too—nagas "grew tired of us", as another fisherman explained, and no longer visited the town.

The invocation of the naga here is no accident. Nagas form a constant figure in the folktales and everyday life of those living on the Mekong (Ngaosrivathana 2009; Sumet 1988). They are of particular interest as they are neither entirely benevolent nor antagonistic to humans—one must know each, individually, even if such beings are beyond a certain point unknowable (Johnson 2020). In the *Phadaeng Nang Ai* (Wajuppa 1990), a local folktale, naga appear as powerful and dangerous beings, separate from humans but who nonetheless live in their own (underwater) city, court human princesses, and thus occupy a kind of middle ground between the landscape, the "natural" world, and the human world—here, too, recall the middle ground that naga occupy for Mon: distant in their alien outlook and non-human characteristics, but able to be dealt with, unlike Chinese state projects. Here, too, Piyawit Moonkham (2021) describes how naga situate Mekong dwellers within this landscape—in so doing, they mete out punishments (i.e., accidents) and blessings (i.e., prosperity) (Hefner 2015, 9). Rather, within the figure of the naga, Mekong dwellers have an interlocutor between the human world and an alter, full of untamed potential: dangerous, but also vital for the preservation and becoming of Mekong lives.

Here, crisis emerges in wave after wave: the river changes, the naga—those beings that make the landscape vital—no longer visit, and fish vanish. While Mon was actively opposing local mining concerns and other development projects sited in the village, the impact of the dams was simply too much, too fast. "How can you fight against the power of China [*suu mahamanaaj jiin*]?" Some along the river have organized resistance; Mon was too aware of his own limited capacity to do so. He waited.

1 Water

Water has become an important source of academic inquiry in recent years. Ballastero (2019) identifies several trends: a focus on rights and access, other-than-human worlds and the meanings of water, and alternate ways of knowing. Here, I am most interested in how these spheres intersect, which is an issue of rights and access that alters water's materiality and the ways of knowing and being (what Baumann 2022 calls "onto-epistemology") around it. Like the material itself, the ways of being and knowing water elide attempts at its containment: water, that "unruly commodity" (Bakker 2010), resists the attempts to control it (see also Edgeworth 2011). It is this fluid resistance that the stories of catfish tunnels recall—despite the dams, water and its slimy denizens slip out of their grasp.

Naveeda Khan, in her *River Life and the Upspring of Nature* (2023), explores what she calls the "entrainment" of cultural worlds by the environments in which they live. She asks us not to see the natural world as either a system of resources to be extracted or utilized by agentive humans, nor as a deterministic template that shapes cultural reactions, but rather as something that shapes how social relations are founded. The Mekong riverbank, with its high hills and rocky substrate, are not the shifting *char* sandbars of Khan's ethnography, but the great river provides a link beyond that of the nation between Lao communities on either bank, one that eludes hardline narratives of Thai national belonging. As Mon's family story indicates, the Mekong is a river that can link communities as much as it is thought to separate them—Mon had a cousin who was a Laotian police officer on the opposite bank, and he would think little of crossing the river for social visits or religious events. More often, Lao villagers would cross on market days or festival days, especially when a *mor lam sing* [Lao country music] band was playing. Further, village heads from either bank would meet to discuss shared concerns—especially fishing plots and their allotment—in ways that did not involve either the Thai or Lao governments.

But such a sense of entrainment on the river is not shared by all—and certainly not by the Jinghong Hydroelectric Power Plant or its kin at Xayaburi and elsewhere. These are sites where the river is a resource—owned by a national government—and destined for extraction.

Anthropological scholarship on resistance to what de la Cadena and Blaser (2017) term "extractivism"—the understanding and exploitation of natural resources *as resources*—takes distinct forms. One of the most charismatic kinds of forms are those ways of being that become incorporated into protest. Consider Marisol de la Cadena's "earth beings" (2015)—animate places in the landscape that demand recognition of indigenous autochthony, respect, and preservation. In her work, she documents carefully the ways that "earth beings" come to stand in uncomfortable ways vis-à-vis Andean politics—derided as "infantile" by politicians (including those on the left) and standing in the way of progress, indigenous activists nonetheless use the *tirakuna* as a means of asserting their own legitimacy to sources as diffuse as the Catholic Church (Blaser 2013).

The political action of animism is what is at stake here—from taking images of animists as a part of the landscape (and, by extension, outside of the realm of politics) towards seeing such claims of a world irreconcilable with an idea of resources as yields to be extracted as political statements (see Blaser and De la

Cadena 2017). Here, Heidegger's work (1977) on "technological" ways of seeing comes to mind: to see a river (or a forest, in Heidegger's case) as a means to an end is to curtail its other potentialities. Mario Blaser, for his part, argues that we need to inhabit such worlds to see how they might be of benefit to a world already vulnerable to exploitation. Such a political ontology (Blaser 2013), a world where alternate ways of seeing arrive at a common political goal—in Blaser's terms, an end to extractivism—allows for ontological diversity at the same time as political action.

Mon clearly fits into some of de la Cadena's notions—he advocates for a river that is a morally informed confederation of humans, nonhumans, and what elsewhere I call "inhuman"—fishermen, fish, and spirits. While the fishermen have their own political groups to advocate for them, the fish have their Lord of the Catfish [*jao pa beuk*], who provides catches if humans are morally upright enough to deserve them. But Mon does not stage a protest in Bangkok's parliament. Indeed, while he was eager for such an opportunity, he was largely alienated from Bangkokian political groups, situated as they are within other concerns far from his own—Mon spoke Lao, and recognized that many Bangkok-based environmentalist groups espoused a royalist doctrine very far from his own democratic inclinations. Laos, too, was a closed door for him. While he proudly *spoke* Lao as his mother tongue, claims to be "*khon lao*" (a Lao person) were too wrapped up in the government of Laos, in relation to its heavy-handed tactics with regard to environmentalist movements, its political repression, and its perceived underdeveloped status with regard to Thailand.

So let us pause here, with de la Cadena and Blaser's enthusiasm. What they describe are politically savvy protest groups, able to speak the language of power and able to be heard. This is not the perspective of most of those in the crosshairs of the environmental changes that the world is at present undergoing. But such figures make for sympathetic heroes.

We could find similar heroes on the Mekong. Kru Tee, of the Rak Mekong Group, is one such. His group on the Thai bank of the river is an organization that lobbies against the blasting of rapids and hydropower proliferation on that great river, and does so in a way that garners national and international acclaim. Kru Tee is the whole picture—Northern-style mo hom shirt, iconic white moustache, long hair, and recognition in multiple presses.

But such heroes presuppose a few things. One is that there is "a public"—a sphere in which all are heard and which can include—or exclude—particular voices. Another is that by speaking up, such voices are made heard. It is a perspective that goes well with a certain kind of student-led activism.

But there is another perspective, perhaps one more attuned to the lives of Mekong fishermen than students and faculty. That, as with Khan's charuras (dwellers of *chars*, alluvial formations in the river), derives from the quiet existential questions that arise when facing the rise or fall of the river, one where the public to whom one could voice rights is indeterminate (in an uncaring Bangkok, or a forcibly silenced Vientiane?), where a political platform seems impossible, and where consensus even within the community is hard to come by. In place of a pluriverse where a defined group, e.g., "indigenous Andeans", is replaced by a multitude of individual voices, questioning what is, and what should be, and arriving at no clear answer. In short, I turn here not to a political movement, but towards one that questions the makeup of the world, one in which the river itself helps to structure what is natural, and what is normal.

By "existential", I point here to not to a personal crisis, but a crisis about the nature of the world and one's place within it. From Rappaport (1979) onwards, environmental anthropology has pointed to the co-creation of ecological worlds and the involvement of people within their surroundings. Work by Ingold (2005) and others (Khan 2023; Strang 2004; Van Schendel 2023), too, furthers this work to show how human labour creates worlds in a partnership—most notably, Raffles (2002) describes how the Amazon, that river thought of as synonymous with nature fundamentally separated from human activity, has been profoundly shaped by human actions both pre- and post-colonial.

Extending the notion of personhood outwards towards the environment, then, reveals new vulnerabilities to this personhood when the surrounding world changes. Here, "existence" denotes more than livelihood, more than simply productive activity, but the way that the world is—its poetics, aesthetics, its affect. I am reminded of the fireflies of my youth in rural Virginia. On summer nights, these would at times be so numerous as to make glowing orange clouds across humid meadows. These fireflies have all but disappeared in recent years, for unclear reasons. Summer nights are still humid, and the air still hangs low and close, but it is lit now only by flickers of heat lightning in the distance. Something has vanished. It has not affected my ability to live my life, nor has it—although other insects will—affected my ability to find food. But something has died. The changes occurring on the Mekong are of many orders—some directly affecting livelihoods, while others have influences on these more intimate, affective levels. It is all of these that I seek to capture here.

2 Mekong Worlds

Mon's fishing plots were centred on the features of the landscape: rushing *kaeng*, dry *hin*, slow-moving *wern*. These are points defined not by a grid or map, but by the behaviour of the water—a slow eddy (*wern*) might expand or contract (or disappear altogether) given the water's distinct flow and level on that day. I spent several years with Mon, disentangling spiny fish from his gill-nets, frightening fish in shallow eddies into the range of his throw-net, or simply riding in the back of his longtail boat from catch to catch.

Fishing varies by the time of day, and it varies by season. These cycles follow the patterns of fish, and also water—murky "red" water means that gill-nets are a more effective trap, as fish cannot see them until it is too late, whereas low, clear water makes for better use of a throw-net or dip-net. Fishing is best done in the early morning or late afternoon, when the Lao sun is lower on the horizon and is less offensive to those in boats—midday is for resting in the shade under stilt-houses, and, for many, drinking. Fishing outside of these hours raises the suspicion that one is using techniques—electric car batteries or bombs—that one would not wish others to know. These new fishing techniques are themselves features of the Anthropocene. They are a kind of shadow "local knowledge"—fishing with such devices requires careful training and knowledge in order to avoid becoming a victim—two fishermen lost their lives to an electric shock during my fieldwork. Each sends out a deadly blast—one concussive, one electric—into the water, causing fish in a wide radius to come to the surface (including many who escape capture and whose bodies float downstream), but fish outside of this radius also fall victim, and Mon and others on the river often reported seeing fish swimming oddly, or fish in the market whose flesh was distorted—signs, they told me, of trauma induced by a shock. Damage to fishes' reproductive organs, too, was often attributed to such trauma. In short, these new techniques are another form of extractivism: as scarcity mounts, such strip-mining of fisheries becomes one of the only ways that fishermen can capture enough to turn a profit.

Fishing, too, is a technique that only some are born into. Mon was unusual in the community for staying in it—many men married into other families, many of them inland, and others came into the community from elsewhere. Lao families are customarily uxorilocal, meaning that they join the household of their wife, and are taught local techniques by their fathers-in-law.

This forms a body of local knowledge, but to reify it as such, to make an argument for "local knowledge" as a singular corpus of wisdom passed down about the natural environment, is to ignore the very real differences within communities, and between individuals, about the river and how to deal with it. And, indeed, how to deal with uncertainty itself—a clear and present danger on a river which, following the logics of hydropower, changes daily.

3 An Existential View

For Heidegger and those like Ingold (2005), inspired by him, dwelling and activity co-create worlds. This has long been an observation within environmental studies more broadly as well—see Richard White's *Organic Machine* (1995), or Timothy Morton's *Ecology Without Nature* (2009). Here, a split between a natural world and a human world is untenable: we are locked into a mutual co-creation.

The same question of human influence over nature dogs the concept of the Anthropocene—do we condemn humans as a whole, or simply capitalism (Moore 2015)? What do we make of the span of geologic epochs with the very short time that we can see Anthropocenic effects (Chakrabarty 2018)? The devastations on the Mekong are certainly a part of a larger crisis that will remake the world: alternations in river channels, extinction of species, and the like will forever alter the way that the world works, although perhaps our influence upon the planet is not an epoch, but an event. But my framing here is more personal and less ecological or political-ecological. Instead, as Mon's story indicates, I seek to understand the kinds of existential alterations that occur on an individual level, within one's own consciousness.

Consider Bachelard's approach to space (1958), and to water (1942)—here is a perspective that brings the poetic qualities of space to the foreground. For Bachelard, the childhood home creates a template that one constantly rediscovers through daydreaming or dreaming. While we may disagree with Bachelard's assumption that his own French home is some kind of universal template, consider the Mekong something similar, for Mon—here is a constantly changing thing that is nonetheless a part of the world of childhood; here is the means for his livelihood.

While a deconstruction of a nature/culture binary has become de rigeur in much anthropological literature on the topic, here I seek to follow an existential take. As Khan argues, the natural world entrains societies in its own logics, bending social and political orders around its own exigencies. But such a world—especially in times of flux—can fragment as much as it creates commonalities. Reactions to the water's change in Ban Beuk are not uniform. Mon himself maintained a hope that the subterranean naga had carved channels deep enough for fish to hide in, that their powers were in the end greater than that of dam-builders. "People string nets up and down this river. You can't go far without running into one. And no one has seen a Mekong Giant Catfish for years. Then, suddenly, here is one. Where did it come from?"

Here is the uncontainable force of the river—the water that slides around attempts to trap it—made manifest. For Mon, the forces of the river were simply too great, and too separate from human activity to fall victim to it. The river is ruled by a triumvirate of powers—human, supernatural, and piscine—and the intransigence of one of these simply means that others will band together. The powers of the river were such that the Mekong Giant Catfish would have a protector, even if this meant the alienation of human powers from its workings.

It is a hopeful narrative, one that speaks to the resilience of nature and an animist co-existence that nonetheless holds space for the challenging of human overreach. But it is not the only perspective in Ban Beuk—nor even the only animist perspective.

Animism in the region is most often invoked via reference to spirit mediumship (Morris 2000; Visisya 2018; Baumann 2022). Nang thiam, "artificial women", post-menopausal or transgender women who are taken over by the spirits of the river or other aquatic features, act as focal points between the powers of the land and those humans who seek to enter into a relationship with them. In Ban Beuk, this most often involved seeking fish (especially the catfish), but also involved questions about law and custom, labour contracts, love, or issues of fate and fortune (e.g., illness or the underground lottery). Were we to see something of a common perspective upon the river, we might expect these to issue stern rebukes to humanity's hubris in channelling the river. But in my research, I just as often discovered rather an embrace of narratives of development even by the spirits—as one medium put it, as her tutelary spirit communed with his Laotian supernatural counterpart, dams are "what is needed for [Lao] national development".

It is a narrative that is quite different from that of Mon's—if Mon posits an animist world that includes humans as a component part, but a lesser one, this is a narrative that posits progress as an ultimate goal. The Laotian tutelary spirit seeks the same notion of progress (i.e., "national development") as the Laotian state—an embrace of change and the dissolution of what had come before. Here, villagers such as Mon are cast by supernatural authorities as stubborn Luddites, much as they are cast by developers. Even Mon's nephew, returned to Ban Beuk from his regular job at a car dealership in Bangkok, put it bluntly: "our parents were fishermen, and now we work in the city. As for change, I think it's a good thing. It's progress. Old ways of life cannot be."

The diversity of views within Ban Beuk points to a similar diversity within the dynamics of dwelling, especially in a state of collapse. With non-human partners changing or disappearing, our relationship with space likewise alters: Mon seeks to understand what has gone wrong, and his nephew understands the collapse as a natural progression. For others, still, the alterations on the river simply opened the door to nihilism. Lo, a fisherman of Mon's generation (and Mon's neighbour), a man whose fortunes largely involved working with the river (fishing, cross-border trade, and, eventually, opening a congee stall for drivers following the river on holiday trips from the regional capital), responded bluntly: "when I see the soldiers from Bangkok, I just want to shoot them all." As each business venture failed—the congee stall was owned and run by his wife (to whom he was incessantly unfaithful), his fishing plots were ruined by floods, and his cross-border trade was shut down by new regulations from the Thai military government (2014–present)—he sat in the market square, and drank all morning, after which he moved to his wife's restaurant and continued to drink. In short, if the co-creation of natureculture can be likened to a marriage (that is, to something that is not reducible to a biography of either partner; see Cohen 2005), what we see here are ways of dealing with divorce.

The construction of villagers or states as owners of a lucrative patrimony seeking to capitalize on it—"seeing like a state" (Scott 1999), in other words—is only relevant here if we take the logic of states and resources as a given. They are not. Neither is de la Cadena's assumption of rights and other logics of state—or protest—power. Rather, in an existentialist framework, one is simply thrown into being. Objects—rivers and dams, or more conceptual things like progress and lordship (i.e., spirits)—exist, but their meaning and trajectory for communities is less clear, especially in times of disruption, such as the present. They may make the same demands upon others as before; or they may reinforce larger, more present demands (e.g., by the economy or state), or individuals like Lo may simply decide that all demands are meaningless and sink into nihilism. In short, what environmental disaster does to communities is present a wealth of possibility, much of it dire.

What, then, does resistance mean here? An existentialist fight for freedom in an absurd world means little here—the stakes are simply too great. The hero of Camus or Sartre who embraces the meaninglessness of the everyday to forge his own possibilities is undercut when one's livelihood is at stake—those philosophers can always return home to a meal and a bed. Similarly, there is little avenue for Mon, Lert, and Lo to adopt the "weapons of the weak" (Scott 1985)—the enemy here is diffuse. Indeed, while Scott's narratives often place a noble anarchist in opposition to a totalizing state, rather, in the ethnographies that Scott himself cites (Jonsson 2005), there is a more nuanced approach—a balance between accommodation and resistance, especially when facing an enemy who may be distant, but who nonetheless can grant possible benefits—in other words, not immediately apprehended as a threat, but as a powerful other being with whom one must live. Taking a step towards a "pluriverse", it is worth noting that strangers and spirits share commonalities—beings outside of the everyday that nonetheless come to impact it. Strangers—spirits, foreigners, and other powers—represent the potential that the world might be other than it is, something attractive when the world as it is seems to be in a state of collapse.

And it is this network—states, spirits, carbon, and hydropower—that come to bear upon the transformation of Mon's backyard from a red-brown muddy river into an unpredictable, algae-clogged clear flow. It is worth, then, listening to voices such as these not for sources for inspiration for a cause that we have already predetermined, but as a way to see fundamentally different spaces as lived spaces, places where people reminisce, hope, and dream.

What, then, do resistance and accommodation look like when they do not speak the language of power? Mon's narrative of spirits' accommodation, especially in the face of the island king's failure, is one. The quiet defeatism of the medium and of Mon's nephew is another. Finally, Lo's nihilism might yet be a third path here. Neither will change the calculus of power on the Mekong, but all point to the reality that the world is otherwise here—so, too, must we become.

References

Bachelard, Gaston. 1958. *La Poetique de L'Éspace.* Presses Universitaires de France.

Bachelard, Gaston. 1942. *L'eau et les rêves.* Presses Universitaires de France.

Bakker, Karen. *Privatizing Water: Governance failure and the World's Urban Water Crisis.* Cornell University Press.

Ballastero, Andrea. 2019. "The Anthropology of Water." In *Annual Review of Anthropology* 48: 405–21.

Baumann, Benjamin. 2022. "Enunciating Ambiguity: Thailand's *phi* and the epistemological decolonization of Thai studies." In *South East Asia Research* 30(2): 161–79.

Biba, Sebastian. 2018. "China's 'old' and 'new' Mekong River politics: the Lancang-Mekong Cooperation from a comparative benefit-sharing perspective." In *Water International* 43(5): 622–41.

Blaser, Mario. 2013. "Ontological Conflicts and the Stories of Peoples in Spite of Europe." In *Current Anthropology* 54(5): 547–68.

Blaser, Mario, and Marisol de la Cadena. 2017. "The Uncommons: An Introduction." In *Anthropogica* 59(2): 185–93.

Chakrabarty, Dipesh. 2018. "Anthropocene Time" in *History & Theory* 58(1): 5–32.

Cohen, Benjamin. 2005. "Escaping the False Binary of Nature and Culture Through Connection." In *Organization and Environment* 18(4): 445–57.

De la Cadena, Marisol. 2015. *Earth Beings: Ecologies of Practice Across Andean Worlds.* Duke University Press.

Edgeworth, Matthew. 2011. *Fluid Pasts: Archaeology of Flow.* Bristol: Bristol Classical Press.

Hefner, Carl J. 2015. "Tempting the Naga: Local Knowledge and Mysteries of the Mekong." In *At Home and in the Field: Ethnographic Encounters in Asia and the Pacific Islands.* University of Hawaii Press.

Heidegger, Martin. 1977. "The Question Concerning Technology." In *The Question Concerning Technology and Other Essays*: 3–35. Harper.

Ingold, Tim. 2005. "Epilogue: Towards a Politics of Dwelling" in *Conservation & Society* 3(2): 501–8.

Khan, Naveeda. 2023. *River Life and the Upspring of Nature.* Duke University Press.

Jakkrit Sangkhamanee. 2007. "Ethnohydrology and Mekong Knowledge in Transition: An Introductory Approach to Mekong Hydraulic Cognition." Conference paper, 'Critical transitions in the Mekong Region'.

Johnson, Andrew Alan. 2020. *Mekong Dreaming: Life and Death along a Changing River.* Duke University Press.

Jonsson, Hjorleifur. 2005. *Mien Relations: Mountain People and State Control in Thailand.* Silkworm Books.

Middleton, Carl, and Jeremy Allouche. 2016. "Watershed or Powershed? Critical Hydropolitics, China and the 'Lancang-Mekong Cooperation Framework'." In *The International Spectator* 51(3): 100–17.

Moore, Jason. 2015. *Capitalism in the Web of Life: Ecology and the Accumulation of Capital.* Verso Books.

Morris, Rosalind. 2000. *In the Place of Origins: Modernity and its Mediums in Northern Thailand.* Duke University Press.

Morton, Timothy. 2009. *Ecology Without Nature: Rethinking Environmental Aesthetics.* Harvard University Press.

Nadasdy, Paul. 1999. "The Politics of Tek: Power and the 'Integration' of Knowledge" in *Arctic Anthropology*" 36(1/2): 1–18.

Nagarjuna. 1995. *The Fundamental Wisdom of the Middle Way: Nagarjuna's Mulamadhyamakakarika.* Jay Garfield, trans. Oxford University Press.

Mayoury Ngaosrivathana and Pheuiphanh Ngaosrivathana. 2009. *The Enduring Sacred Landscape of the Naga.* Silkworm Press.

Piwat Moonkham. 2021. "Etnohistorical Archaeology and the Mythscape of the Naga in the Chiang Saen Basin, Thailand." In *Trans: Transnational and Regional Studies of Southeast Asia.* 9(2): 185–202.

Raffles, Hugh. 2002. *In Amazonia: A Natural History.* Princeton University Press.

Rappaport, Roy. 1979. *Pigs for the Ancestors: Ritual in the Ecology of a New Guinea People.* Yale University Press.

Scott, James. 1999. *Seeing Like a State: How Certain Schemes to Improve the Human Condition Have Failed.* Yale University Press.

Scott, James. 1985. *Weapons of the Weak: Everyday Forms of Peasant Resistance.* Yale University Press.

Strang, Veroica. 2004. *The Meaning of Water.* Routledge.

Sumet Jumsai. 1988. *Naga: Cultural Origins in Siam and the West Pacific.* Oxford University Press.

Van Schendel, Willem. 2023. "Non-Human Labor History? Three Short Questions." In *Notebooks: the Journal for Studies on Power.* Published online: https://brill.com/view/journals/powr/2/2/article-p189_006.xml?language=en.

Visisya Pinthongvijayakul. 2018. "Personhood and Political Subjectivity Through Ritual Enactment in Isan (Northeast Thailand)." In *Journal of Southeast Asian Studies* 49(1): 63–83.

Wajuppa Tossa. 1990. *Phadaeng Nang Ai: A Translation of a Thai-Isan Folk Epic in Verse.* Bucknell University Press.

White, Richard. 1995. *Organic Machine: The Remaking of the Colombia River.* Hill and Wang.

Yos Santasombat. 2012. *The River of Life: Changing Ecosystems of the Mekong.* Silkworm Books.

CHAPTER 16

The Fluidity of 'Tradition' in Japanese River Improvement: Evolving Interpretations of Technology and Craftsmanship

Ryoji Soda

Characterized by steep terrain and a fragile geological structure dominated by granite, Japan's fast-flowing rivers transport large amounts of sediment downstream. In Japan, approximately 50% of the total population lives in the alluvial plains, which cover only about 10% of the national territory, and 75% of the country's assets are concentrated there. These lowlands are prone to flooding due to the presence of raised-bed rivers, exposing the population to a constant risk of flooding. With an annual rainfall of 1,718 mm, concentrated during the rainy and typhoon seasons, flooding is common not only in major rivers but also in smaller ones. In this natural environment, the Japanese have been engaged in river management for centuries. As a result, Japan, known as a country prone to water-related disasters, has maintained a high level of flood control technology worldwide. Also, even today, some "traditional" river engineering methods still exist alongside modern high-tech approaches.

This chapter focuses on river engineering methods (techniques/skills/craftsmanship) related to flood control, discussing the topological relationship between people and rivers and its changes. In particular, it examines the "traditional" engineering methods of flood control, which, according to Tomino (2002, 11), are not about controlling nature, but rather about skilfully harnessing the regenerative power of nature and the energy of floods to improve river ecosystems and the rivers themselves.

These river engineering methods declined rapidly in the 1970s with the advent of the 'age of iron and concrete' in civil engineering. Since the 1990s, however, they have regained attention and recognition as environment-friendly "traditional" methods amid growing environmental awareness. Many structures built using "traditional" methods begin to accumulate sediment on their surfaces and in their surroundings shortly after installation. Grasses, shrubs, and algae begin to thrive, providing habitat, resting, and spawning areas for aquatic life. Over time, they blend into the landscape of the river basin.

In Japan, virtually all aspects of river landscapes have been modified in some way, including not only large dams and concrete revetments but also

 | HTTPS://DOI.ORG/10.1163/9789004758100-018

riverbanks that appear to be lush with natural vegetation and riverbeds that appear composed of natural stone formations. In addition, human-altered landscapes are constantly being reshaped by the actions of water, wind, and aquatic organisms. Rivers can therefore be seen as prime examples of hybridity, encompassing 'social nature' or 'anthropogenic nature' (Whatmore 2002). In this chapter, I aim to explore how river engineering methods and practices have evolved within such 'hybrid spaces.' In doing so, I will pay particular attention to how the place-based and context-dependent nature of river engineering influences the development of river improvement.

1 Technology and Craftsmanship in River Improvement Work

The concepts of "craftsmanship" (わざ, "*waza*" in Japanese) and "technology" (技術, "*gijutsu*" in Japanese) are often qualified by descriptors such as "traditional" or "modern", which tends to set them in opposition to each other. However, the demarcation between these concepts is not clear-cut; they are intertwined with contextual subtleties related to specific places and periods, resulting in a nuanced and complex relationship.

I argue that there is no fundamental difference between "*gijutsu*" (technology) and "*waza*" (craftsmanship). Moreover, it is overly simplistic to categorize them into a binary where technology equals modernity and craftsmanship equals tradition. This perspective arises from the recognition that, in the context of river engineering, the boundaries between 'tradition' and 'modernity' are particularly blurred, as will be discussed below. Typically, 'technology' is associated with a quest for universal application and quantification, while 'craftsmanship' tends to be seen as based on empirical knowledge, often shrouded in 'black boxes' of unquantified expertise. While it is true that certain 'traditional' river engineering methods have not been scientifically quantified for their effectiveness, this does not mean that they are purely empirical or artisanal in nature.

In this chapter, "craftsmanship" is not simply characterized as an intangible aptitude exclusive to individual craftspeople. Instead, it is conceptualized as a dynamic process or orientation toward ingenuity within a set of constraints. The ability to develop an approach to construction—or an amalgam of approaches—that is tailored to the specific circumstances of a given time and place is indeed empirical, especially when resources such as materials, labour, funding, and time are limited. When appropriate construction methods and materials are elusive, practitioners may resort to integrating several techniques or inventing novel applications for machines, tools, and materials that deviate from their intended use during on-site work. This adaptability can be likened to a form of 'bricolage'. However, it is not limited to the conventional dichotomy of traditional versus modern paradigms.

Unlike the 'craftsmanship' approach, which emphasizes adaptability and creativity, 'technology' tends toward a trajectory of standardization and universal applicability. This trend is characterized by the validation of a method—originally designed and tested in one context—as broadly effective when applied to another setting with analogous characteristics, thereby contributing to a cumulative body of knowledge. In the process, some constraints are either mitigated or reinterpreted as 'fixed parameters.'

In any construction project, constraints are inevitable. Both technology and craftsmanship are used to deal with these constraints; however, the difference between them is not in their inherent nature but in their approach to overcoming these challenges. Craftsmanship is characterized by its ability to adapt and transform methods of dealing with constraints, demonstrating a dynamic response. In contrast, technology is characterized by its ability to operate within constraints, seeking flexibility and establishing a standardized pattern of operation that minimizes the impact of such constraints.

The vectors of 'craftsmanship' and 'technology' are not diametrically opposed but can coexist harmoniously within the same construction project. The trajectory of the construction is shaped by the synergistic interplay of these two forces (see Figure 16.1). To elaborate, the 'technology' vector typically prioritizes the adaptability of construction techniques, aiming for universal

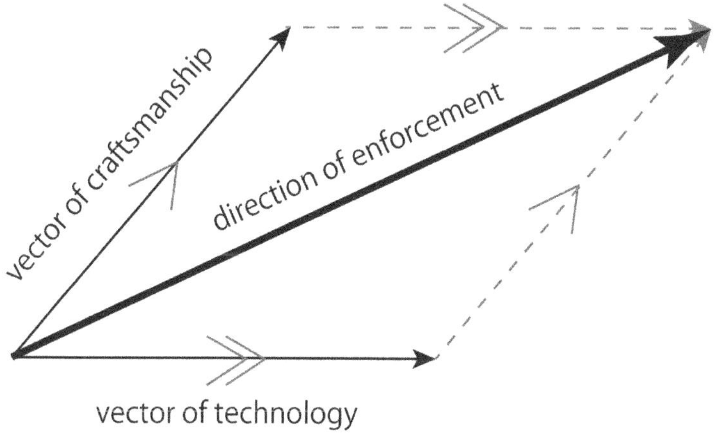

FIGURE 16.1 The vectors of 'technology' and 'craftsmanship'

application. Conversely, the 'craftsmanship' vector values the uniqueness of the immediate context, often resulting in construction practices that have strong site-specific and bespoke characteristics, tailored to the nuances of the site.

Until the 1970s, the methods used in river engineering were often a product of the interplay between the various stakeholders involved in such projects. These construction approaches and strategic orientations emerged from collaborative efforts rather than from the opaque empirical knowledge or proprietary techniques of individual engineers and craftsmen. In many cases, this collective genesis of methods was the norm. The following section explores the evolution of these 'collaborative' skills and works, using specific cases to illustrate the developments.

2 Interactions between Human and River

2.1 *Agency of River*

When considering traditional river construction methods, it is important to note the characteristic interaction between people and rivers. The history of river rehabilitation has been a process of mutual response between people and rivers.[1] The flow of water vividly reflects the responses to anthropogenic changes. In this regard, river engineering requires a heightened level of consideration for human–nature interaction compared to other infrastructure development, fostering a fluid relationship between the two.[2]

Installing artificial structures on riverbanks and riverbeds will undoubtedly alter the flow of water. For example, Figure 16.2 shows a continuous stone-piled

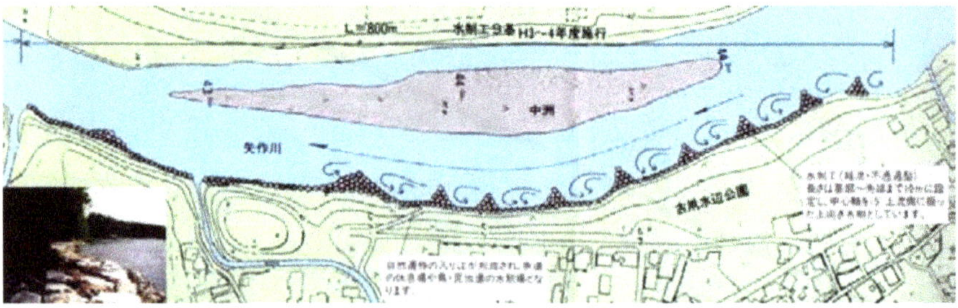

FIGURE 16.2 Plan of *itachi-suisei* river groynes (Nakamura 2006).

1 As for the question of subjectivity and agency of rivers, refer to Hasan (2025), this book.
2 Refer to the Introduction to this volume.

groyne, known locally as '*itachi-suisei*', installed on the Yahagi River in Aichi Prefecture. Nine clusters of natural stones are positioned along the meandering outer bank, serving as water impellers. This causes water turbulence around them, resulting in a decrease in flow velocity near the riverbank. As a result, erosion of the riverbank is suppressed, while the flow channel is pushed back (fixed) toward the centre of the river (toward the sandbar).

This water control system was installed in 1991–1992 drawing inspiration from the masonry water control systems of Switzerland and Germany, which were pioneers in "near-nature construction" methods. This system was adapted to both the masonry techniques developed in Japan and the unique characteristics of Japanese rivers (Nakamura 2006). The method has been recognized for its effectiveness in capturing water flow and regenerating riverbanks, including riparian forests, and was awarded the JSCE (Japan Society of Civil Engineers) Design Prize in 2007.

To accurately predict the changes in river channels, it is crucial to understand not only the flow of water, but also the movement of sediment. Japan, which is located in a tectonically active zone and is mainly underlain by weather-prone granite layers, experiences a high incidence of slope failures and erosion in its mountainous areas. This results in a significant supply of sediment to the rivers. As a result, Japan has gained worldwide recognition for its advanced sediment control techniques. In the field of river engineering, the Japanese term '*sabo*' (砂防), when written in Roman letters, is universally understood to refer to sediment or erosion control, which recognizes Japan's expertise in this area.

In Japan, it is essential to consider sand behaviour in addition to water behaviour when protecting riverbanks and riverbeds. Among the traditional methods, the '*soda* mattress' (see Figure. 16.3) is a technique for bank protection that effectively utilizes sediment. For example, Figure 16.4 shows a schematic diagram of the scour process at the base of a riverbank slope. By placing a submerged woven mat of thin hardwoods at this location, the mat provides resistance to the flowing water, thereby reducing the velocity of the water near the bank. In addition, the woven mats contain many gaps and voids that trap and accumulate the sediment carried from upstream. This accumulation of sediment strengthens the base of the bank, reducing the risk of erosion and collapse and providing a revetment effect. Thus, the ability to predict and assess water flow and sand movement is of great importance in flood control measures.

But understanding the behaviour of water is a complex task, even for engineers. Consider the following exercise: pour water onto an acrylic plate. Varying the volume of water or adjusting the angle of the plate will change the flow. But

FIGURE 16.3 Soda mattress settled in riverbed

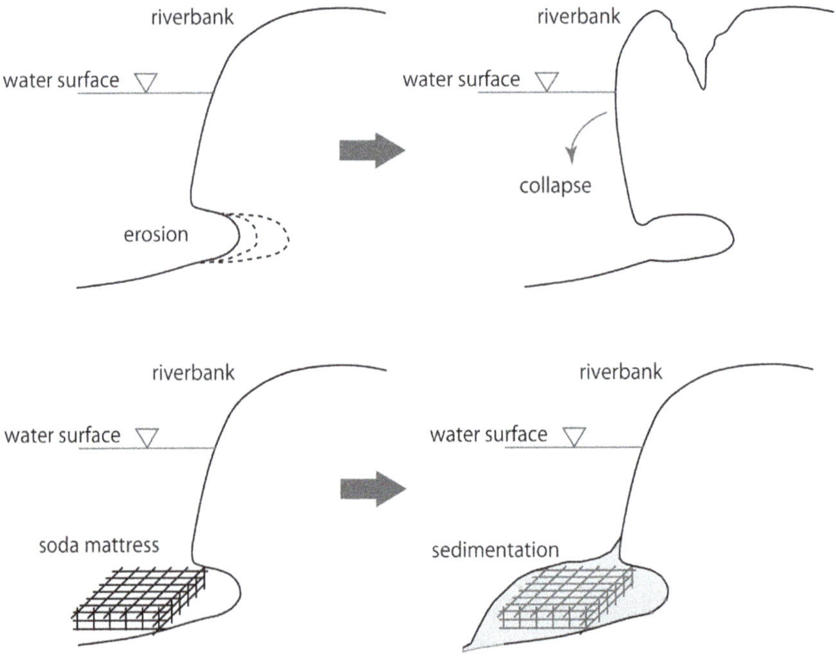

FIGURE 16.4 The effect of a soda mattress

instead of maintaining a straight path, the water winds and branches. To date, no theoretical principle fully explains why water behaves in such an unpredictable way. Understanding the movement of water, even on low-resistance acrylic panels, is not easy; therefore, predicting it in real river channels with various obstacles is a very challenging task, even for experienced engineers.

FIGURE 16.5 Each step in the creation of a soda mattress.

The complexity and uniqueness of river engineering stems from this very aspect. While the interplay between humans and nature has been extensively discussed, dealing with natural elements such as water requires the consideration of not only sediment flow, but also the biological communities that depend on the wat er. The uncertainty and unpredictability of their behaviour and the resulting chain of events can be extraordinarily intricate.

2.2 *Human–River Interactions and Connections*

When considering the relationship between humans and rivers, it is important to keep in mind their topological relationship (Whatmore 2002). As discussed earlier, the history of river improvement is a story of reciprocal responses between humans and rivers. Water often responds strongly and directly to human interventions. For example, when the right bank is fortified, the water cannot dissipate its energy there. Instead, it redirects its accumulated energy to the left bank (the opposite side), weakening it and increasing the likelihood of bank failure or overtopping. Similarly, strengthening downstream structures can increase risks upstream, and strengthening the main channel can affect tributaries. The reverse is also true.

In essence, artificial changes in certain locations can have unintended consequences in other sites. Unless the total volume of water is changed, the potential energy contained or diverted at one point will be released at another, increasing the risk of flooding. Consequently, areas where the risk has escalated may require additional retrofitting to mitigate these new vulnerabilities.

In addition, engineered structures inevitably degrade over time, resulting in a shift in the nature and magnitude of risk. As a result, the level of risk may eventually reverse between older and more recently upgraded sites. When this occurs, the embankments at the older sites will require maintenance, repair, or complete reconstruction to ensure continued effectiveness.

Namba (2017), who argues for an 'infrastructuralist turn', emphasizes that the construction of infrastructure is inherently unpredictable and never truly complete. River improvement work exemplifies this characteristic of infrastructure. Looking at the human–river dynamic in terms of energy absorption, reflection, and dissipation, it becomes clear that an ongoing cycle of action and reaction—akin to a perpetual game of cat and mouse—unfolds at the site and its surroundings. The intensity of this interaction fluctuates with time and place, and the indefinite mutual influence remains almost permanent. In this sense, the temporality of infrastructure is important to the relationship between humans and rivers.

The relationship between humans and rivers is extremely ambiguous, even compared to other civil engineering projects. This is also true for urban rivers, where human influence is more pronounced. Even in rivers where all three sides of the banks and the riverbed are lined with concrete, or where waterways have been culverted and turned into underground channels that are usually hidden from view, water can easily overflow from underground channels during prolonged heavy rains, causing internal flooding and significant impacts on human life. In other words, the relationship between humans and rivers continues to involve a constant push and pull, even as the visible forms of rivers change.

With this relationship between humans and rivers in mind, the following discussion focuses specifically on river improvements known as 'traditional' methods, examining various aspects of the techniques and skills observed in them.

3 Localization of River Engineering

3.1 *River Theory and Localization*

River engineering in Japan is considered to have undergone a modernization beginning in the early Meiji era (1870s). This perspective attributes the 'modernization' to the adoption of innovative river improvement techniques introduced by Dutch engineers recruited for their expertise. Indeed, within a year of his arrival in Japan in 1872, the Dutch engineer, van Doorn, authored a technical manual entitled "Ideas for the Improvement of Rivers in Japan". In this work, he presented a number of novel construction methods that would later influence the field (Iguchi 1995).

However, in the field of civil engineering, it is sometimes said that the difference between tradition and modernity is small. Tetsuro Nakaoka, a renowned figure in the field of the history of technology, points out in his book "The Formation of Modern Japanese Technology: Dynamics of 'Tradition' and 'Modernity'" (2006) that there is a relatively small qualitative difference between ancient and medieval civil engineering techniques and modern ones. He also notes that among the Western technologies transferred to Japan in the early Meiji period, civil engineering and metallurgical techniques showed the least difference from indigenous technologies (Nakaoka 2006, 448).

Furthermore, when considering infrastructure development that targets 'natural entities' such as river engineering, it may be reasonable to expect a stronger qualitative proximity between indigenous and modern technologies. This is because river engineering often reflects a strong element of 'localization' compared to infrastructure projects such as road building, railway construction, or urban planning, where technology transfer is (almost) directly feasible. Drawing on Larkin's (2013) discussion, Namba (2018) emphasizes the "replicability" of infrastructure planning in modernity, but in the case of river civil engineering, simple replication is not necessarily straightforward.

The flow of each river is unique—a seemingly obvious fact—but it was Aki Koichi's '*kasō* (river phase) theory' that formally established this concept (Aki 1944, 1952). In his book "River Engineering" (1952), Aki, who served as an engineer in the Ministry of Internal Affairs before becoming a professor at the University of Tokyo, stated: "The rivers of reality can only be understood in terms of their unique characteristics. The rivers we encounter are not 'universal' rivers, but rather are characterized by their specific locations." He further argued that river engineering methods that have been successful on one river cannot simply be applied to another, emphasizing the critical need to understand the individual characteristics of each river and adapt technology accordingly to improve it. In essence, although 'modern' technology tends to strive for universality and standardization, the need for localized approaches to river engineering was equally emphasized.

The discussion of localization among Japanese engineers began soon after the introduction of various Dutch methods to Japan. One of the most significant methods introduced during the Meiji period was the 'soda mattress' (Iguchi 1995, 171), which, according to Kusakabe's 1885 paper, was exhibited at the Domestic Industrial Exposition by the Public Works Bureau of the Ministry of Internal Affairs (Kusakabe 1885, 31). This detail underscores the innovation and importance of the soda mattress as a method of river construction and suggests that it was already beginning to take root in Japan. However, Tsujimura (1889), in discussing river engineering methods imported from the

Netherlands, asserts that it is imperative to eventually develop a 'Japanese-style' approach. He emphasizes that this is particularly true for the adaptation of the soda mattress, as the technology must be tailored to the specific conditions of each place and time.

The topography, geology, and climate are naturally different between the Netherlands and Japan, as are the materials available. In this context, the methods of technology transfer must be different from those used for railways and roads. From the point of view of design and construction, these differences also constitute 'constraints' that impede generalization and universalization, requiring a significant degree of 'technology localization' compared to other types of infrastructure construction.

3.2 *Between 'tradition' and 'modernity' in River Technology*

As suggested earlier, the question arises: did the Dutch really introduce Western 'modern' technology to Japan? As Nakaoka (2006) points out, the gap between 'tradition' and 'modernity' is narrow in the field of civil engineering, and moreover, construction projects involving natural elements such as rivers require a high degree of localization. In other words, the universal application of river engineering is not easy.

It is reasonable to assume that various river improvement techniques introduced by Dutch engineers were quite empirical in nature, and it is also noted that similar methods already existed in various parts of Japan (Ukisu and Igarashi 2002). There are records of the use of techniques similar to the soda mattress in the pre-Meiji renovation of the Kitakami river.

It is probable that Dutch engineers, when implementing sediment-bed techniques in Japan, modified what was used in their home country, while also referring to precedents in Japan, and carried out reorganization and standardization, thus spreading these techniques to river projects throughout Japan. It can be inferred that this process involved the selection and creation of adaptable river engineering methods. Such adaptability was demonstrated at various stages of river engineering.

For example, de Reijke and Escher, commissioned to improve the lower Yodo river for flood control and to secure a shipping route, advised the Meiji government to prioritize mountain management. They identified the influx of sediment from the barren mountains upstream as a key problem, but their advice was not immediately implemented. To manage sediment more directly, several sediment control dams were constructed around the Kizu river, a tributary of the Yodo river and a major source of sandy soil (located about 40 kilometres east-northeast in a straight line from the river mouth). In addition, wooden water control structures were built downstream to stabilize the channel. In this way, they approached the river and mountains with flexibility, executing site-specific construction despite various constraints.

The construction practices of these Dutch engineers are often highly praised for revitalizing rivers throughout Japan (Kanbayashi 1999). It is interesting to note, however, that some key members of the Meiji government at the time questioned the wisdom of inviting Dutch engineers. The scepticism was rooted in the belief that engineers from the low-lying Netherlands, a country characterized by land reclamation, might not be able to manage Japan's fast-flowing rivers (Matsuura 1990).

Consequently, Dutch engineers did not directly transplant their country's river engineering methods to Japan. While flood control techniques from the low-lying Netherlands undoubtedly played an important role in Japan, where the population is concentrated in the alluvial plains of the lowlands, river engineering methods that considered the entire basin of a fast-flowing river, including sediment transport, were actually based on the experience of the Dutch colony of Java at the time.

Thus, while concepts and technologies from the Netherlands influenced the river design in Japan, their direct application was limited due to the different geography and climate. Clearly, while drawing on the experience of the Dutch colony of Java—which shares similar geographic features with Japan—engineers also considered the unique landscapes of each Japanese river basin, as well as the social and political context. They developed their 'technology and craftsmanship' as far as possible, seeking solutions that were appropriate to Japan's milieu (environmental characteristics). In this process, 'Western modernity = new technology' was not simply copied, but an 'adjusted modernity' introduced from Java was fused with 'Japanese tradition'. In other words, the 'modern' river engineering methods developed in Japan during the Meiji era were highly site-specific, localized 'technologies/craftsmanship'.

4 Locality of River Improvement Technology

4.1 *The Characteristics of the Soda Mattress: Local Resource Recycling Method*

In this section, I delve into the details of the soda mattress, considered one of the most important techniques introduced from the Netherlands, with an emphasis on its dependence on place. Therefore, I will focus on the local resource recycling aspect of the soda mattress, which is a large, flexible mat constructed from thin trunks and branches of hardwood trees. This mat is submerged onto the riverbed to reduce erosion.

A key aspect of soda mattress construction is the sustainable use of local resources (Nakagawa 2006). Consider the Hokuriku region, where the use of soda mattresses is still widespread. The wood for these mattresses is harvested from nearby deciduous forests. The optimal time to cut the trees is in late autumn or early winter, when there is less rain and the trees have the lowest water content. In the rural areas of the Hokuriku region, where rice farming is the main livelihood, this wood harvesting is an important off-season job after the rice harvest.

After harvesting, it is desirable to allow a drying period of several months to half a year. This is to sufficiently reduce the moisture content in the wood; otherwise, the products may be more susceptible to rotting. The harvested forests are left as they are. Due to the predominance of deciduous vegetation, the forests regenerate (semi-) naturally by sprouting (coppicing). The rate of vegetation recovery varies depending on sunlight exposure, soil fertility, slope angle, etc., but typically the forest becomes reusable within several to a dozen years. When this wood is used to construct riverbeds, considerable labour is required for tasks such as making long bundles, weaving the bundles into mats, and stacking them in layers. This labour is most efficiently provided by surplus labour during the agricultural off-season.

After weaving, the soda mattress is submerged onto the riverbed. The natural stones used as weights for this immersion process were collected from the banks of the middle and upper rivers until about 1970. Alternatively, stones were sometimes collected from the gravel beds formed after floods or landslides in the surrounding areas of the basin. While this practice of local resource recycling—or 'local production for local consumption'—may have been recognized as a rational method, it is likely that this rationale was applied retrospectively. In reality, it was probably more a matter of using what was readily available within the constraints of material procurement.

The procurement of construction resources according to site conditions probably reflected a search for rationality in the face of various constraints. This approach was, in a sense, site-dependent and, conversely, had the potential to influence the choice of construction methods based on resource availability. Below I explore the relationship between local constraints and 'technology/craftsmanship' through specific examples from the field.

4.2 *Interaction between Design Engineers and Construction Workers*

Until the late 1960s, the utilization and constraints of local resources were also observed in the procurement of labour. When a site for a major river improvement project was selected, a construction office was established nearby. In the case of the Hokuriku Regional Construction Bureau, the bureau would send a construction office manager and engineers to the site. These engineers recruited temporary workers through personal contacts. The number of workers hired and their wages were at the discretion of the construction office manager.

Young engineers often learned construction/improvement methods directly from lower-level workers/labourers in the field. Engineers often determined the locations and methods of improvement, taking into account local opinions gathered through these workers (as noted by Mr. MK, a former engineer). Conversely, if there were landowners who were opposed to the rehabilitation, negotiations could extend into the morning over drinks and sometimes even escalate into fights (as reported by Mr. TR, a former engineer). It was also a crucial role of the engineers to find a 'compromise' in the midst of such disputes. Former worker, Mr. KS, also mentioned that the success of the construction work depended heavily on the trust between the engineers and the workers.

Although many workers of the time were skilled in what are now called 'traditional' construction methods, there were no craftsmen or skill groups dedicated to a single method. Instead, they became adept at different methods by working on multiple construction/renovation projects. Some of these skilled craftsmen were so respected that the engineering staff referred to them as 'gods of the site.'

When repairing embankments damaged by heavy rains or floods, the relationship between engineers and workers is also important. When partial damage to the embankment is detected, engineers immediately go to the site and urgently recruit workers through trusted foremen or crew leaders. According to Mr. TR, "workers quickly gathered under the leadership of respected foremen". Furthermore, due to the urgency of the repairs at the damaged site, "no drawings were even made". Although this statement may be exaggerated, Mr. TR mentioned, "For the restoration work on the special embankment in Matsuhama that was damaged by the flood, it was basically as if the plans were made on site by surveying and investigating, and then constantly making changes accordingly. We had to do both field work and administrative work at the same time."

For example, the placement of a soda mattress could potentially alter the flow of water and sediment. However, relying on this method alone may not provide sufficient risk mitigation. According to Mr. KS, an experienced construction worker, the approach extended beyond soda mattresses to include structures such as *jakago* (gabion mattress) or *kawakura* (river crib). This adaptive strategy involved closely observing the water dynamics and making adjustments accordingly. In essence, the decisions made by engineers and site managers were dynamically influenced by changes in the field driven by water/sand behaviour and the collaborative interactions among the various actors/actants involved in the project.

The experiences and keen observations of local workers (craftsmen and labourers) regarding their "own dike" were key elements in advancing such construction projects. Engineers determined the direction of repair and restoration while being influenced by the experiences of local residents, such as "This area is always prone to erosion" and the intuition of craftsmen working at the water's edge, such as "Maybe we should use a *jakago* (gabion mattress) here". At the same time, the engineers considered the possibilities and limitations of material procurement and redrew the plans accordingly.

In this way, river projects determined their strategies on a case-by-case basis, and the choice of techniques/methods to be employed varied depending on the situation. The implementation of technology and the demonstration of craftsmanship for river management occurred in the midst of complex interactions among various factors, including engineers, workers, materials, water dynamics, sediment transport, and local communities. The application of 'technology/craftsmanship' to river control can be characterized not so much as residing in individual engineers or craftsmen, but as being collectively formed and activated on the ground in each specific context.

By interacting with local workers and residents, and by considering the constraints and opportunities of local resources, the engineers honed their skills in the process of iterating on situational and one-off construction practices. They searched for more versatile 'technology/craftsmanship' and learned how to apply them to other sites. Until about 1970, dense, multi-directional interactions among actors in local settings were observed in this way on construction sites across the country. In the 1970s, however, 'contracting out' (outsourcing to the private sector) became mainstream in design and construction. At the same time, river construction methods using natural wood and stone, such as the soda mattress, rapidly declined. This decline was significantly influenced by the manualization of design and changes in the relationship between engineers and contractors.

4.3 *Advancement of Manualization: Separation of Planning, Design, and Construction*

Since the mid-1960s, public works projects have gradually shifted from a system in which the national and prefectural governments directly managed construction, known as 'direct management', to a system in which construction consultants were hired to design and construction companies, especially general contractors, were responsible for execution, known as 'contracting out'. This change was the result of various factors, including the expansion of construction projects, the increase in the size of heavy machinery, the promotion of private enterprises, reforms in the civil service system, and changes in the materials used. As a result, civil engineers employed by the national and prefectural governments were transferred to consulting firms, heavy machinery owned by the government was sold off to private companies, and government offices shifted their primary functions to planning. Although the shift from direct management to contracting out took place gradually over a decade, it brought about significant changes in various aspects. Of particular note are the trends toward specialization (division of labour) and manualization.

The planning, design, and construction phases eventually became completely separate over time, but what is crucial here is that the process of generating 'technology/craftsmanship', which had previously been collective, mixed, and interactive, began to change. The simultaneous progression of design and construction became completely separated, with a clear sequence established, where design by consultants preceded, followed by construction companies carrying out the work accordingly. This marked a shift toward 'design-centric' principles, meaning that construction sites became subordinate to planned designs. In other words, the once interactive relationship between engineers, construction workers, material suppliers, and the river (including water, sand, living organisms, etc.) was severed.

Another important aspect is the advance of manualization/standardization. The 'technology/craftsmanship' collectively generated in response to the conditions of the river or the agency exercised among the actors has been inherently localized, emerging on a situational basis from the relationships among the actors, embodying a one-time aspect. Manualization/standardization, however, requires a more universal technique that separates the process of generating the integrated and interactive 'technology/craftsmanship' into 'design technologies' and 'construction craftsmanship', with an emphasis on reinforcing the former.

The specialization and unidirectional approach from planning (administration) to design (consultants) to construction (contractors) severed the interactive network between the actors. As a result, there was a need to create 'standards' that could be shared among the disconnected actors. In particular, the introduction of the tendering system required prior cost calculations, including methods for calculating material and labour costs. Without the establishment of predetermined unit rates or 'standard quantities', competition would not be possible.

In order to ensure 'fairness' among the design consultants and construction companies, as well as to facilitate efficient inspection after the completion of construction projects, a unified set of standards became necessary. In fact, in 1976, the "River and Erosion Control Technical Standards (Draft)" (National Edition) was issued by the Ministry of Construction as the first nationwide unified standard, marking a step toward the 'universalization of technology'. It also included the re-expression of location-dependent limiting factors as 'constants'. In other words, progress was made in establishing numerical standards for streambed gradients, cross-sectional shapes, bank slopes, flow rates, and velocities. Design engineers were thus constrained within these frameworks and became complicit in reinforcing these standards, perpetuating the 'universalized technology'.

5 Reimagining 'Traditional' Construction Methods

5.1 *Questioning Extreme River Control: the Process towards Creating "near-nature river restoration"*

In Japan in the 1970s and 1980s, rivers were generally not considered natural, especially in urban areas. The rivers that flowed through the downtown areas of Tokyo and Osaka were notorious for their thick sludge and foul smell. It was a common sight to see people covering their noses with handkerchiefs while crossing bridges or boarding ferries (Yoshikawa 2007).

For those who grew up in the 1970s and 1980s, urban rivers were often perceived as black, dirty, and smelly. In particular, rivers lined with concrete on three sides were nothing more than canals (or drainage ditches, sewers, and gutters). This was also a consequence of the standardization of river construction. When the artificial alteration and pollution of rivers became so severe, it reached a state where it was difficult to call them 'natural', regardless of whether they were described with adjectives such as 'secondary', 'anthropogenic', 'social', 'semi', or 'pseudo.'

Rivers in Japan have been shaped by human intervention for thousands of years. It is estimated that the oldest existing irrigation canal in Japan dates back to the first century AD. In this sense, Japanese rivers have remained anthropogenic in nature, exhibiting characteristics of a secondary nature, semi-natural elements, and hybridity since ancient times. However, since the period of rapid economic growth, extreme artificialization and socialization have progressed, leading to a state where rivers have become so heavily engineered and socialized that their hybridity and ambiguity are no longer perceptible.

As seen in the above case from the Hokuriku region, rivers that had been artificially altered until the 1960s through interactions with nature and local residents underwent significant changes in both construction methods and landscapes in the 1970s and early 1980s. In terms of construction methods, the use of only iron and concrete became the norm, and the demand for natural wood and stone rapidly declined. As a result, from a scenic perspective, river-bank vegetation was eliminated and aquatic life was drastically reduced.

The 1980s marked the peak of the artificialization or socialization of Japan's rivers, and it was also a period when the extreme centralization of river management was thoroughly implemented. However, as the excessive management and artificialization progressed, doubts and concerns about Japan's river management accumulated from various quarters. These became the driving force that prepared the ground for a major policy shift in river administration in the 1990s (Soda and Yuhora 2013).

As a result of these movements, a new directive was issued in 1990 by the Ministry of Construction to regional construction bureaus, promoting the so-called 'near-nature river restoration' approach. This approach emphasized environmental considerations and community participation as important factors, which were later codified into law with the River Law Amendment of 1997 (Soda and Yuhora 2012; Kado 2020).

5.2 *'Traditionalized' River Engineering Techniques*

One might wonder when river construction methods that have been partially but continuously used in various regions, such as the soda mattress, stone masonry groynes, gabion mattress, and wooden riverbed construction, came to be considered 'traditional'. Confirmation with several experienced river engineers suggests that this recognition probably occurred after 1990. In fact, it is reasonable to assume that within the movement to promote 'near-nature' river engineering, 'traditional' methods have been significantly (re)emphasized. The circumstances are as follows.

The Ministry of Construction did not explicitly define what 'near-nature' meant, but rather aimed to encourage river engineers to think for themselves about what kind of designs should be applied to each river (Seki 1994). In the context of the progress of 'technological universalization' since the 1970s, the importance of the interaction between humans and rivers was forgotten among those involved in river improvement, and river engineers who had become overly dependent on numerical calculations without understanding the actual sites were urged to revive their imagination and creativity in dealing with nature. It can be said that this was an attempt to redirect rigid 'technology', which had become dependent on manuals, towards the vector of 'craftsmanship', which is oriented towards the site. However, among the river engineers who had already embraced the notion that flood control and environmental protection were in a trade-off relationship, 'near-natural' river improvement was perceived as having a completely opposite orientation to their conventional work. In addition, the lack of clear guidance on the near-natural method left many engineers confused.

In the midst of this confusion, trial and error were conducted. One aspect was the adjustment of the landscape. Attempts were made to achieve various visual effects, ranging from forcibly bending straightened rivers back to their original curves to developing concrete that mimicked the appearance of natural stone. However, questions have been raised about how to ensure safety and whether it is appropriate to call something 'near-nature' based solely on its appearance. On the other hand, there was an accelerated movement to install fish ladders in weirs and dams, in consideration of ecosystems. However, large dams required quite long concrete fish ladders, and many of these did not take into account the downstream migration of fish, although they could provide some effectiveness for upstream migration (Multi-nature River Creation Review Committee 2006).

In the midst of this confusion, one solution considered by engineers was to revive the construction methods using natural wood and stone. While this was a desperate measure for engineers, the use of natural materials in construction was perceived as environmentally friendly and in keeping with the 'near-natural' approach. In particular, methods such as the soda mattress, the masonry groyne, and the gabion mattress were likely to provide ecological benefits, as aquatic organisms could easily inhabit the interstices of the structures. In addition, these methods could be combined with iron and concrete construction to create structures similar to conventional ones. Such use of natural materials was a common construction practice until the early 1970s, but as mentioned above, the technical knowledge had been lost in many regions, and for the new generation of engineers these methods were considered 'traditional' techniques of a bygone era.

In this way, within a few decades, construction methods that did not rely on iron and concrete came to be called 'traditional'. In other words, 'traditional' river engineering methods were rediscovered and recreated by engineers in the 1990s, driven by the need for 'near-nature' river restoration.

6 The Collectivity and Constraints of Techniques/Craftsmanship

6.1 *The Hybrid Nature of 'Traditional' Construction Methods*

Whether observing the regional resource recycling construction method or the emergency embankment repair process in disaster situations in the Hokuriku region, it has been found that 'traditional' construction methods are established through the presence of various actors and that the interactivity among them is very strong. The landscapes created by 'traditional' construction methods may appear natural, but they are in fact highly interactive and hybrid forms of an anthropogenic nature.

The following factors contributed to the high interactivity of 'traditional' construction methods: 1) Compared to iron or concrete, alternative materials are not easy to obtain. One has to consider the availability of materials, while constantly keeping an eye on the condition of the forests as a source of supply—how many trees and branches were cut in the early winter of the previous year, whether the cut trees/branches had sufficient time to dry, whether the flexibility of the trees/branches was maintained, etc. 2) Among the various natural elements, the behaviour of water is highly uncertain and unpredictable. While water always responds to human intervention, predicting its behaviour is not easy. Moreover, water may exhibit movements that affect people without human intervention. In other words, the active–passive relationship between humans and rivers is more 'symmetrical', and construction methods are forced to be chosen within such relationships. 3) Even when other factors such as wood procurement, labour acquisition, and local perceptions of water movement were taken into account, the interconnectedness at the local level was strong. Wood procurement and labour acquisition were always considered in relation to the workers' primary livelihoods (mainly agriculture and forestry). In terms of countermeasures to water movement, the knowledge of local people was sometimes given priority over that of engineers. Water, materials, engineers, workers, and local residents, despite any hierarchy or disparity in job rank or wages, exhibited more flattened agency and significant relational effects in the realization of the collective 'technology/craftsmanship' tested on the site.

In the midst of such arrangements and interrelationships, technology/craftsmanship tailored to the situation on the ground emerged and manifested as something collective (or as a result of it). Such collective and hybrid technology/craftsmanship has a sense of locality, and Shinohara (2018) refers to this as "the traditional construction methods of each river", while noting that the shape of the river is "determined by the engineers and the river itself." In reality, however, it is not only the engineers who are deeply involved, but also workers, local residents, forest owners, stone material suppliers, and others. In more extreme terms, the use of technology/craftsmanship to control rivers is something that is 'collectively' produced in the moment, on site, within the relationships between these different actors/actants, and in this sense has an aspect of 'site-specific contextuality'. The practice of going to the site "without even drawing a plan" is thought to be influenced by this background.

As noted earlier, the behaviour of water can appear to be unrelated to human intervention, or it can exhibit movements that even river engineers and hydrologists cannot understand. Consequently, there are times (or almost always) when humans are forced into a passive role. The more mysterious the water's behaviour, and the more constrained by resources and time, the stronger the vector of 'craftsmanship' becomes in the work done there, while in the opposite situation the vector of 'technology' may be strengthened. It is within these relationships that the river construction methods were chosen.

6.2 *Temporal and Spatial Constraints of Technology/Craftsmanship*

Choosing which tree to cut from which mountain forest and at what time is critical to ensuring the flexibility and durability of the soda mattress. Even when bundling cut trees, there are specific ways to tie ropes or wire to maintain flexibility. Twisting the corners during fencing creates a tightening effect to ensure that the coarse weave does not unravel or break (see Figure 16.5). Where the constructed soda mattresses are submerged depends on the experience of the engineers and local workers. When sinking the mattress onto the riverbed, one must adjust the timing of the cobblestone drop while predicting the direction and distance that the water current will carry the mattress before it settles on the riverbed. Each of these tasks does not necessarily require a high level of skill and is not difficult to learn. They are not esoteric skills shrouded in mystery within the individual craftsman, but could be considered 'small crafts' in which many actors are involved, and these small crafts chain together to create a collective technology/craftsmanship. While this may be shared among those involved in the work on site, it is also a one-time construction method that embodies local constraints.

Morita (2012) discusses the lack of blueprints/design plans in the manufacturing and repair processes of tractors in rural Thailand. For Japanese engineers who provide financial and technical assistance, blueprints are essential. However, tractors operating in Thailand's different soils, climates, and crops require constant repair and improvement. The shape of each tractor varies depending on how the limited parts available in Thai villages are used. These machines are 'fluid' in nature, and their creation involves a dialog with the 'material', a process called 'non-design-centric production.'

It could be said that the use of natural materials in river improvement projects has a similar 'fluid' aspect. However, this does not mean that blueprints are never made. Artificial structures that are installed with the assumption that they will eventually fail require historical blueprints for future repairs. In the case of emergency repairs, where the urgency is high, the blueprints are made at the same time as the construction work is being done in the field. In a sense, this is a form of post-hoc blueprinting. This can be said to be the task of recording the physical aspects of the collective 'technology/craftsmanship' formed by various actors, while embodying both locality and uniqueness.

As engineers accumulate such experience in the field, they move on to other locations. The process of designing blueprints probably has the effect of finding some commonality and versatility in the accumulation of these experiences. In fact, from the late 1960s to the early 1970s, each Regional Construction Bureau under the Ministry of Construction was engaged in the task of consolidating these local 'techniques/skills' and extracting those with high versatility (according to Mr. US, a former engineer). This accumulation led to the creation of the aforementioned "River and Erosion Control Technical Standards" (standard manuals). However, with the systemic shift from direct management to contracting out, it became difficult for engineers and contractors to share field experience, and as a result, the standard manuals began to take on a life of their own.

7 The Future of 'traditional' Construction Methods

It is inherently difficult to distinguish between 'technology' and 'craftsmanship', and this is especially true in civil engineering, which deals with nature. This is because the types of constraints faced by engineers and contractors are diverse, and depending on the situation, the vectors of 'technology' and 'craftsmanship' can easily change. Both 'technology' and 'craft' must have always oscillated between universalization and localization (Raj 2007; Livingstone 2003). However, in the process of separating design from construction and establishing the superiority of design over construction, the vector towards localization was inevitably greatly reduced. This is because under the so-called 'design-centric' approach that followed the shift to contracting, constructors began to take on a subordinate role, forced to follow the finalized design plans, and the scope for improvisation in the face of the unpredictable flow of river on site was significantly reduced. In other words, 'non-design-centric production', which involves building while testing, became virtually impossible.

Mr. WS, an active construction consultant in the Hokuriku area, emphasizes that "Technology is meaningless unless it is tested in the field", but notes that this has become increasingly difficult. As seen in the previous section, 'trying things out' cannot be done unless engineers and contractors share the construction site at the same time. In other words, traditional construction methods using natural materials have seen the light of day again since the 1990s in terms of rediscovery or re-creation, but this has not involved a re-enforcement of the vector of 'craftsmanship'. Many parts of the network between different actors/actants, including materials, have been severed, and even when new relationships have been established, they have been transformed into something with a different phase from the previous relationships. Here, 'technology/craftsmanship' has been divided into 'design technology' and 'construction craftsmanship', and the degree of freedom on the construction site has been significantly reduced.

Mr. WS also said the following: "Traditional construction methods, such as the soda mattress, are empirically known to be effective for river improvement, but they have not been numerically proven from the perspective of fluid dynamics or structural mechanics." At present, it is becoming increasingly difficult to incorporate into designs those methods that cannot be explained numerically, and the future of the 'traditional' construction methods that were 'revived' in the 1990s is again becoming unclear. Conversely, if the usefulness of a method can be proven numerically and theoretically, it cannot be denied that an alternative to the 'traditional' construction methods could be created with 'scientific' backing. If this happens, traditional methods may become unnecessary and 'craftsmanship' may be completely absorbed by 'technology'. What aspect the hybrid nature of rivers may show in such a scenario is still anyone's guess.

Acknowledgement

The images in Figures 16.3 and 16.5 were provided by the Hokuriku Soda Industry Promotion Association. I express my sincere gratitude for their contribution. This chapter is a modified and translated version of an article originally published in the edited volume "Anthropology of Techniques" (Kyoto University, 2022) edited by Ikuya Tokoro.

References

Aki, Koichi. 1944. *Kaso-ron* (River phase theory). Tokiwa Shobo. (in Japanese)

Aki, Koichi. 1952. *River engineering*. Kyoritu Shuppan. (in Japanese)

Iguchi, Shohei. 1995. "Introduction of soda mattress and karreppu water management." In River Traditional Construction Techniques Study Group ed. *River traditional construction methods*. Regional Development Consultants: 166–94. (in Japanese)

Kado, Kazunori. 2020. "Potential of process analysis through environment control system theory: the transformation process of postwar Japanese river administration and environment control systems." In Kayano, T. and Yuasa, Y. eds. *Sociology of environmental problems: environment control system theory and its application.* Toshindo: 199–225. (in Japanese)

Kanbayashi, Yoshiyuki. 1999. Engineer De Reijke, who revitalized the rivers of Japan. *Soshisha.* (in Japanese)

Kusakabe, Benjiro. 1885. "The improvement of the Kitakami River: an overview of soda mattress." *Annual Report of Society of Engineers* 2: 15–39. (in Japanese)

Larkin, Brian. 2013. "The Politic and poetics of infrastructure." *Annual Review of Anthropology* 42: 327–43.

Livingstone, David. 2003. *Putting science in its place: geographies of scientific knowledge.* University of Chicago Press.

Matsuura, Shigeki. 1990. "Circumstances that the hireling Dutch engineers came to Japan early in the Meiji Era." *Water Science* 34(3): 1–12. (in Japanese)

Morita, Atsushi. 2012. *Savage engineering: Anthropology of humans and materials in Thai small and medium industries.* Sekai Shisosha. (in Japanese)

Multi-nature river construction review committee 2006. The development of multi-nature river construction: the direction river construction should aim for and policies for promotion moving forward. (Report of multi-nature river construction review committee). (in Japanese)

Nakagawa, Takeo. 2006. "Technical succession of soda mattress." *Rivers* 2006(11): 47–50.

Nakamura, Shunroku. 2006. "Hydraulic model experiments for the succession of traditional techniques in stone masonry water control." (Report on the River Improvement Fund Subsidy Project for Fiscal Year 2005). (in Japanese)

Nakaoka, Tetsuro. 2006. "The formation of modern technology in Japanese technology: the dynamics of 'tradition' and 'modernity.'" Asahi Shinbunsha. (in Japanese)

Namba, Miki. 2017. "Infrastructure as conceptual analytical and empirical ethnographic tool." In Ueda, T. ed. *The politics of materiality and 'infrastructural turn.'* The Institute for Research in Humanities and Social Sciences, Aichi University, 53–63. (in Japanese)

Namba, Miki. 2018. "Visible infrastructure and the 'pseudo-modern' in Vientiane, capital of the Lao PDR." *Japanese Journal of Cultural Anthropology* 83(3): 404–22. (in Japanese)

Raj, Kapil. 2007. *Relocating modern science: circulation and the construction of knowledge in South Asia and Europe, 1650–1900.* Palgrave MacMillan.

Seki, Masakazu. 1994. *Rivers of the earth: revitalizing the rivers in Japan.* Soshisha. (in Japanese)

Shinohara, Osamu. 2018. *How three generations of river engineers viewed rivers: Aki Koichi, Takahashi Yutaka, Okuma Takashi and 150 years of modern river administration.* Nosangyoson Bunka Kyokai. (in Japanese)

Soda, Ryoji and Yuhora Kazuhiro. 2012. "What was the aim of 'nature-oriented river work'?" *E-journal GEO* 7(2): 147–57. (in Japanese)

Soda, Ryoji and Yuhora Kazuhiro. 2013. "River improvement works and changes in human-nature relations: from the case of Agano river in Japan." *Studies in the Humanities* 64: 197–212. (in Japanese).

Tomino, Akira. 2006. *Traditional river works in Japan [I]*. Shinzansha. (in Japanese)

Tsujimura Yokichi. 1889. "Sedimentation control work and soda methods." *Seizan Anpo*. (in Japanese)

Ukisu, Shuei and Igarashi Shinji. 2002. "On the modularization of soda mattress." *Rivers* 2002(10): 33–36. (in Japanese)

Whatmore, Sarah. 2002. *Hybrid Geographies: Natures Cultures Spaces*. Sage Publications.

Yoshikawa, Katsuhide. 2007. *Beyond Multi-Nature River Construction*. Gakugei Shuppansha.

CHAPTER 17

Riverine Lifelines of the Brunei Sultanate

Gabriel Yong, Izni Azrien bin Noor Azalie and Noor Hasharina Hassan

The nation-state of Brunei Darussalam was once a powerful Sultanate with territories that spanned the entire northern coast of Borneo for most of its history. At its zenith (1500–1578), this included parts of the Philippines. The seat of power was a city built over the water in the Brunei estuary. The Magellan chronicler, Antonio Pigafetta, compared its splendour to the European city of Venice following a visit in 1521 (Nicholl 2007). The Sultanate's fortunes began to wane after the Castilian War of 1578, and were exacerbated by a civil war that lasted 13 years from 1660 to 1673. The arrival of the British and the establishment of Sarawak under the White Rajahs led to the loss of much of its territories in the 19th Century. By the 20th Century, the Sultanate had become a British protectorate, essentially governed by Britain until the Second World War (WW2). The British administration began a systematic transformation of the Sultanate, shifting settlement, development and economy onto land, and installed political and education systems modelled after its own. Brunei transitioned from a Malay sultanate into a modern nation-state following WW2, gaining full independence in 1984.

In stark contrast to its pre-20th-century existence, present-day Brunei Darussalam is a small country in terms of land area and population. Its citizenry, however, enjoys a high standard of living due to an economy powered by oil and gas resources. Its capital has moved from the estuary onto the north bank of Sungai (Sg, 'river') Brunei. In 1970, Brunei Town was renamed Bandar Seri Begawan (BSB) after the 28th Sultan, widely regarded as the architect of modern Brunei (Ibrahim 1971). Meanwhile, the water city has devolved into an unexceptional settlement complex known as Kampong Ayer (KA, 'water village'), which is bereft of economy and industry. Presently, KA is a vestigial feature of the water city of the old empire. However, accounts of early to mid-20th-century KA still depicted a vibrant populace and economy, despite the Sultanate having lost most of its territory and power (Ibrahim 1971; Bakar 2018; Azalie & Yong 2022). A British engineer who worked in the *kachu* ('cutch') factory in the early part of the 20th Century described KA as a "city of many waters", where the population lived "a semi aquatic life, and their methods of living, household arrangements, family life, and town government, adapted as they had been to life over the water" (Blundell 1923, 201). He also described

© 2026 THE AUTHOR(S) | HTTPS://DOI.ORG/10.1163/9789004758100-019

Brunei as a country with no roads (ibid, 45), denoting a fluvial characteristic of Borneo before the arrival of the Europeans.

This chapter however raises the question regarding the survival of a 'water city' for over 500 years. In particular, where and how it obtained its daily supply of food, water and materials? Without land to grow crops or raise livestock, these resources would have to be brought to the city from elsewhere. In a humid tropical environment, rivers would be the natural transit highways. This chapter, therefore, aims to trace the riverine network that connected KA to the supply of its daily needs and shed light on the cultural manifestation consequent of commerce, trade, migration, inter-marriages and settlement. It will also examine how these riverine lifelines faded as Brunei transitioned into a modern nation state on land, but also how they could regain prominence as the impact of climate change becomes more pronounced. The paper also compares its treatment of the river with that of Ludden (Chapter 3) and Rashid (Chapter 12), and discusses the findings about the concept of the Anthropocene.

1 The Need for a Dynamic Research Framework

According to the Oxford Dictionary, a lifeline is anything on which someone depends for continued existence (Cowie 1989, 721). Its usage is apt in this study because (i) the riverine network provided the ancient water city of the Brunei Sultanate the means to exist for five centuries in an estuary, and (ii) its nautical root serves as an appropriate metaphor. As rivers were the primary facilitators of trade and commerce, they served as lifelines that sustained the city of the Sultanate. However, the fact that Brunei underwent drastic transformation in the 20th Century complicates the research framework, as both the cultural and physical contexts over the Sultanate's history are unrecognizable from the recent times, which were written mainly by outsiders, primarily those who share a common perspective with the western colonizers.

Figure 17.1 shows the research framework of the study. It separated four major periods relevant to understanding the history and development of Brunei into their corresponding cultural and physical realms. This will be used as the research context.

1.1 *Drastic Changes in the Cultural Realm*

The Brunei sultanate was a thalassocracy that rose to prominence on the northern coast of Borneo in the 16th Century (Saunders 2003; de Vienne 2015). It shared similar characteristics with the coastal Malay kingdoms in the region in terms of external trade (Wolters 1967). Commerce in the city of the Sultanate,

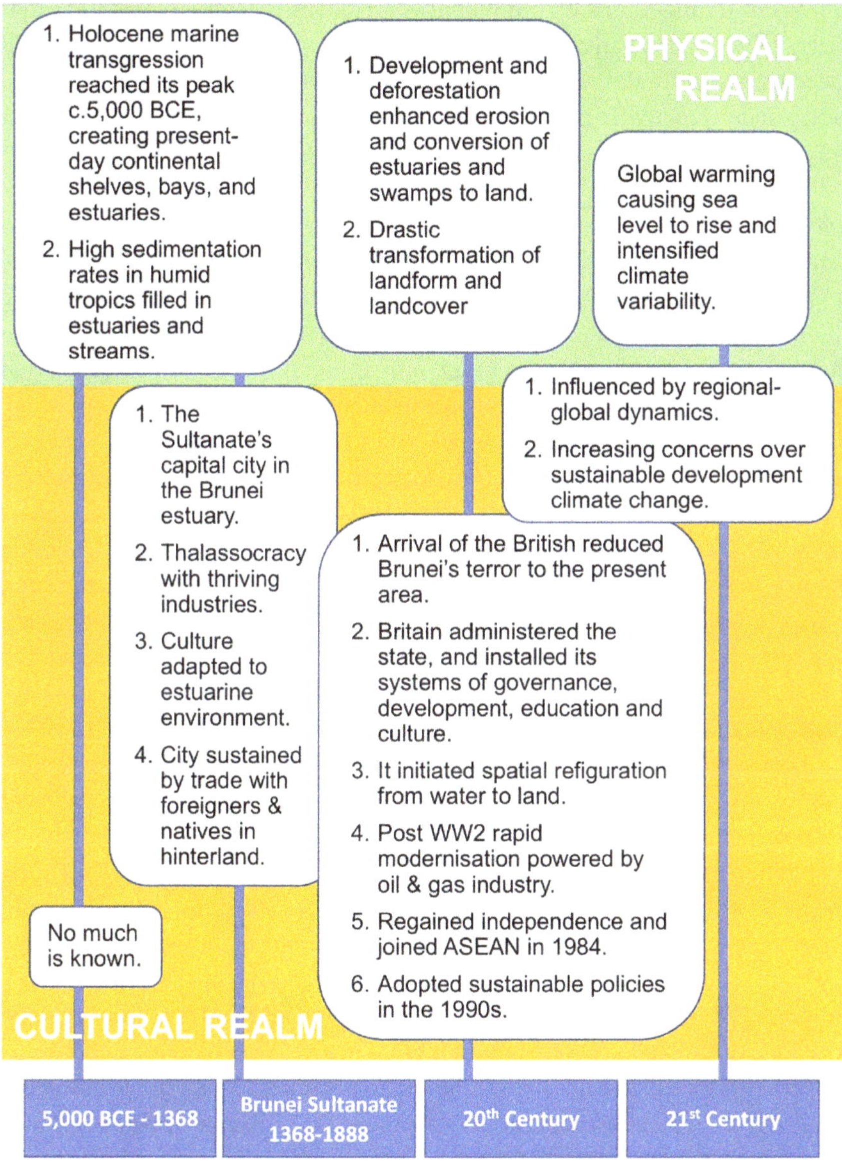

FIGURE 17.1 Research framework

however, was conducted on the estuary by women selling daily necessities from house to house in small boats called *gubang*, *bidar* or *perahu*. Both the sellers, identifiable by their large hats (*siraung*), and the floating mobile market are known as *padian* (Ibrahim 1970). This unique market was recorded as early as 1521 by Pigafetta (Nicholl 2007) and prevailed until the latter half of the

20th century, together with the *pengalu* ('merchants'), who supplied the goods (e.g., Bakar 2018). As many have direct or indirect[1] knowledge of *padian* and *pengalu*, there are therefore quite a number of articles about them, published and on social media, in both English and Malay. An article in *Sukan Kitani* (Our Sports), a football blog, is an example of an unlikely but insightful source.[2]

The Daily Resource,[3] which is penned by a former senior government official with vast knowledge of and interest in the country's development and history, gave good details of the role played by the *pengalu* in facilitating trade in fine camphor, sandalwood, tortoise shells, wild bee honey and others (The Daily Resources 2007b). It was the *pengalus* that made the journey upstream to trade with indigenous tribes in the hinterland. It is generally assumed that *pengalus* were part of the KA populace, which although the majority of which is Malay, is also not homogenous; KA was a centre of trade. Unverified personal knowledge of people who were formerly *pengalus* suggests that they are not always KA Malay. What proportion of *pengalus* and their crew members or family are of other ethnicities or even nationalities? Whom did they trade with, and what ethnicities were they? Although the location of trade routes, markets, and boat designs are common knowledge to many locals who share the same life-world, there are conflicting details, such as the name or location of lesser-known trading bases (*pengkalan*), items traded, duration of the journey, ethnicity of people involved, and so on.

Another pertinent question concerns the rules or laws governing the activities of the *pengalu*. In the Brunei sultanate, there were three categories of land ownership (Mail and Bakar 2016): (i) *Sungai Kerajaan*, land belonging to the Sultan as the ruler; (ii) *Sungai Kuripan*, land belonging to the *Wazir*, the highest-level officials in the sultanate; and (iii) *Sungai Tulin*, land that could be inherited or bequeathed to others. Land ownership is therefore determined by river basins, commonly referred to only by the name of the river (*sungai*). In pre-modern Brunei, KA would have been a Sungai Kerajaan, i.e., owned by the Sultan. As *pengalus* operated from bases in KA, this would mean they must operate under the auspices of the Sultan and perhaps pay a tax to the monarch. In some river basins, native Bornean leaders were given titles and power to act on the Sultan's behalf. An example is a Kiput (a Bornean ethnic) in the Tutoh Valley (just south of present-day Brunei), who was bestowed the title

1 Directly, where they have familial relationships with pengalus or have interacted with them in the past; indirectly, by way of stories shared to the younger generations by the former group.

2 http://subuk-saja.blogspot.com/2009/01/padian-pengalu-in-brunei.html.

3 http://bruneiresources.blogspot.com/2007/07/role-of-pengalu-in-brunei-history.html.

Orang Kaya Pemancha (Abdul Rahman 2013), and therefore, land ownership associated with the office/title.

A *Nakhoda* is the captain of vessels engaging in maritime trade. However, they were also active in river basins, such as the Baram, which was once part of the sultanate's territory (Godfrey 2014). Although there is similarity in their scope of work with *pengalus*, the latter appears to refer to those who operate in KA and its riverine network. Of particular relevance to this study is a blog that provides details of exchanges between *pengalus* and local communities as well as of the trade routes in the Kulapis area (Rais 2009). Suryani (2011) also offers insights into trade in the Limbang Valley. Ismail et al. (2019) provide detailed insights into people, place and cultural activities during the transition period. Shariffuddin (1969) is the main information source on the Kedayan people,[4] who are the main suppliers of food to KA from the surrounding land. This study could be treated as a complement to the article by Md Mamunar Rasid (Chapter 12) because they shared a common region, but this article provides an insider's perspective of the ethno-commercial network and dynamics at the ground level.

1.2 *From Water to Land*

David Ludden (Chapter 6, 139) elucidated the spatial history, power and precarity of rivers, and highlighted their role in toponyms, myths, and how they are inextricably intertwined with social life, culture, economy and indeed, civilisation. This is echoed in Kathrina Daud's (Chapter 3) fascinating article connecting rivers to contemporary imaginaries. Our study, however, shed further light on the physical features of the river itself and its environment, because they are dynamic and continually evolving. The change occurs at a rate that is imperceptible to the casual observer. The differences, over a substantial period (decades, centuries) are, however, quite stark. The present warm geological epoch (Holocene) began some 11,700 years ago (Chung 2024, March 5). Sea level rose as glaciers melted, reaching its maximum around 5000 BCE. This inundated coastal lowlands and river valleys, creating continental shelves and estuaries. This includes the valley of the Brunei river, which were submerged beneath the estuary separating KA into a northern and southern segment. The KA area 5000 BCE was a vast expanse of water, based on soil and geomorphic data (Yong and Hassan 2022).

In the humid tropical region, hydrometeorological processes are continually eroding the land and transporting sediments to river channels and topographic

4 One of the seven ethnic groups identified as indigenous to Brunei and classified as Malay under the Nationality Act of 1961, revised in 2002.

lows, infilling them over time. Estuaries, because of their connection to the sea at one end and fluvial discharges at the other, are depositional environments, i.e., they will be filled in by sediments, a process known as sedimentation. Swamps develop in the shallow intertidal estuarine mudflats, eventually transforming the area into forested land as the leading edge of mangroves advances into deeper water. Due to the low relief of much of the river basins in the western region, rivers in Brunei have low gradients, which allows estuaries to penetrate deep into the middle reaches of river systems. Consequently, they filled in rapidly, and modernisation accelerated the process, because of deforestation, which exposed bare soil to storm runoff.

The rainforest is the natural land cover, which intercepts the heavy rainfall, and therefore slows down erosion and sedimentation. However, large areas of rainforest were removed as part of Brunei's modern development in the 20th century. Land surfaces became increasingly impervious due to urbanisation, where soil was replaced by concrete, asphalt and other construction materials. This amplified storm runoff and its erosive power, and promoted sedimentation. The process explains why many mapped rivers are little more than depressions or relatively small artificial drains today. During this period, the environment in the vicinity of KA transformed from largely watery to dry land crossed by rapidly narrowing and vanishing rivers. In the 21st Century, however, sea level is expected to rise due to climate change. Since the Earth Summit at Rio de Janeiro in 1992, all nations have become party to the UNFCCC to address rising temperatures due to anthropogenic greenhouse gases (GHG) and associated climate-related risks. Scientists monitoring the Earth system can observe and measure concentrations of various GHG, surface temperatures, glacial melt and changing sea levels. Estimates for this region range from 3–4 mm/year to about double the rate along the coast, suggesting that sea level could be higher by 30 cm in the ocean or 60 cm along the coast by the end of the century. The effect in Sg Brunei will be pronounced due to its narrow form (Hassan and Yong 2023).

2 Earth Surface Processes and Human-Environment Relationships: towards a Dynamic Methodological Framework

This study combines two core traditions in geography: (a) earth surface processes or geomorphology; and (b) human-environment inter-relationships. It therefore combines physical and human geographic disciplines to examine how the riverine network sustained KA for over 500 years and gave rise to the distinctive Brunei culture and peoples. It will focus on the period of the

20th century when the sultanate underwent drastic transition, where riverine trade survived until the 1970s. It will also examine the relationship between KA, traders and communities along the rivers and surrounding river valleys, and the cultures that emerged as outcomes of these relationships, and how modernisation altered them. Finally, it will discuss the findings in relation to the Anthropocene.

The first part of this study traced the riverine connections between KA and the hinterland. River networks are readily observable from aerial imagery and topographic maps. The latter also contains information on cultural features, which reflect the influence of trade with KA. The following are key sources of information used:

- Brunei Town (1962), Sheet 4/114/4, 1:50,000, published by Director of Overseas Survey, Surrey, England. The map is based on aerial photography by British Royal Air Force (B.R.A.F.) in 1958 and fieldwork completed in 1961.
- Soil Map of Sarawak, Sheet B (1968), Sarawak Series No. 8, published by the Directorate of National Mapping, Malaysia. Ground surveys were carried out by the Sarawak Soil Survey Team between 1955 and 1968, while aerial surveys were carried out by the B.R.A.F. between 1947 and 1967.
- Soil maps of Brunei, Hunting Technical Services (1969). This was a study to determine the potential for the development of Brunei's land resources. It involved drilling 1,401 boreholes (depth < 1.2 m), 58 soil profile pits of representative soil units, and over 280 km of trace lines cut through the forest. As Brunei and Sarawak share the same regional landform and soil types, the survey used the Sarawak soils classification by Soil Survey Staff (SSS) (1966).
- Bandar Seri Begawan (1981, 1987), 1:50,000, published by Surveyor General, Brunei.

The second part shed light on the socioeconomic activities and cultural landscape along these trade routes. Blogs and articles provided useful information of the interactions between *pengalu* and the community.[5] Interviews were also conducted with two informants. The first (INF–1) is a female herbalist, age 91 years old, who has lived in the Kupang area along Sg Damuan most of her life. The second (INF–2) is a young male government officer, age 33 years old, whose family owns a large area of land in the Sengkurong area. The matriarch of his family (INF–2) kept detailed records of business activities. Both are known to the authors, who also have insider knowledge of the study area. The informants are of Kedayan stock. Interviews were conducted informally in a

5 As pointed out above, that many have direct and indirect knowledge of pengalus and activities, blogs become important source of information when individuals use them to share their personal knowledge.

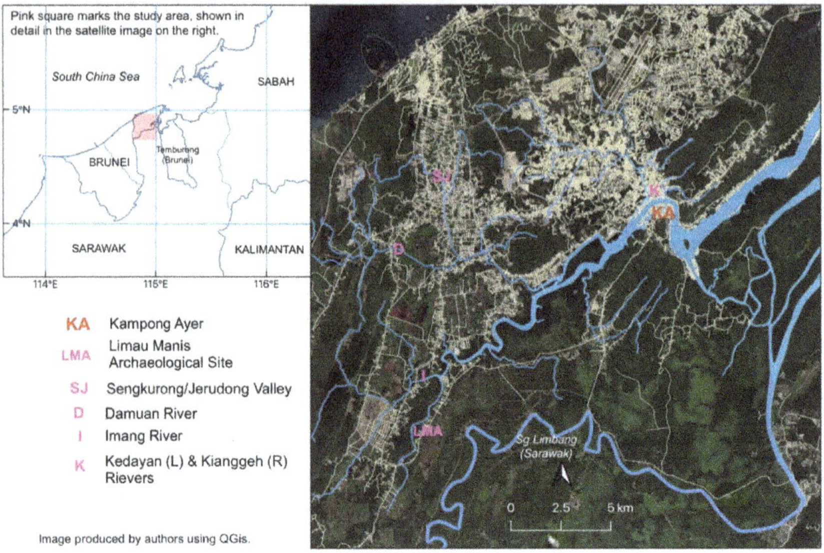

FIGURE 17.2 Location and environment of the study area.

conversational manner to acquire answers to a specific set of questions. The questions revolve around details of the Damuan-Sengkurong area before the area was altered beyond recognition by urban development after the 1970s.

Figure 17.2 shows the study area, which covers most of the Brunei-Muara District of Brunei, and a small part of northern Limbang, Sarawak, which included the northern section of the large and once important Sg. Limbang. It can be seen that most of the land area, from north to west of KA, is highly urbanised and has incorporated parts of the river network into the urban structure (engineered drainage system).

3 From Landscape Physicality to the Social: a Renewed Approach Towards Human-Environment Relationship and Network

The result of the study is elaborated in two sections: (a) riverine network and environment, and (b) cultural manifestations.

3.1 *Riverine Network and Environment*

The Brunei estuary in which KA is located was formed due to the Holocene marine transgression, which inundated the valley. The submerged river exists as the deep channel that separates KA into two clusters, a smaller one to the north, and the main cluster to the south. A number of relatively small river

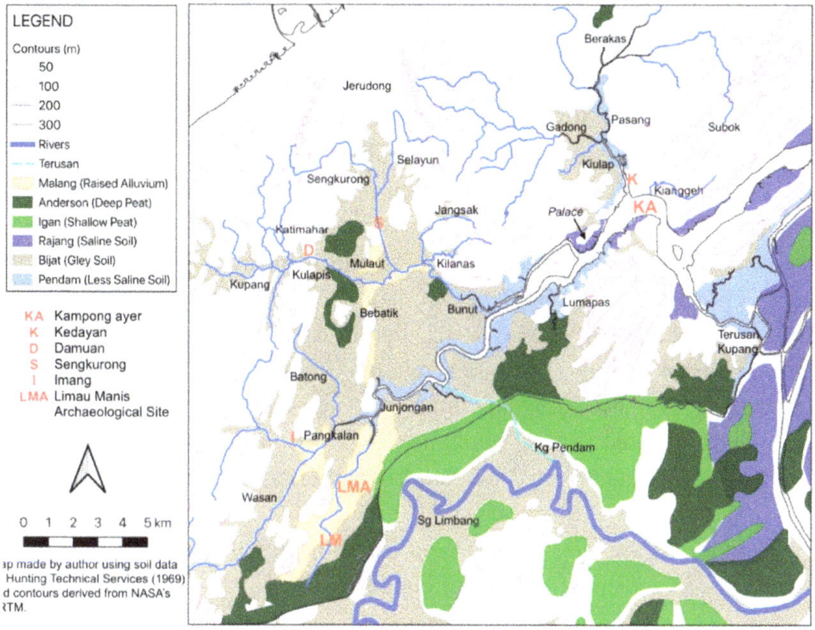

FIGURE 17.3 River networks, soils and environment.

systems drain into this section of the estuary from the north, west and southwest. From the south, Sg Limbang, draining a catchment area $(3{,}950\text{km}^2)$[6] seven times larger than the entire Brunei-Muara District, is connected to KA via the Brunei Bay and Terusan Kupang, where a port had existed in the past (Osman 2004). The riverine network of KA is linked to four main areas: (i) Imang and Limau Manis; (ii) Damuan and Sengkurong; (iii) Kedayan; and (iv) Limbang. Figure 17.3 shows this network, plotted with soil data[7] and contours. The rivers in the Brunei-Muara District all drain from low ridge systems, which are part of synclines that have been uplifted and tilted to the northwest (Sandal 1996). Geological faults gave rise to the Kedayan, Limbang and Sengkurong rivers.

3.1.1 The Imang—Limau Manis Route

The Imang river flows from a narrow ridge that is part of the Belait Syncline. Streams draining along strata converged to join Sg Brunei at Junjongan. Soil data revealed that much of the area is low-lying, flood-prone and water-logged (gley soils). A small patch of raised alluvium at the confluence of streams flowing

6 National Register of River Basins (2003), Malaysian Department of Drainage and Sewerage.

7 Names of soil shown in the legend are based on the Sarawak Soil Series with types of soil in brackets.

from Wasan, Imang and Batong suggests an area of cultivable land existed in the midst of swamps. A much larger similar deposit is found a kilometre to the east, where the archaeological remains of an ancient city-port were discovered in 2002 (Osman 2004). Similar alluvial deposits could be found northwards at Bebatik and Mulaut. They would have been like the tropical swamp equivalent to oases in the desert when the estuary was more extensive during pre-modern times. These features are most likely to be sand bars, which typically form at the mouth or intersections of rivers (Yong and Hassan 2022). The pattern of saline soil further reveals that river channels in the past were 2–5 times wider at different sections before they were filled in by sedimentation. Sections of the river network has also been incorporated into the urban structure. Pangkalan Batu[8] and Limau Manis[9] are important places in this network. Yong and Hassan (2022) posited that Sg Limbang flowed through this area in the past before its mouth was diverted eastward through natural sedimentation and fluvial processes. Prior to its diversion, Sg Limbang would have discharged a large amount of freshwater into the area. This is corroborated by the extensive gley and less saline Pendam soils (see Figure. 17.3).

3.1.2 The Damuan—Sengkurong Route

The Damuan river drains from the same ridge as the Imang, but reaches further west into the Tutong District. Although it is a small channel that connects to Sg Brunei just west of the palace of the reigning sultan, it was a much larger river in the living memories of the few who grew up in the various settlements that developed along its length, such as INF-1. According to the soil data, its channel was once 5 times wider at Bunut, where Sg Damuan branched off from Sg Brunei in the past, some 4 km further upstream from the present-day connection. Elongated raised alluvium found along its length between Mulaut and Kilanas indicates considerable discharge and sediment supply. Sg Sengkurong, flowing along the Jerudong valley,[10] joins the Damuan channel here and contributed to the development of the sand bars. The deep peat deposits in this region (Figure 17.3) were former lakes, which eventually transformed into peat bogs naturally over time. They would therefore have appeared as deep waters,

8 Labelled as Kg Pangkalan in the 1962 map.

9 Different maps have different locations for Kg Limau Manis along the same alluvial area, with the present-day village at the southern end. Limau Manis most is more likely to be an area than a specific site, such as a village or port.

10 Sg Limbang is believed to have flowed through the area in recent geological times and created the valley, before tectonic uplift altered its course.

which would be brackish due to their unobstructed connection to the Brunei Bay. Indeed, the name Mulaut means 'surrounded by sea.' Note that a number of isolated ridges with the same orientation as the ridge to the west and the axis of the Jerudong valley can be found here. They would have appeared like islands in the swampy environment in pre-modern times, particularly after rain storms and during high tides.

3.1.3 The Kedayan Route

The Kedayan system drains from the uplifted strata of the Berakas syncline. Its main flow direction runs counter to the general gradient of the land, which is tilted downward towards the northwest. This is because of the BSB Fault, which also gave rise to the sharp kink in the shape of Sg Brunei at KA and the island of Pulau Berambang.[11] As with the other rivers, the main channel was wider in the past (refer to saline soil, Figure 17.3). The Kedayan has two main sets of tributaries. The western set gave rise to the swampy depressions in Gadong and Kiulap, while the northern set drains from ridges at Subok to the east and undulating lands to the north. Sg Kianggeh was a small tributary that joins the Kedayan system close to its mouth. It was later separated by reclamation works in the construction of Brunei Town. An interesting feature is a constricted section at Pasang. This is caused by a build-up of sediments from the Gadong river, which produced a pronounced bend. This place is known as *Kumbang Pasang* ('beetle' 'flood tide'), which was said to refer to swirling water currents observed in the area;[12] such eddy currents could also form when tidal currents enter or exit a narrow passage. The place is also called *Kambang pasang* ('rising' 'flood tide'), probably referring to a pronounced tidal effect, which would form when rising tide squeezes through a narrow passage, as well as when storm discharge is encounters rising tide. A similar bend is observed at Kiulap, which appears to be caused by Sg Kiulap. The main channel here was however much wider and therefore facilitates river discharge better. The Kedayan system comprised of many small rivers and river valleys in close proximity to KA. As such, they naturally became important sources of food and materials for KA, as they are readily reachable by small boats within an hour.

---

11 The fault severed from the ridge from the mainland, turning it into an island.

12 In the past, before modification of the channel by engineering works. Source of Kumbang Pasang explanation: http://bruneiresources.blogspot.com/2006/11/more-origin-of-place-names-in-brunei.html.

3.1.4 The Changing Limbang Route

Sg Limbang's connection to Limau Manis and the Jerudong Valley is facilitated by the Sengkurong fault system (Sandal 1996; Yong and Hassan 2022). Although it has changed course due to natural processes, the implication of its connection to KA from the Limau Manis area on trade and mobility is highly significant because Sg Limbang is a massive river and therefore a major highway for many tribes and villages along the way to its headwaters in the Kelabit Highlands, some 130 km south of Limau Manis. The process that diverted its mouth from the Limau Manis area to present-day Limbang town to the east gave rise to much of the land and swamps in its lower reach (Yong and Hassan 2022). As the river changed course, a *terusan* ('canal', 'passage') was cut through the swamps that had developed to maintain a link between Sg Brunei and Sg Limbang. The defunct canal is still marked in topographic maps today. A 1906 map of British North Borneo[13] clearly showed that it was still actively used during that period. The diversion of Sg Limbang's mouth altered its connected to KA from west to east, via Terusan Kupang and the Brunei Bay.

3.2 *Salient Cultural Features*

Land use and cultural activities along the river networks were derived from 1962, 1981 and 1987 topographic maps and cross-referenced with information obtained through interviews, reports and blogs. The results are presented in four sections.

3.2.1 The Pengalu

The *pengalu* is one of the main actors in the local economy of the riverine world of the Brunei sultanate. In the early-to-mid 20th Century, they operated from two bases in KA: (i) Kg Ujong Kelinik[14] and (ii) Limbongan[15] (see Figure 17.4). The first group traded with settlements to the east of KA and with Limbang to the south, while the second with villages and *pengkalan* ('trading posts') to the west. This chapter will, however, only concentrate on the river networks to the west and south because they are the main source of the staple food, rice. Furthermore, most of the well-established trading centres are located upstream of Sg Brunei. The *pengalu* played a crucial role in the viability of KA, supplying its population with their daily needs. They transported goods along the river networks in 'house boats', essentially a large *sampan* (wooden boat) with a hut to accommodate the *pengalu* and his crew, which usually

13 British North Borneo, https://www.bmarchives.org/items/show/100205135.

14 At the eastern end of KA, part of Sg Siamas.

15 Within Sg Kedayan, close to its mouth.

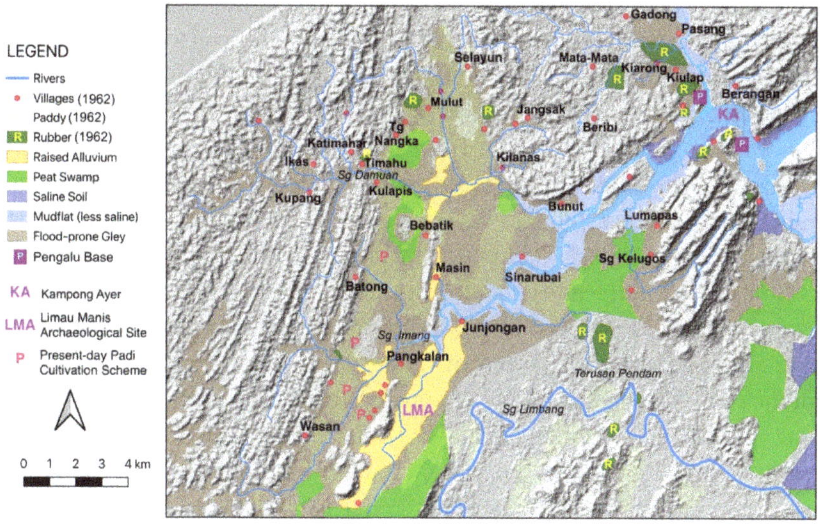

FIGURE 17.4 Villages and cultivation along riverine network.

comprised of his family (Mail and Bakar 2016). On board accommodation is necessary as each trip could take a few days before returning to KA. The boats were propelled manually using large oars that are fixed to the boat's structure. These merchants connected people and facilitated exchanges among communities of various ethnicity at different locations along the riverine network and KA. They 'modernise' the hinterland in the process, spreading products, imported or produced in KA, and its culture, to communities along the trade routes and more remote regions. *Pengalus* were active until as recently as the 1970s, decades into the modern era, where development and economy had shifted onto land. They eventually ceased operations when they, as businessmen, saw and took the opportunities presented by modernisation on land.

3.2.2 Bartering at the 'Tamu'

Sg Damuan is one of the two main trade routes west of KA. It provides connectivity to the wide area of swampy lowland and the ridge that separates the region from Tutong, the adjacent river basin. Within a gap in the ridge, a large number of settlements flourished (see Figure 17.4). They include Kulapis, Katimahar and Kupang, where INF-1 has lived for 91 years. According to her, between 20–30 boats would gather there to exchange goods. Commerce was done through bartering. *Pengalus* would bring dry and salted seafood, brass and iron wares, porcelain ware, textiles, traditional biscuits (*kuih-muih*) and other manufactured or processed products from KA in exchange for fresh vegetables, fruits, jungle products, cured meat (buffalo, deer) and rice from the communities here, the most prominent of which are the Kedayans, who are believed to have originated from Java and brought paddy cultivation to the sultanate (Shariffuddin 1969, 17).

According to the INF-1, *ambulong* (sago[16]) from farther afield in the Tutong basin, was also traded at Kupang. This centre of commerce appears to have a reach beyond the extent of the river network.[17] A closer examination of the terrain from maps revealed that the watershed is generally 60–100 m high, rising to 200 m only in isolated areas. This physical barrier is therefore readily surmountable by indigenous people. INF-1 recalled the excitement with every arrival of a *pengalu*, whether from KA or from further inland. Commerce was always vibrant. According to her, the name of the river, *Damuan*, is derived from the term, *betamu* ('meeting'), as it was the meeting place between *orang Brunei* (people of KA) and "paadian"[sic], referring to the Kedayan[18] villagers. She also said that the 'nearby' river, Sg. Matambu was similarly named after the word, *tamu*, which is commonly used to refer to a bazaar in the present day. Sg Matambu is a minor tributary that connects to Sg Damuan (7 km downstream), at the eastern edge of the Jerudong valley, near Kilanas.

The information from INF-1 suggests that the main meeting place for trade and commerce could stretch from Kilanas to Kupang. This section of the river traverses the width of the Mulaut lowland, where paddy cultivation was carried out extensively until the mid-1980s. Indeed, paddy cultivation also extended northwards into the Jerudong valley and south-southwest to Wasan (see Figure 17.4), where paddy cultivation has continued to the present day, as a government scheme. The extent of paddy cultivation and vibrancy of commerce was corroborated by INF-2, whose family owns large expanse of land in Jerudong Valley, including at Jangsak. As the sultanate transitioned to a modern nation state, land ownership was institutionalised under the Land Code in 1909. The Kedayans would become wealthy landowners.[19] INF-2 shared stories

16 This is a Bornean staple because the environment is not conducive to wet paddy cultivation on a large scale, due to the relatively small size of river catchment and hence, saline or acidic swamps, which were more suitable for sago palms. The swamp paddy (*padi paya*) grown by the Kedayans probably evolved from wet paddy varieties brought to the region by their Javanese ancestors. This hypothesis however requires further research.

17 This is corroborated by personal communications with residents from Tutong, who said that their forefathers used to travel a day or two to exchange goods with pengalus at Kupang.

18 It should be noted that this 'paadian' differs from the "padians" at KA, which refers to the sellers on boats or the floating mobile market, itself. This refers to the people who grows paddy.

19 This is evident in a 1967 map of the Brunei-Muara District published by the Survey Department, which shows land alienation of the paddy cultivation area.

of trade with KA passed down via oral traditions. Figure 17.4 also shows cultivation of rubber trees on hill slopes. They were introduced by the British (they are not endemic to Borneo) as part of their economic transition initiatives.

Within the ridge, the main settlements that have developed along Sg Damuan include Ikas, Lugu, Katimahar on the north bank, and Kupang and Kulapis on the south bank. Rais (2009), who hailed from Kulapis, provided detailed accounts of commerce and culture during the days of the *pengalu*. He recounted how villagers looked forward to the arrival of *pengalu*s and the commerce that would ensue. In INF-1's account, *pengalu*s have at times supplied goods on 'credit'. She related this to a situation in which one village suffered from a poor harvest and therefore had nothing with which to barter. The *pengalu* allowed the villagers to take what they needed with an agreement to pay back in his next or later visits; this trust economy is still practiced in present-day KA between informal merchants and low-income individuals (Hassan et al. 2022). The *pengalu*, therefore, played an important role in the sustainability of villages along the trade route, in addition to its primary role in sustaining the KA population. According to Rais, some chose to haul their produce 10–12 km across the swamps directly to KA using a *gubang* (small boat) pulled by buffaloes, bypassing the *pengalu*.

3.2.3 The Stone Trading Post and the Limbang Connection

The upper reach of Sg Brunei is the other main trade route, which continues into Sg Imang. The main trading post here is Pangkalan Batu ('stone trading base'), 2 km along Sg Imang from its confluence. The area is, however, often referred to as Limau Manis, the name of the other much smaller 'river.'[20] Accounts (Ismail n.d.) from as far away as Mata-Mata, Beribi and Kiarong refers to regular trade with Limau Manis. The archaeological artefacts discovered at Limau Manis (Osman 2004), 2 km south of Pangkalan Batu, support the idea that it was an important place of trade during its heyday.[21] Additionally, the area was known until the 1980s for paddy cultivation. Presently, the government scheme is confronted with limited water supply. However, when Sg Limbang flowed through the area in the past, it would have created the ideal environment for wet paddy cultivation. The change in environment due to diversion of Sg Limbang might

20 Sg Limau Manis does not have a catchment, like a normal river. It appears to be a vestige of the time when Sg Limbang flowed through the area or a man-made feature (a canal).

21 It is not possible to determine this period without further evidence. Yong and Hassan (2022) believe it could be as recent as the 19th Century while Osman (2014) thought it would be 10th–14th century.

explain the hitherto unresearched swamp variety of wet paddy (*padi paya*) cultivated by villagers in Brunei.

The connection of Sg Limbang to Pangkalan Batu and Limau Manis was of immense importance, as it provided access to the resource-rich valley. It was a major highway for riverine trade. This is evident in a statement attributed to Sultan Hashim in response to the annexation of the Limbang valley by Sarawak in 1890, when he referred to it as the *periok nasi* ('rice pot') of the Sultanate (Suryani 2011).

3.2.4 Extensions of the Water City

Sg Kedayan and Sg Kianggeh are essentially extensions of KA due to their close proximity. The small and narrow valleys are amenable to the cultivation of fruits and vegetables, and the young sedimentary rocks and intense weathering contribute to the ease of working the land. Kedayan communities settled in many nooks and crannies, producing food through swidden cultivation and providing KA with a lifeline of fresh fruits, vegetables, fowls, deer, as well as fish, shellfish, and crustaceans from the estuary. Some of the main settlements were at Mata-Mata, Beribi, Kiarong and Jangsak (Ismail n.d.). Dry hilly paddy (*padi tugal*) was also grown, albeit in smaller amounts compared to Mulaut. As the resettlement program initiated by the British Resident moved people out of KA, some settle in the nearby valleys, cultivating rubber in Gadong and paddy in Kiulap, before paddy gave way to rubber estates (ibid.). At KA, residents of Sg Kebun cultivated rubber in the adjacent hill slope, on the south bank (Ismail 2019). The whole character of KA and the surrounding land were transformed by the actions of the British locally as well as globally, which also affected KA.

Traditional industries in KA, such as iron, silver and gold smithing, production of brass wares, textiles and artisanal fish traps and boats, were replaced by cutch and rubber processing, and dockyards for shipbuilding and repairs (Forrest 1776, in Di Vienne 2015). Eventually, several shipyards were operating in the Subok area. Forested ridges were converted into plantations for rubber, coffee, tea and coconuts, as roads and bridges were built to traverse the land. KA extended into Sg Kianggeh, as Malay *bangsawan* ('entourage of the queen') and Chinese merchants settled in the area while they collaborated in a reclamation project to establish a fixed centre of commerce in KA, when the mobile *padian* market was still the norm. The constructed island (known as a *bakut*) became the first 'town', recognised as such when it was renamed *Pekan Lama* ('old town') (Hassan and Yong 2019). A hybrid (Malay-Chinese) culture had emerged, which preferred imported wares, use of kerosine lamps, duck rearing in the estuary, salted eggs, and eventually, water taxis with outboard motors that have become an integral part of modern KA. While *pengalus* were still active, a modern port was developed in a reclaimed area between Sg Kedayan and Sg Kianggeh, which grew into Brunei Town. After WW2, there was a period where *padians* and *pengalus* co-existed with the modern world before the eventual disappearance of the old world, leaving KA as merely one of many settlements in the country. The Kedayan and Kianggeh river valleys became highly urbanised.

4 Post-Second World War Modernisation, Globalisation and Climate Change: Its Transformative Effects on the Lifelines

Modernisation transformed the landscape and culture of Brunei so drastically that most of its past has been obliterated. The generations that grew up since the late 1990s are oblivious to the fact that the capital city was confined to a small section of the Sg Brunei estuary and how riverine networks, many of which are now largely storm drains or small inconspicuous channels, played a pivotal role in the life and prosperity of the city built over water. Indeed, modern Brunei prospered on land with an economy largely fuelled by oil and gas exports. The *pengalus*, so crucial as the mediator of life and resources in pre-modern Brunei, has assimilated into the modern world, with many in the upper echelons of government and industry. Meanwhile, highly mobile migrants from the Chinese diaspora, as well as the indigenous population of Sabah and Sarawak, also added to the fabric of the new nation state on land. As time progressed, life in the old world is but a memory in fewer and fewer of the surviving grey population.

In the interviews conducted, the subjects recounted with nostalgia life along the rivers, the lively *tamu* that followed the arrival of *pengalu*s, and the variety of food and wares that can no longer be found. They still retain memories of the old landscape, for example, of a once large and important river and trade route along Sg Damuan at Kupang, which is now a small unremarkable channel crossed by a trunk road, inconspicuous in the cacophony of urban modernity. However, while the majority of present-day society is unaware of past realities, the few that have such memories appear to be increasingly compelled to share their life-world, as evident in the many posts on social media; they are important sources of information for this chapter. As insiders, we share, to varying degrees, the subjective collective history and culture of the old world, largely through stories shared among friend and family in communities across the land, even when dispersed and assimilated into the modern world through education and life in a globalised world. As such, we are able to access this life-world somewhat effortlessly when exposed to new details. They revealed to us the human-environment dynamics across the two worlds—as academics in a discipline that examines human-environment interrelationships and natural processes, and as locals who are connected to the past via life-worlds transcending time evoked by stories shared. We were therefore able to recognise the riverine networks of KA as lifelines of the old world.

Furthermore, it is also apparent that whilst these rivers still exist, albeit much smaller in dimension and altered in parts by urban development, the subjective notion of lifelines has taken on a new turn. The oil and gas industry, which had given KA a new lease of life in the late 1960s with modern infrastructure and amenities, and saw an increase in its population to the highest by the 1980s, replaced the riverine trade as lifelines of modern Brunei. It facilitated and sustained the national resettlement program initiated by the British, but has refined its purpose to ensure that every family has a home on land (Hassan and Yong 2018). It was oil and gas that allowed Brunei to remain an independent state instead of having to become part of Malaysia, as it provided the fledging modern nation with a secure lifeline to stand on its own. Indeed, the 'petro' lifeline gave Brunei the means to systematically modernise the country through 5-year NDPs and participate in regional and global communities. However, the rapid industrialisation and development following WW2 in Brunei and across the globe has led some to question whether the planet has entered into a new age by the end of the century.

The Anthropocene is a geological epoch proposed by Nobel Laurette Paul Crutzen in an article in Nature in 2002, "Geology of Mankind". Although it has received extensive attention in the scientific community, and established an Anthropocene Working Group (AWG) in 2009 within the International Commission on Stratigraphy (ICS), the proposal was never likely to be accepted because the Anthropocene does not meet the criteria for inclusion as a formal unit of geological time (Finney 2015). Firstly, its stratigraphic record is minimal, given that the period spans only one human lifespan, and secondly, its evidence, which is based on observation rather than stratigraphic record, cannot be attributed solely to humankind. The proposal and the concerted push for its recognition as a geological epoch, despite not meeting established criteria, therefore appears to be motivated by political rather than scientific interest. It is therefore not surprising that after 15 years of debate, the ICS decided that the Anthropocene will not be officially recognised as a geological epoch (Chung 2024, March 5; Witze 2024 March 6). Although widely supported, the Anthropocene has its share of critics. Arons (2023), for example, pointed to the uneven treatment of the costs and benefits of the 'Age of Man', while Horváth and Lovász (2024) provided a broader discussion on it.

Our study suggests that modernisation has drastically and irreversibly altered the natural and cultural landscapes of Brunei. However, we concur that it cannot be attributed to humankind as a whole because they have existed before modernisation, which deliberately and systematically exploited the planet during the period of western colonialization. We are more inclined to use the term Capitalocene proposed by some (ibid.), as it narrowed the plethora of environmental issues to a specific root cause. This is the economic system founded on capitalism, which consumes resources and generates wastes at unsustainable rates, caused widespread environmental degradation, exacerbating inequality, and giving rise to the climate crisis that is confronting humanity. Regarding the latter, the International Panel on Climate Change's latest assessment report (2021) stated unequivocally that anthropogenic climate change is unfolding. The United Nations had responded to the impacts of global capitalism with a conference on environment and development ("Earth Summit") in 1992, which resulted in several global initiatives, the current stage of which is the 2030 Agenda for Sustainable Development with 17 Sustainable Development Goals (SDG).

Over the past decade, concerns over climate change have increased in urgency, with warnings of a global catastrophe, should mean global temperature rises 2 °C above 'pre-industrial' levels. In Brunei, the government has adopted a 'whole of nations' approach to combating climate change. As a coastal country in a humid tropical region, rising sea levels and increased rainfall are the main concerns. Satellite altimetry data indicate that sea levels in the region are rising slowly but surely at 3–4 mm/year since the turn of the Millennium (Fu et al., 2019). If the trend continues, this would raise the levels of estuaries to inundate the riparian zones, which have become highly urbanised. Ironically, this would resemble a step back into the old world of living in a water environment connected by river networks. Rising sea levels would increasing disrupt road networks and activities in low-lying areas. In Brunei, this includes the oil and gas industry and most of its industry and settlement, seriously threatening its present-day lifeline. Rivers might play important roles once again as connective highways and drivers of blue urban development. As for the Anthropocene, it is never likely to be accepted as a geological epoch, simply because it cannot meet the criteria for being defined as such. It should be regarded as a concept, if it were to achieve its goals of alerting humanity to the unsustainable trends caused by human activity. However, as pointed out above, it would be more prudent to focus attention on the root cause than to point the finger on humanity as a whole. Capitalism thrives on exploitation and therefore inequality, which is antithetical to concept of sustainable development (WCED 1987). Nevertheless, advocated of the Anthropocene might be jumping the gun because the impacts caused by modernisation represents only the beginning stage of heightened anthropogenic impacts. It does not account for the fact that humankind is sentient and capable of reasoning and innovation. Among the scenarios presented by the IPCC's Working Group 2 is one where humanity worked towards achieving the global SDGs, fostering inclusion, enhancing knowledge diversity and ecosystem stewardship, and achieving equity and justice. This could reverse the damage done by capitalistic industrial society.

5 River Lifelines and Change: a Conclusion

Rivers are a familiar part of the landscape to most people in Brunei and the region. They are associated with water transport, fishing and recreation. In Brunei, some rivers are sources of potable water supply. In pre-modern Brunei, rivers were the lifelines of the sultanate, as it was the connections to food and material to sustain a city built over water. However, these lifelines would be useless without the *pengalu*, who traverse these waterways to facilitate trade, and in the process, sustains both the population of KA as well as the hinterland. However, in an environment where change is constant, these facilitators must continually adapt to changing routes, population centres and demands. Ultimately, in the case of Brunei, the change caused by external agency attained such magnitude that the Sultanate itself together with its lifelines were completely altered and replaced with new realities. The rivers that used to sustain the sultanate now only exists in the life-world memory and stories of some and in articles such as this chapter. In a future of a warming climate and rising sea levels, however, rivers might once again take on prominence as drivers of blue urban development, where society must adapt to life over water yet again. The constant interplay between humanity and the natural environment continues. Man, in Brunei and other places, will continue to alter its environment locally, but it also responds to environmental change, which is planetary in scale.

References

Abdul Rahman, Abdul Karim. 2013. *Lipan Baram: Hubungan Brunei dengan Orang Ulu di Sungai Baram, Belaga dan Bintulu pade Abad Ke-19*. Brunei History Centre.

Abu Bakar, Nani Suryani. 2007. "A Historical Overview of Brunei's Economy Before the Discovery of Oil and Subsequent Issues". *Southeast Asia: A Multidisciplinary Journal* 7 (1): 89- 103.

Arons, Wendy. 2023. We Should Be Talking about the Capitalocene. TDR: The Drama Review. 67(1):35-40. doi:10.1017/S1054204322000697.

Azalie, Izni Azrein, and Yong, Gabriel Yit Vui. 2022. "Heritage Industry as Conservation Strategy for Kampong Ayer (Brunei Darussalam)." *Brunei Museum Journal*: 26–50.

Bakar, Haji Tassim Abu and Ariff, Mohammad Raduan Mohd. 2012. "Dari Padian ke Tamu Kianggeh: Perubahan Corak Perniagaan Peniaga-Peniaga Kecil Bandar Seri Begawan, Brunei". *Borneo Research Journal* 6: 18- 28.

Bakar, Haji Tassim Abu. 2018. "The Residents of Kampong Ayer, 1906–1940: The Activities of *Pengalu, Padian* and Handiwork". *Tawarikh: International Journal for Historical Studies* 9 (2): 151–158.

Blundell, Peter. 1923. *A City of Many Waters*. J.W. Arrowsmith.

Cowie, Anthony, ed. 1989. *Oxford Advanced Learner's Dictionary* (*Fourth Edition*). Oxford University Press.

Chung, Emily. 2024, March 5. "Goodbye, Anthropocene? Scientists Vote Against New Epoch." CBC News. https://www.cbc.ca/news/science/anthropocene-epoch-scientists-vote-1.7134449.

Daud, Kathrina. 2025. "Monsters and Mysteries: Rivers in the Contemporary Bruneian Imaginary." In *The River and the Anthropocene Life world in the Asia-Pacific,* edited by Iftekhar Iqbal, Noor Hasharina Hassan and Asiyah Kumpoh. Brill.

De Vienne, Marie Sybille. 2015. *Brunei: From the Age of Commerce to the 21st Century.* NUS Press.

Godfrey, Helen L. 2014. Trade and transformation on the Borneo frontier: towards a study of the Baram region of Sarawak in the late nineteenth century. *Borneo Research Bulletin*. ISSN 0006–7806. Accessed 22 January,2019. https://www.thefreelibrary.com/_/print/PrintArticle.aspx?id=420435224.

Horváth, Márk and Lovász, Ádám. 2024. "From the Anthropocene to the Capitalocene and beyond." The Anthropocene Review. https://doi.org/10.1177/20530196241270703.

Finney, Stanley C. 2015. "The "Anthropocene" epoch: Scientific decision or political statement?" *GSA Today*, v. 26, no. 3–4, doi: 10.1130/GSATG270A.1.

Fu, Y., Zhou, X., Zhou, D., Sun, W. and Jiang, C. 2019. "Sea Level Trend and Variability in the South China Sea". *ISPRS Annals of the Photogrammetry, Remote Sensing and Spatial Information Sciences*, IV-2/W5: 589–593. https://doi.org/10.5194/isprs-annals-IV-2-W5-589-2019, 2019.

Hassan, Noor Hasharina, Rigg, Jonathan, Azalie, Izni Azrein, Yong, Gabriel Yit Vui, Zainuddin, Nurul Hazirah Haji, Muhammad Shamsul, Mohammad Addy Shahril. 2022. "Making do and staying poor: The poverty context of Urban Brunei". *Geoforum* 136 (November): 132–141. https://doi.org/10.1016/j.geoforum.2022.09.012.

Hassan, Noor Hasharina and Yong, Gabriel Yit Vui. 2023. "Climate Change and Sustainable Development of Kampong Ayer: Its History, People, and Environment".

In *Public Policy's Role in Achieving Sustainable Development Goals*, edited by Mahani Hamdan, Muhammad Anshari, Norainie Ahmad & Emil Ali. ICI Global.

Hassan, Noor Hasharina and Yong, Gabriel Yit Vui. 2019. "The cultural value of bakuts in Kampong Ayer, Brunei Darussalam." *Southeast Asia: A Multidisciplinary Journal* 19: 47–63.

Hassan, Noor Hasharina and Yong, Gabriel Yit Vui. 2018. "A Vision in Which Every Family Has Basic Shelter". In *Sustainable Development Goals in Southeast Asia and ASEAN, edited by* Ronald Holzhacker & Dafri Agussalim. Brill. https://doi.org/10.1163/9789004391949_010.

Ibrahim, Haji Abdul Latif Haji. 1970. "Padian: Its Market and the Women Vendors". *Brunei Museum Journal* 2 (1): 39–51.

Ibrahim, Haji Abdul Latif Haji. 1971. "Variations and Changes in the Names and Locations of the Wards of Kampong Ayer over the Last Century". *Brunei Museum Journal* 2 (3): 56–73.

Intergovernmental Panel on Climate Change (IPCC). 2021. Climate Change 2021: The Physical Science Basis. Contribution of Working Group I to the Sixth Assessment Report of the Intergovernmental Panel on Climate Change. Edited by Masson-Delmotte, V., P. Zhai, A. Pirani, S.L. Connors, C. Péan, S. Berger, N. Caud, Y. Chen, L. Goldfarb, M.I. Gomis, M. Huang, K. Leitzell, E. Lonnoy, J.B.R. Matthews, T.K. Maycock, T. Waterfield, O. Yelekçi, R. Yu, and B. Zhou. Cambridge: Cambridge University Press. https://doi.org/10.1017/9781009157896.

Ismail, Rosidah, Mohd Akir, Zawiyatun Ni'mah, and Abd Rahman, Abu Bakar. 2019. *Mukim Sungai Kebun Sejarah Kampung*. Government of Brunei (Information Department).

Ismail, Rosidah, Mohd Akir, Zawiyatun Ni'mah, and Abd Rahman, Abu Bakar. n.d. *Mukim Gadong 'B' Sejarah Kampung*. Government of Brunei (Information Department).

Ludden, David. 2025. "River Territory: Spatial History, Power, and Precarity." In *The River and the Anthropocene Life world in the Asia-Pacific,* edited by Iftekhar Iqbal, Noor Hasharina Hassan and Asiyah Kumpoh. Brill.

Mail, Asbol and Bakar, Tassim Abu. 2016. "Live on Water: The Lifestyle of Kampong Ayer Community during the British Residency Period, 1906–1941". In *Brunei: History, Islam, Society and Contemporary Issues,* edited by Ooi Keat Gin. Routledge.

Nicholl, Robert, ed. 2007. *European Sources for the History of the Sultanate of Brunei in the Sixteenth Century.* 3rd edition. Brunei Museums.

Osman, Karim, ed. 2004. *Sungai Limau Manis: Tapak Arkeologi Abad Ke-10-13 Masihi.* Brunei Darussalam Museums Department.

Rais, Mohd Jaffar. 2009. "Blog Kpg Kulapis – Maklumat dan Info Terkini". Accessed January20, 2023. kpgkulapis.blogspot.com/2009/10/sejarah-kampong-kulapis.html.

Rashid, Md Mamunur. 2025. "Ethno-commercial Networks in Northwest Borneo: Perspective from the River." In *The River and the Anthropocene Life world in the Asia-Pacific,* edited by Iftekhar Igbal, Noor Hasharina Hassan and Asiyah Kumpoh. Brill.

Sandal, T. Svein, ed. 1996. *The Geology and Hydrocarbon Resources of Negara Brunei Darussalam.* Brunei Shell Petroleum & Brunei Museums.

Saunders, Graham E. 1994. *A History of Brunei.* Oxford University Press.

Shariffuddin, P.M. 1969. "The Kedayans". *Brunei Museum Journals* 1 (1): 15–23.

Suryani, Nani. 2011. "Migration of the People from Brunei to Limbang after the Annexation of Limbang by Charles Brooke of Sarawak in 1890". Paper presented at Merantau: Imagining Migration in the Malay World. International Seminar in Honour of Prof. E.U. Kratz. University of Frankfurt, Germany, 30–31 March. https://www.academia.edu/8446318/Migration_of_the_people_from_Brunei_to_Limbang_after_the_annexation_of_Limbang_by_Charles_Brooke_of_Sarawak_in_1890.

The Daily Brunei Resources. 2007a. "Bandar Seri Begawan – A New City, An Ancient Capital". http://bruneiresources.blogspot.com/2007/09/bandar-seri-begawan-new-city-ancient.html.

The Daily Resources. 2007b. "The Role of "Pengalu" in Brunei History". http://bruneiresources.blogspot.com/2007/07/role-of-pengalu-in-brunei-history.html.

Witze, Alexandra. 2024, March 6. "Geologists Reject the Anthropocene as Earth's New Epoch – After 15 Years of Debate." *Nature*, 627, 249–250. https://doi.org/10.1038/d41586-024-00675-8.

World Commission for Environement and Development (WCED). 1987. *Our Common Future.* Oxford Universiti Press.

Yong, Gabriel Yit Vui and Hassan, Noor Hasharina. 2022. "Historical Geography of the Limau Manis Archaeological Site, Brunei Darussalam." *Brunei Museum Journal*: 51–70.

PART 4

Epilogue

CHAPTER 18

Rivers and Planetary Politics in the Anthropocene

Rohan D'Souza

In his 'Short Introduction' to the twentieth century, the celebrated British historian Eric Hobsbawm characterized the decades between 1914–1991 in world history as the 'Age of Extremes.' This 'Age of Extremes,' was chiefly marked by the unruly march of capitalism alongside many a dogged struggle for socialism and communism. By the cusp of the twenty-first century, however, free market enthusiasts appeared victorious. Dramatized in the collapse of the Berlin Wall and the stunning dissolution of the Soviet Union. Hobsbawm, nonetheless, remained grudging and unconvinced. Though neoliberalism seemed triumphant, he reasoned, as an ideological force it offered no meaningful resolution to the innumerable bitter political conflicts and social contradictions that continued to haunt much of the world (Hobsbawm 1995). In the words of an equally celebrated contemporary Eugene Genovese, Hobsbawm's twentieth century seemed to have ended in 'a bang and a whimper', without a 'blueprint for the coming century' (Genovese 2012).

In hindsight, one cannot help but wonder if Hobsbawm, inadvertently or otherwise, naively chose only a limited set of colours and a few broad-brush strokes to draw out his claims. Notably, by narrowing the twentieth century to only being a battlefield between capitalism and Marxist critiques. A limitation that became all the more pronounced when barely a few decades after its publication Hobsbawm's bestseller was laid bare over how little it offered on the question of global warming and the dire possibility of the planet being engulfed in a mass extinction event. Clearly, in terms of a framework, Hobsbawm trained his lens exclusively on issues of political economy. His conceptual orientation, in other words, focused on the human and the social, while sidestepping their deeper entanglements with the nonhuman and the natural world. In a landmark paper published in 2009, Dipesh Chakrabarty, a leading historian and postcolonial theorist, by way of a frank realization offers a probable insight into why the current 'planetary conjuncture' might have caught many a scholar by surprise.

 | HTTPS://DOI.ORG/10.1163/9789004758100-020

> As the crisis [global warming] gathered momentum in the last few years, I realized that all my readings in theories of globalization, Marxist analysis of capital, subaltern studies, and postcolonial criticism over the last twenty-five years, while enormously useful in studying globalization, had not really prepared me for making sense of this planetary conjuncture within which humanity finds itself today. (Chakrabarty 2009, 199)

In effect, for Chakrabarty, the concentration on globalization in the social sciences and the humanities might have blindsided many from grasping the implications of global warming, which increasingly calls for radical shifts in existing perspectives within academia, especially in how we grapple with the 'epoch of the Anthropocene.' The Anthropocene, simply stated, announces that we are in a new geological epoch in which the previous climatic variability that defined the Holocene epoch (preceding 11,500 years) has been radically disrupted. To be living in the Anthropocene, however, is not only to declare that the previous Earth System—the coupled 'ecosphere' (atmosphere, biosphere, and cryosphere) and the 'anthroposphere (all human activity)—has been irreversibly shifted but that this 'altered earth' has been primarily brought about by humans, who, by releasing vast amounts of Green House Gases (GHS) into the earth's atmosphere, have ended up profoundly transforming the chemical composition of the latter (Thomas 2022). Humans, in other words, by becoming the main drivers of global warming can potentially bring about the sixth mass extinction of life on the planet (Kolbert 2015).

The Anthropocene as a notion is not, however, free of friction. For one, the scientific and institutional bodies of geologists are yet to officially accept the claim that planet Earth has indeed undergone an epoch-level change (Witze 2024). Second, if we are indeed in the Anthropocene, what should be the appropriate scale for environmental action: should it be an environmentalism that involves the local, regional and the national or should policies and interventions be directed at the planetary level. Christophe Bonneuil and Jean-Baptiste Fressoz (2016), in their *Shock of the Anthropocene* explain why the scale for environmental action has become the source for intense debate:

> A few decades ago, the "environment" was still understood as that which surrounds us, the place where humans went to extract resources, deposit waste, or even that in certain places was to be left virgin ... The concept of the Anthropocene challenges this separation and ... [in] place of the "environment" there is now the Earth system ... Instead of "master and possessors of nature", we find ourselves each day a bit more entangled in the immense feedback loops of the Earth system (20).

That is, while Holocene-centered environmentalism responded to how humans and their different social formations modified or degraded ecosystems, in the epoch of the Anthropocene the emphasis for humans will be focused on monitoring and staying within the planetary boundaries of the earth system (Rockstrom and Klum 2015). This shift towards a planetary perspective, however, is not merely a call for connecting the local, regional and the national with the planetary but, more critically, being able to grasp in a single frame not only the entanglements of nature and culture but also their innumerable webs and linkages within the larger feedback loops of the earth system—understanding how biological possibilities are entangled and enmeshed within larger planetary level physio-chemical and geological processes. In other words, ensuring human survivability and flourishing in the Anthropocene requires us to not only go beyond conventional environmentalism such as achieving sustainable development or conservation agendas but to critically rethink what environmental conditions enable human survivability in conditions of an altered earth system.

It is in the context of the turn away from the conventional environmentalism of the Holocene and growing alarms over extreme weather events linked to climate change that this edited collection makes a persuasive case for reconsidering a range of existing attitudes to rivers in general and those flowing in the Asia-Pacific region in particular. The compelling introduction and the seventeen essays that make up the edited collection are, in fact, focused on arguing for a profound conceptual shift. Central to which is the claim that rivers need to be grasped as a complex interweave of physical flows and social 'life-worlds'—that is, flows are entanglements of 'more-than-human' entities with 'human engagements' that comprise a mix of environmental imaginations, livelihood strategies and discourses on development and governance. While, undoubtedly, the emphasis on describing and understanding rivers principally as 'riverine worlds' is conceptually novel, the edited collection must also be commended for further deepening current critiques of the reigning paradigm in which rivers are treated and referred to as a modern resource—mere adjunct to the economy of land and input to agrarian production.[1]

In the modern resource framework, rivers are principally viewed as a colourless homogenous volume in motion and with flows contained between the banks of a channel. In great measure, this overtly technical and engineering

1 For how the notion of 'resource' marked a conceptual break with the advent of modern colonialism and industrialism by treating Nature as an input for industrial production and colonial trade see Vandana Shiva, "Resources." In Wolfang Sachs, *The Development Dictionary: A Guide to Knowledge as Power* (Orient Longman 2000), 276–92.

imagination of the river was arguably first assembled sometime in 16th century Europe, a period marked by agricultural 'improvements' or the push towards agrarian capitalism alongside the emergent economic and technical stirrings of what came later to be described as the 'Industrial Revolution.'[2] Throughout the century, as several studies have shown, rural Europe and vast swathes of its hinterlands witnessed a burst in drainage, reclamation and embankment campaigns that gathered speed in northeast Italy, the lower Loire, East Anglican Fenlands and the coastal deltas in the low countries (Netherlands) (Getzler 2004; Ash 2017). Pumps, dredging devices, locks and sluices were enthusiastically deployed by impatient land speculators and innovators who were seeking to transform once soluble and precarious waterscapes such as marshes and swamps into agricultural lands. In effect, trying to create more land to produce agricultural commodities through a process that the Venetian hydrologist Nicolo Zen (1515–65) described as the 'divinely decreed quest' for effecting 'separations between water and land, between fresh and salt waters, between clear and turbid flows, between individual channels, between lagoon water and river water and between city and *terraferma*' (Cosgrove 2016). By the 17th and early 19th century in step with rapid industrialization and agricultural growth, governments and a range of entrepreneurs began to aggressively pursue projects for canalizing rivers into a network of navigable waterways for carrying out a robust trade in manufactures and commodities.[3]

The systematic transformation of rivers into resources in South Asia was initiated by British colonial rule sometime in the middle half of the nineteenth century. Unlike the quest in Europe to transform rivers into navigable watercourses, the hydraulic rupture was initiated in British India with the construction of the *Ganges Canal* (1854), which was the first 'perennial canal irrigation' scheme built exclusively by colonial engineers in the sub-continent. Perennial canal irrigation schemes were meant to irrigate fields during the 'dry season'

2 The literature and debates on the issue of agrarian capitalism is vast. For a useful, accessible and concise introduction please see Ellen Meiksins Wood, *The Origins of Capitalism* (VERSO, 2002 [1999]). Denis Cosgrove and Geoff Petts, eds, *Water, Engineering and Landscape: Water Control and Landscape Transformation in the Modern Period* (Belhaven press, 1990). Also see Stefania Barca, *Enclosing Water: Nature and Political Economy in a Mediterranean Valley, 1796–1916* (White Horse Press, 2010).

3 See Reed G. Geiger, *Planning the French Canals: Bureaucracy, Politics and Enterprise Under the Restoration* (University of Delaware Press, 1994); L.T.C. Holt, *Navigable Waterways* (Longmans, 1963); Pierre Claude Reynard, "Charting Environmental Concerns: reactions to Hydraulic Public Works in Eighteenth-Century France," *Environment and History* 9, no. 3 (2003): 251–273 and Marc Chioc, *The Rhine; An Eco-Biography, 1815–2000* (University of Washington Press, 2002); Chandra Mukerji, *Impossible Engineering: Technology and Territoriality on the Canal du Midi* (Princeton University Press, 2009).

(the non-monsoon months of November –April), which was in sharp contrast to earlier strategies, when the local populace overwhelmingly chose to tap rivers for their fields only in the monsoon season (June–September) and during low floods.[4]

The perennial-canal-irrigation complex, however, as I have argued elsewhere, announced the emergence of an unprecedented hydraulic intervention in South Asia. The construction of 'permanent headworks' such as weirs and barrages to build canal networks went on to spur the colonial authorities to profoundly recast the notion of the river itself. That is, rivers were no longer to be grasped through cultural contexts and local knowledges but were chiefly conceptualized as 'calculable resources' and meant to be apprehended only through the colonial engineer's expertise in applying universal methods such as mathematical abstraction, physical laws and statistical quantification (D'Souza, Forthcoming). This view that fluvial behaviour could be predicted and tracked through universal principles, however, drew mainly from the British colonial government's strong belief that rivers were to be exclusively harnessed for irrigation. In effect, while land came under the purview of property, revenue collection, administration and political economy, river control and management fell exclusively under the concerns of colonial engineering (D'Souza 2015).

The attempts at maintaining a hard distinction—land as political and the river as technical—came in for sharp scholarly scrutiny by the latter decades of the twentieth century. The emergence of the field of environmental history in South Asia was particularly insightful and telling in exploring the interweaves between political economy and natural worlds. In arguably the first debate amongst historians over colonial irrigation policies that was carried out between Whitcombe (1972) and Stone (1985), the varied environmental impacts spawned by the Ganges Canal were discussed not principally as technical complications but explored in the main as colonial efforts to transform semi-arid landscapes into stable revenue yielding irrigated tracts.

Subsequent to the Whitcombe-Stone debate, a number of publications followed, which took as their main focus the transformation of rivers systems and

---

4 On the Ganges Canal and the debate on its consequences see Elizabeth Whitcombe, *Agrarian Conditions in Northern India, Vol. 1: The United Provinces under British Rule (1800–1900)* (University of California Press, 1972) and Ian Stone, *Canal Irrigation in British India* (Cambridge University Press, 1985). For a review of the literature that studied the colonial impact on river control and water management in British India see Rohan D'Souza, "Water in British India: The Making of a Colonial Hydrology," *History Compass* 4 no. 4 (2006): 621–8.

their ecologies in South Asia under British colonial rule.[5] This burst in scholarship was also striking for the manner in which they wove the cultural, natural, political and technical for formulating a range of fresh narratives and themes around rivers: notably, flood control, riverine communities, navigation, large dams, rivers and urbanization, fisheries, 'biographies' of rivers, geo-politics and even how the need for river control ended up sparking interest in developing the meteorological sciences in South Asia.[6]

At the core of this scholarly interest in the fates of the rivers was a sharp critique of the modern resource framework. Writings on the environmental histories of South Asia, in particular, were able to persuasively argue that rivers were more than volumes in motion and that flows were, in fact, interwoven biological and geological forces. Rivers in South Asia, moreover, needed to be explored as histories of how local communities had crafted a range of possibilities for livelihoods that spanned the spectrum from fishing to flood recession agriculture. In other words, the modern resource framework by treating rivers merely as a sum total of abstract volumes, physical laws and statistical quantities had ended up ignoring a range of local contexts, knowledges and varied cultural- ecological relationships. The rivers, hence, in such a community-centered imagination appeared more as an 'endowment' that needed to be sustained for livelihood and cultural possibilities, while the restricted engineering vision predominantly treated flows as abstract resources that could be harnessed and utilized for short-term profits. In effect, environmental histories on rivers in South Asia, and the Asia-Pacific in general, have re-oriented the

5 The monographs that focused on the northern canal system in British India would be Imran Ali, *The Punjab under Imperialism, 1885–1947* (Princeton University Press, 1988) and David Gilmartin, *Blood and Water: The Indus River Basin in Modern History* (University of California Press, 2015).

6 A set of these writings that could broadly be termed as environmental Christopher V. Hill, *River of Sorrow: Environment and Social Control in Riparian North India, 1770–1994* (Association for Asian Studies, 1997); Iftekhar Iqbal, *The Bengal Delta: Ecology, State and Social Change, 1840–1943* (Palgrave Macmillan, 2010); Rohan D'Souza, *Drowned and Dammed: Colonial Capitalism and flood control in Eastern India* (Oxford University Press, 2006); Daniel Klingensmith, *One Valley and a Thousand': Dams, Nationalism and Development* (Oxford University Press, 2007); Daniel Haines, *Rivers Divided: Indus Basin Waters in the Making of India and Pakistan* (Oxford University, 2016); Sudipta Sen, *Ganges: the many pasts of an Indian river* (Yale University Press, 2018); Debjani Bhattacharyya, *Empire and Ecology in the Bengal Delta: The Making of Calcutta* (Cambridge University Press, 2018); Sunil Amrith, *Unruly Waters: how rains, rivers, coasts and seas have shaped Asia's History* (Basic Books, 2018); Arupjyoti Saikia, *The unquiet river: A biography of the Brahmaputra* (Oxford University Press, 2019). On the 'diara' or people of the 'char' see Kuntala Lahiri-Dutt and Gopa Samanta, *Dancing with the River: People and Life on the Chars of South Asia* (Yale University Press, 2013).

previous emphasis on studying rivers as technical objects and abstract quantifiable flows to instead aiming to grasp the complex relationships of flows between culture, nature, and power through time.

Deepening the paradigm change in how rivers are understood—from resource to endowment—is, in fact, what this edited collection of insightful essays have as their central aim and given the growing alarms about global warming and the probable shift in the Earth System, it has become all the more urgent and compelling for rivers to be grasped and sustained as 'life-worlds.'

References

Ash, Eric H. 2017. *The Draining of the Fens: Projectors, Popular Politics and State Building in Early and Modern England.* John Hopkins University Press.

Bonneuil, Christophe, and Jean-Baptiste Fressoz. 2016. *The shock of the Anthropocene: The earth, history and us.* Translated by David Fernbach. Verso Books.

Chakrabarty, Dipesh. 2009. "The climate of history: Four theses." *Critical inquiry* 35 (2): 197–222.

Cosgrove, Denis. 1990. "Platonism and Practicality: Hydrology, Engineering and Landscape in Sixteenth-century Venice." In *Water, Engineering and Landscape: Water Control and Landscape Transformation in the Modern Period,* edited by Denis Cosgrove and Geoff Petts. Belhaven press: 35–53.

D'Souza, Rohan. "Fins in the inland Ocean and the River as Pulse in Monsoon Asia" in *History and Society in the Hydrosphere of Monsoon Asia,* edited by Prasenjit Duara and James Westcoat. (Forthcoming).

D'Souza, Rohan. 2015. "Drainage, River Erosion and Chaurs: an environmental history of land in Colonial Eastern India." *Working Paper Series, History and Society.* New Series 84: 3–24, Nehru Memorial Museum and Library.

Genovese, Eugene. 2012. "From the Archives: Eugene Genovese on Eric Hobsbawm." *The New Republic,* October 1. Accessed July 1, 2024. https://newrepublic.com/article/107966/eugene-genovese-eric-hobsbawm-age-of-extremes.

Getzler, Joshua. 2004. *A History of Water Rights at Common Law.* Oxford University Press.

Hobsbawm, Eric J. 1995. *The Age of Extremes: short history of the twentieth century, 1914–1991.* Abacus.

Kolbert, Elizabeth. 2015. *The Sixth Extinction: An Unnatural History.* Picador.

Rockstrom, Johan and Mattias Klum. 2015. *Big world, small planet: Abundance within Planetary Boundaries.* Yale University Press.

Stone, Ian. 1985. *Canal Irrigation in British India.* Cambridge University Press.

Thomas, J. Adeney, ed. 2022. *Altered Earth: Getting the Anthropocene Right*. Cambridge University Press.

Whitcombe, Elizabeth. 1972. *Agrarian Conditions in Northern India, Vol. 1: The United Provinces under British Rule (1800–1900)*. University of California Press.

Witze, Alexandra. 2024. "Geologists Reject the Anthropocene as Earth's New Epoch—after 15 Years of Debate." *Nature* 627: 249–250.

Index

www.ingramcontent.com/pod-product-compliance
Lightning Source LLC
Chambersburg PA
CBHW071335230926
51849CB00002B/402
9789004758094